Discrete Signals and Inverse Problems

Discrete Signals and Inverse Problems

An Introduction for Engineers and Scientists

J. Carlos Santamarina
Georgia Institute of Technology, USA

Dante Fratta
University of Wisconsin-Madison, USA

John Wiley & Sons, Ltd

Telephone (+44) 1243 779777

Email (for orders and customer service enquiries): cs-books@wiley.co.uk
Visit our Home Page on www.wiley.com

Other Wiley Editorial Offices

John Wiley & Sons Inc., 111 River Street, Hoboken, NJ 07030, USA

Jossey-Bass, 989 Market Street, San Francisco, CA 94103-1741, USA

Wiley-VCH Verlag GmbH, Boschstr. 12, D-69469 Weinheim, Germany

John Wiley & Sons Australia Ltd, 42 McDougall Street, Milton, Queensland 4064, Australia

John Wiley & Sons (Asia) Pte Ltd, 2 Clementi Loop #02-01, Jin Xing Distripark, Singapore 129809

John Wiley & Sons Canada Ltd, 22 Worcester Road, Etobicoke, Ontario, Canada M9W 1L1

Wiley also publishes its books in a variety of electronic formats. Some content that appears in print may not be available in electronic books.

Library of Congress Cataloging-in-Publication Data

Santamarina, J. Carlos.
Discrete signals and inverse problems : an introduction for engineers and scientists / J. Carlos Santamarina, Dante Fratta.
p. cm.
Includes bibliographical references and index.
ISBN 0-470-02187-X (cloth : alk.paper)
1. Civil engineering—Mathematics. 2. Signal processing—Mathematics. 3. Inverse problems (Differential equations) I. Fratta, Dante. II. Title.
TA331.S33 2005
621.382′2—dc22 2005005805

British Library Cataloguing in Publication Data

A catalogue record for this book is available from the British Library

ISBN-13 978-0-470-02187-3 (HB)
ISBN-10 0-470-02187-X (HB)

Typeset in 10/12pt Times by Integra Software Services Pvt. Ltd, Pondicherry, India
Printed and bound in Great Britain by TJ International, Padstow, Cornwall
This book is printed on acid-free paper responsibly manufactured from sustainable forestry in which at least two trees are planted for each one used for paper production.

To our families

Contents

Preface

The purpose of this book is to introduce procedures for the analysis of signals and for the solution of inverse problems in engineering and science. The literature on these subjects seldom combines both; however, signal processing and system analysis are intimately interconnected in all real applications. Furthermore, many mathematical techniques are common to both signal processing and inverse problem solving.

Signals and inverse problems are captured in *discrete* form. The discrete representation is compatible with current instrumentation and computer technology, and brings both signal processing and inverse problem solving to the same mathematical framework of arrays.

Publications on signal processing and inverse problem solving tend to be mathematically involved. This is an introductory book. Its depth and breadth reflect our wish to present clearly and concisely the essential concepts that underlie the most useful procedures readers can implement to address their needs.

Equations and algorithms are introduced in a conceptual manner, often following logical rather than formal mathematical derivations. The mathematically minded or the computer programmer will readily identify analytical derivations or computer-efficient implementations. Our intent is to highlight the intuitive nature of procedures and to emphasize the *physical interpretation* of all solutions.

The information presented in the text is reviewed in parallel formats. The numerous figures are designed to facilitate the understanding of main concepts. Step-by-step implementation procedures outline computation algorithms. Examples and solved problems demonstrate the application of those procedures. Finally, the summary at the end of each chapter highlights the most important ideas and concepts.

Problem solving in engineering and science is hands-on. As you read each chapter, consider specific problems of your interest. Identify or simulate typical signals, implement equations and algorithms, study their potential and limitations, search the web for similar implementations, explore creative applications . . . , and have fun!

First edition. The first edition of this manuscript was published by the American Society of Civil Engineers in 1998. While the present edition follows a similar structure, it incorporates new information, corrections, and applications.

Acknowledgments. We have benefited from the work of numerous authors who contributed to the body of knowledge and affected our understanding. The list of suggested reading at the end of each chapter acknowledges their contributions.

Procedures and techniques discussed in this text allowed us to solve research and application problems funded by the: National Science Foundation, US Army, Louisiana Board of Regents, Goizueta Foundation, mining companies in Georgia and petroleum companies worldwide. We are grateful for their support.

Throughout the years, numerous colleagues and students have shared their knowledge with us and stimulated our understanding of discrete signals and inverse problems. We are also thankful to L. Rosenstein who meticulously edited the manuscript, to G. Narsilio for early cover designs, and to W. Hunter and her team at John Wiley & Sons. Views presented in this manuscript do not necessarily reflect the views of these individuals and organizations. Errors are definitely our own.

Finally, we are most thankful to our families!

J. Carlos Santamarina
Georgia Institute of Technology, USA

Dante Fratta
University of Wisconsin-Madison, USA

Brief Comments on Notation

The notation selected in this text is intended to facilitate the interpretation of operations and the encoding of procedures in mathematical software. A brief review of the notation follows:

Letter:	a, k, α	scalar
Single-underlined letter:	$\underline{a}, \underline{x}, \underline{y}, \underline{h}$	one-dimensional array or vector
Double-underlined letter:	$\underline{\underline{a}}, \underline{\underline{x}}, \underline{\underline{y}}, \underline{\underline{h}}$	two-dimensional array or matrix
Capital letter:	$A, \underline{X}, \underline{\underline{F}}$	a capital letter is used to represent a quantity in the frequency domain, which is complex in most cases; it could be a scalar or an array
Bar over capital letter:	$\overline{X}$	complex conjugate of X
Indices (sequence of data points in an array):	i, k	indices in the time domain
	u, v	indices in the frequency domain
Indexed letters:	x_i or $z_{i,k}$	a specific value within arrays $\underline{x}$ or $\underline{\underline{z}}$
Imaginary component:	$a + j \cdot b$	$j^2 = -1$ indicates the imaginary component
Magnitude:	$\lvert a + j \cdot b \rvert$	$\sqrt{a^2 + b^2}$ Pythagorean length
Additional information:	$CC^{<x,y>}$	superscripts in angular brackets are used to provide additional information on the quantity
Point-by-point operations:	$x_i \cdot h_i$	point-by-point product; the operation is defined between specific elements in the arrays
"time":		the term "time" designates the independent variable, such as time, space, or any other independent parameter

1

Introduction

This chapter begins with a brief discussion of signals, systems, and the types of problems encountered in engineering and science. Then, selected applications are described to begin exploring the potential of signal processing and inverse problem solving. Exercises at the end of the chapter invite the reader to extend this preview to other areas of interest, and to gather simple hardware components to obtain discrete signals in different applications.

1.1 SIGNALS, SYSTEMS, AND PROBLEMS

Listen. . . . Touch. . . . See. . . ! Our senses detect signals that convey important information we use for survival. We hear the variation of pressure with time, our fingers feel the spatial variation of surface roughness, and we see the time-varying spatial distribution of color. Clearly, each <u>signal</u> *is the variation of a parameter with respect to one or more independent variables.*

We take these stimuli (input signals) and respond accordingly (output signal). Therefore, each of us is a <u>system</u> that *transforms an input signal into an output signal.* In fact, our response to a given stimulus reveals important information about us. Likewise, a time-varying wind load (input signal) acts on a building (system) causing it to oscillate (output signal), and these oscillations can be used to infer the mechanical characteristics of the building.

A system may transform the input energy into another form of energy. For example, metals dilate (mechanical output) when heated (thermal input). Most transducers are energy-transforming systems: accelerometers produce an electrical output from a mechanical input, and photovoltaic cells convert light energy into electrical energy.

The input signal, the output signal or the system characteristics may be unknown. Our level of knowledge permits classifying *<u>problems in engineering</u>*

Discrete Signals and Inverse Problems J. C. Santamarina and D. Fratta

Table 1.1 Forward and inverse problems in engineering and science

PROBLEMS IN ENGINEERING AND SCIENCE

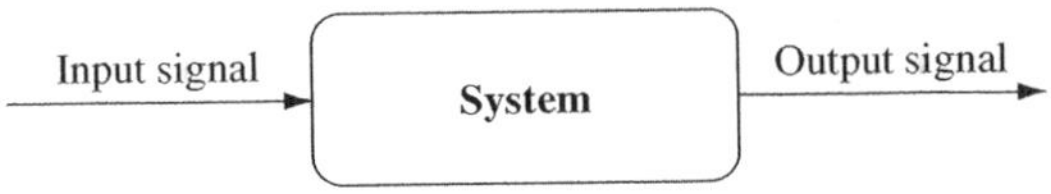

Forward Problems		Inverse Problems	
System design[a]	*Convolution*	*System identification*	*Deconvolution*
Input: Known	Input: Known	Input: Known	Input: Unknown
System: To be designed	System: Known	System: Unknown	System: Known
Output: Predefined	Output: Unknown	Output: Known	Output: Known
Classical training	*Chapters 3–7*	*Chapters 8–11*	

[a] The system is designed to satisfy performance criteria: controlled output for estimated input.

and science, as shown in Table 1.1. Typically, engineers are trained to solve *forward problems*. Emphasis has been placed on the *design of systems* to satisfy predefined performance criteria, based on an estimated design load. Typical examples include the design of a reactor or a transportation system. The other form of forward problems is estimating the response of a system of known characteristics given a known input. This second class of forward problems is a *convolution* of the input with the characteristic system response, such as computing the signal coming out of an amplifier, the flood discharge after a rainfall, or numerical simulations in general.

A wide range of scientific problems – by definition – and many engineering tasks are *inverse problems* whereby the output is known, but either the input or the system characteristics are unknown (Table 1.1). In *system identification* the input and output signals are known, and the task is to determine the characteristics of the system. For example, a bone specimen is loaded and its deformation is measured to determine material properties such as Young's modulus and Poisson ratio. The other type of inverse problems involves the determination of the input signal knowing the system characteristics and the output signal. This is called *deconvolution*, as opposed to the forward problem of convolution. In all measurements, the true signature is computed by deconvolution with the characteristics of the transducer: the earthquake signature is obtained by deconvolving the recorded signal from the characteristics of the seismograph. Inferring the speed of a vehicle before collision is another example of deconvolution in the context of forensic engineering.

Many inverse problems are complex and involve partial knowledge of the system and signals. Hence, it may not be possible to identify a unique solution. For example, we are still puzzled by multiple plausible hypotheses related to the extinction of dinosaurs, the catastrophic failure of Teton dam, and the initiation of various deadly diseases. Even extensive scrutiny may not render enough information to falsify hypotheses, particularly when information may have been lost in the event itself.

1.2 SIGNALS AND SIGNAL PROCESSING – APPLICATION EXAMPLES

Signal processing is an integral part of a wide range of devices used in all areas of science and technology. The following examples introduce common concepts in signal processing within the contexts of our own daily experiences and lead us towards the development of devices and procedures that can have important practical impact. Cases include active and passive systems. Other examples are listed in Table 1.2.

1.2.1 Nondestructive Testing by Echolocation (Active)

Echolocation consists of emitting a sound and detecting the reflected signal. The time difference between sound emission and echo detection is proportional to the distance to the reflecting surface. Differences between the frequency content in the reflected signal with respect to the emitted signal are used to discern characteristics of the object such as its size.

Bats and dolphins are able to use echolocation to enhance their ability to comprehend their surroundings. (People have some echolocation capability, but it is less developed because of our refined vision.) The sound made by bats varies among species. Some bats emit a sine sweep signal or chirp like the one shown in Figure 1.1. This input signal has two important advantages: first, it leads to improved accuracy in travel time determination, and second, it permits assessing the size of the potential prey (Chapters 3–7).

The same technique is used in nondestructive evaluation methods, from medical diagnosis to geophysical prospecting for resource identification (Figure 1.2a; see suggested exercises at the end of this chapter). While the input signal can resemble the signal emitted by bats, the frequency content is selected to optimize the trade-off between penetration depth and resolution (Figure 1.2b).

Table 1.2 Examples of signals

Time and spatial variations in one dimension (1D)

- Acoustics: sonar signals; echolocation by bats and dolphins
- Electrical engineering: signal emitted by a transmission antenna
- Chemistry – material science: temperature history in a chemical reaction
- Finance: the stock market historical record
- Medicine: electrocardiogram and electroencephalogram

Two-dimensional (2D) spatial variations

- Agricultural engineering: vegetation, evaporation and infiltration in a watershed
- Geography – climatology: surface temperature and pressure maps; GIS maps
- Socioeconomics: world distribution of population density and income
- Mechanics – tribology: surface roughness; contact pressure distribution
- Physics: AFM image of a polymer surface
- Traffic engineering: accident rate at intersections across the city

Three-dimensional (3D) volumetric variations

- Physics: porous network in a particulate medium
- Fluid mechanics: flow–velocity profile around airplane wing
- Geotechnology: pore fluid pressure underneath a dam
- Biology: CO_2 distribution in a bioreactor

Note:
The graphical representation of a signal can be simplified if a plane or axis of symmetry is identified. For example, the 4D variation of subsurface temperature in space and time can be captured as a 2D signal in depth–time coordinates if the subsurface is horizontally homogeneous.

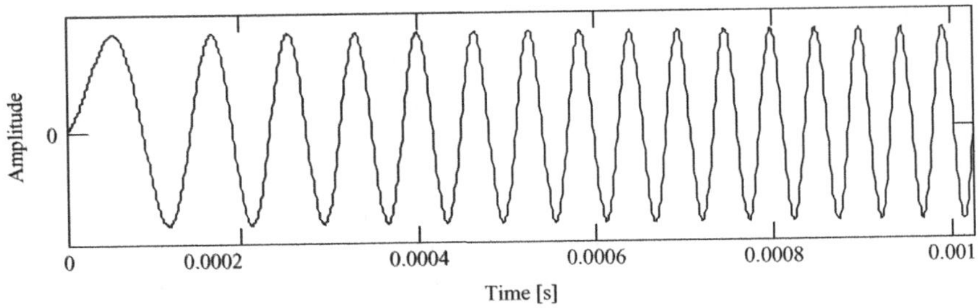

Figure 1.1 A sine sweep signal. The frequency increases with time

1.2.2 *Listening and Understanding Emissions (Passive)*

Many signals are generated without our direct or explicit involvement. In most cases, "passive" signals are unwanted and treated as noise. However, passive signals when carefully analyzed may provide valuable information about the system.

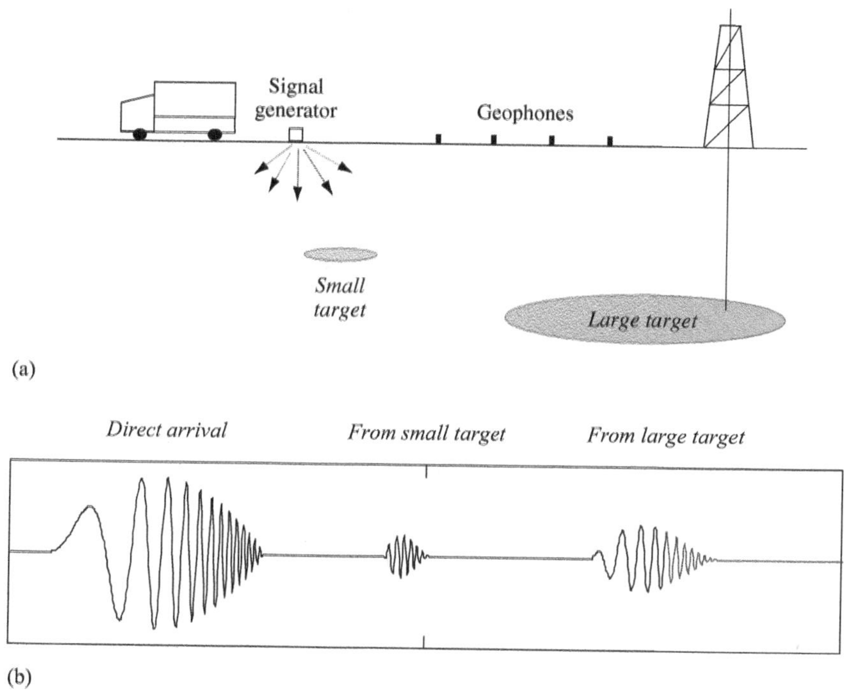

Figure 1.2 The frequency sweep signal is used in geophysical and nondestructive applications. Low frequencies are not reflected by small objects, whereas large objects reflect both low and high frequencies

A stethoscope used by a trained physician to listen to the passive emissions generated by the heart and the lungs remains a valuable diagnostic technique 200 years after its development. Forensic investigators can analyze the sound track recorded when a gun was fired, extract time delays and intensities corresponding to the various sound reflections and constrain the location of the sniper. Likewise, there is information encoded in earthquakes, in changes exhibited by bacterial communities, in economic indicators, and in the distribution of air pollution above a city. We just need to observe and learn how to decode the message.

1.2.3 Feedback and Self-calibration

Organisms are particularly adept at accommodating to changes. Likewise, adaptive systems are engineered to attain optimal vibration control of airplane wings or to minimize traffic congestion by means of intelligent traffic signals.

Natural or computerized adaptive/learning systems include feedback, and when the feedback loop is interrupted, adaptation stops. For example, deaf individuals (the adaptive system in this example) can learn to speak only when alternative feedback is provided to counteract their inability to hear themselves or others. Imagine a visual feedback device that permits trainer and trainee to speak into a microphone and displays their signals on the screen of an oscilloscope as a variation of sound pressure versus time: this is the *time domain* representation (Chapters 3 and 4). This device may also analyze their signals and show the amount of energy in different frequencies: this is the *frequency domain* representation (Chapters 5 and 6). Figure 1.3 presents simple sounds in the time and frequency domains. The trainee's goal is to learn how to emit sounds that match the time domain traces, using frequency domain information to identify needed emphasis on either high-pitch notes or low-pitch sounds.

1.2.4 Digital Image Processing

We seldom pause to assess the extent of our natural abilities to process signals. However, when researchers in artificial intelligence began studying vision, they were confronted with a highly sophisticated process. Only the fact that we do see stopped researchers from concluding that vision as we know it is impossible.

The advent of digital photography has opened important possibilities for a wide range of techniques that were not envisioned a generation ago. A digital image is a matrix of numbers. For example, the pixel value $p_{i,j}$ at location (i, j) in a black-and-white image is a number in a matrix (Figure 1.4). The resolution of digital images is selected to optimize application needs and storage considerations. Resolution is restricted by the pixel size in the computer screen – the grain size in conventional photographic prints is much smaller.

Captured images are displayed on a screen, processed, analyzed, and stored. Image processing includes operations such as smoothing and contrasting, edge detection, and recoloring. Image analysis and data extraction can range from measuring areas and perimeters of objects to the more advanced task of pattern recognition. Digital image analyzers are complementary components to a wide range of devices, such as microscopes, tomographers, and video cameras. These systems are increasingly being used in engineering and science, from materials research to automated quality control in manufacturing processes.

1.2.5 Signals and Noise

Noise is an unwanted signal superimposed on the signal of interest. Eventually, the signal of interest may become indistinguishable when the signal-to-noise ratio is low; yet its presence may still have important consequences on the system

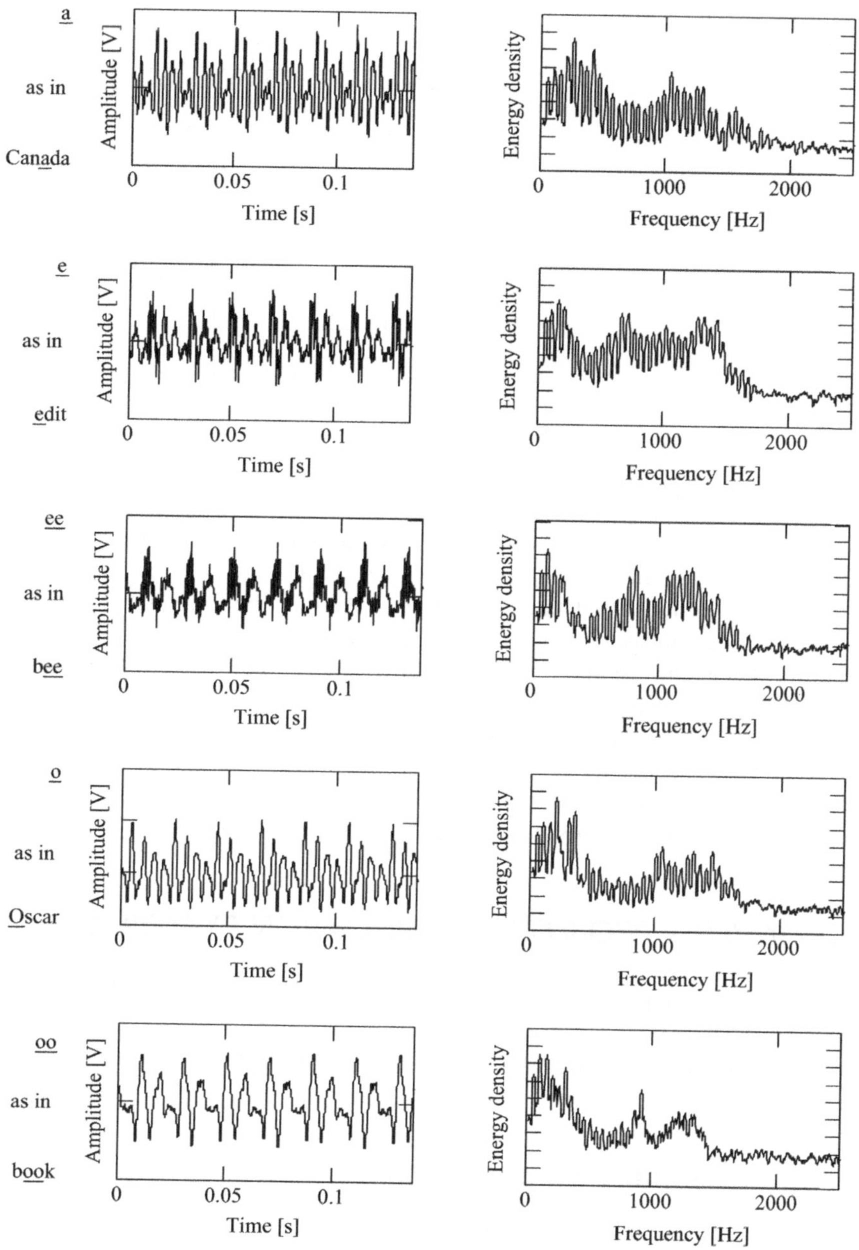

Figure 1.3 Simple sounds in the time and frequency domains

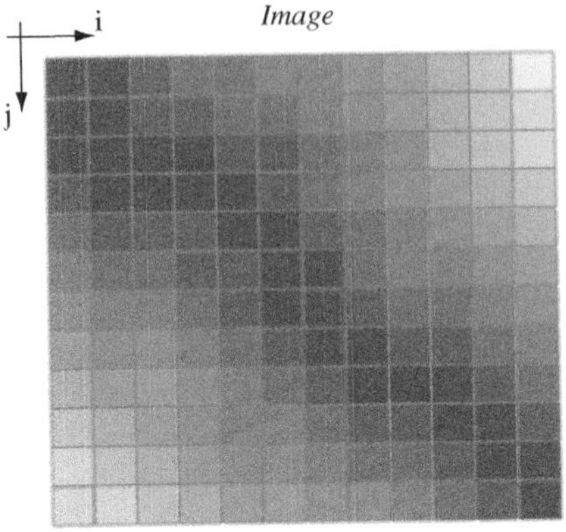

Pixel values $P_{i,j}$

0	0	1	2	2	3	3	4	5	6	7	9
0	0	1	1	2	2	3	4	5	6	7	8
0	0	0	0	1	1	2	3	4	6	7	8
1	0	0	0	0	1	2	3	4	5	6	7
2	1	1	1	0	0	1	2	3	4	5	6
3	2	2	1	1	0	0	2	3	3	4	5
4	3	3	2	1	0	0	1	2	2	3	4
5	4	4	3	2	1	0	0	1	1	2	3
6	5	5	4	3	2	1	0	0	0	1	2
7	6	5	4	3	3	2	1	1	0	0	1
8	7	6	5	4	4	3	2	2	1	0	0
8	8	7	6	5	5	4	4	3	2	1	0

Figure 1.4 A gray scale image and the stored matrix of pixel values

response. For example, it is difficult to recognize the small waves caused by an earthquake in Chile as they propagate across the Pacific Ocean; however, they can produce devastating tsunamis when they reach Hawaii or Japan.

The first goal in every data collection exercise must be to reduce the level of noise that affects measurements. Sometimes, simple "tricks" in the design of the experiment can render major improvements in signal-to-noise ratio. For instance, a work bench made of a massive marble slab sitting on rubber pads can be designed to low-pass filter the mechanical noise in buildings, whereas grounded aluminum foil wrapped around experimental devices and instrumentation is an effective filter of electromagnetic noise. Once the signal is stored, a number of postprocessing techniques are available to separate signal from noise (Chapters 4–6).

1.3 INVERSE PROBLEMS – APPLICATION EXAMPLES

The goal of inverse problem solving is to infer the unknown input or the unknown system characteristics (Table 1.1). Instances of deconvolution and system identification are described next. Other examples in engineering and science are listed in Table 1.3.

1.3.1 Profilometry (Deconvolution)

Many research and application tasks require proper assessment of surface topography, including the following: research on crystal growth, scanning probe microscopy, study of friction, quality assessment of paints and coatings, light

Table 1.3 Examples of inverse problems

System identification
- Constitutive modeling: material properties from experimental data
- Experimental research: transducers' frequency response from calibration data
- Medicine and NDT: tomographic imaging
- Earth science: earth's mantle structure from earthquake data
- Astronomy: origin of the universe from rate of expansion and redshift
- Structural engineering: bridge condition from deformation during load testing

Deconvolution
- Experimental research: variable true time history from the measured time series
- Geophysics: detection of gravity anomaly from surface measurements
- Forensic engineering: gunman location from sound recordings in newscasts
- Environmental monitoring: source characterization from remote measurements

scattering control, rock joints and the stability of rock masses. Measured 1D or 2D surface profiles are analyzed to identify spatial scales or wavelengths that are important to the problem under consideration (see Chapters 5 and 7).

Consider the case of tire–pavement interaction: the short wavelength roughness is important for friction and hydroplaning, whereas long wavelength components affect riding comfort. Furthermore, surface topography also denounces pavement distress; therefore, optimal pavement management benefits from frequent pavement profilometry that can be effectively implemented by mounting an accelerometer on the axis of a wheel riding on the pavement. The measured acceleration vs. distance signal is the response of the wheel–accelerometer system to the input surface topography. Therefore, the surface topography is obtained by deconvolving the characteristic response of the wheel–accelerometer system from the measured signal.

1.3.2 Model Calibration (System Identification)

The analysis of systems always takes place within the framework of assumed models. Hence, biomechanicians interpret the stress–strain response of biological tissue from the perspective of elasto-visco-plastic constitutive models; physicists analyze the electronic polarization of molecules assuming a single degree of freedom system; and structural engineers probe the seismic response of water tanks using an inverted pendulum model. Each model has associated model parameters, such as the mass, damping, and spring constant in vibrating systems.

Model calibration is an inverse problem. It consists of identifying the model parameters that minimize the difference between the observed system response and the model response for the same input. A poor match suggests either an inappropriate model and/or measurement errors. Once calibrated, models are used to represent the system in subsequent analyses.

1.3.3 Tomographic Imaging (System Identification)

Great advances in noninvasive imaging technology have revolutionized medical diagnosis in the twentieth century. Current imaging systems include computerized axial tomography (CAT) scan, positron emission tomography (PET) scan, and magnetic resonance imaging (MRI). In these techniques, *boundary measurements* obtained with transducers placed on the periphery of the body are mathematically processed to compute internal *local values* of material parameters. For example, boundary measurements of total X-ray absorption across the chest are "inverted" to determine the attenuation at different points within the body, and these local values are displayed on a screen using a selected color palette; the resulting picture is the tomographic image. By contrast, the classical X-ray plate collapses the 3D body onto a 2D image that displays the cumulative absorption in the body along each ray path. Similar tomographic techniques are used to explore materials from the micron scale to the planet scale!

1.4 HISTORY – DISCRETE MATHEMATICAL REPRESENTATION

The fields of signal processing and inverse problem solving are relatively young. While the needed mathematical tools were available before the twentieth century, several decisive developments in the last 100 years stimulated revolutions in discrete data processing, in particular (Table 1.4): consumer electronics (1920s), digital processing (1940s), computers (1960s), and single-chip digital signal processors (1980s).

The scope of this book is restricted to the analysis of discrete signals and to the solution of inverse problems that are expressed in discrete form. Consequently, classical definitions in continuous form are restated in discrete form (e.g. impulse – Chapter 3), operations that integrate the product of two functions become matrix multiplications (e.g. cross-correlation – Chapter 4), and integrals are replaced by summations (e.g. Fourier transform – Chapter 5). While the analysis of discrete data can be more intuitive than the mathematics of continuous functions, peculiar effects arise in discrete data analysis and must be carefully understood to avoid misinterpretations.

Table 1.4 Brief history of discrete signals and inverse problem solving

Year	Event
1300	The philosopher and theologian W. Ockham states the rule of parsimony: "*Plurality should not be assumed without necessity.*"
1800s	The main themes are thermodynamics, mechanics, hydrodynamics, acoustics, and electromagnetics; their solution requires new mathematical tools and concepts. J. B. J. Fourier (1768–1830) uses the representation of a function as a series of sinusoids to solve heat flow problems. J. M. C. Duhamel (1797–1872) uses convolution to solve the problem of heat conduction with time-varying boundary conditions. V. Volterra (1860–1940) investigates on integral equations. Analog recorders are invented at the end of the century
1910s	I. Fredholm introduces the concept of generalized inverse for an integral operator (1903). Generalized inverses for differential operators are implied in D. Hilbert's discussion of generalized Green's functions (1904)
1920s	E. H. Moore presents the generalized inverse of matrices (1920). The field of consumer electronics starts with the sale of radios and electronic phonographs. Sound is added to motion pictures
1930s	Car radios and portable radios become common
1940s	N. Wiener develops statistical methods for linear filters and prediction. Correlation techniques develop to recover weak signals in the presence of noise. The Singleton's digital correlator rapidly performs storage, multiplication, and integration by a binary digital process (1949)
1950s	The transistor is invented by J. Bardeen, W. Brattain, and W. Shockley (1947–48 – Nobel Prize in 1956). Sony brings it to mass production and develops pocket-size transistor radio. R. Penrose shows that the Moore's inverse is the unique matrix satisfying four matrix equations (1955). Shannon theorizes that a message can be encoded and transmitted in "bits" (1956)
1960s	Computers emerge and there is a rapid growth in the new field of digital signal processing. Integrated circuits lead to new technology. The development of signal processing starts having a strong impact in consumer electronics related to voice, music and images. J. Tukey and J. Cooley introduce the fast Fourier transform algorithm (1965)
1970s	Microprocessors are developed (1971) and the size of computers decreases to a chip. Consumer electronics begin their transition to digital. A. M. Cormack and G. Hounsfield receive the Nobel prize in 1979 for computerized tomography
1980s	CD players are introduced in 1982. Record players vanish from the market in less than a decade. Texas Instrument brings single-chip digital signal processor into mass production. Commercial cellular phone service starts
1990s	Very few analog consumer electronics remain in the market. There is a rapid growth in digital memory and storage capabilities

1.5 SUMMARY

- Signal characterization, decoding, and interpretation are important components of engineering and science tasks.
- There are forward problems (system design and response computation) and inverse problems (system identification and input estimation).
- The fields of discrete signal processing and inverse problem solving are relatively new. Their growth has been intimately associated with revolutions in computer technology and digital electronics.
- Today, discrete signal processing and inverse problem-solving techniques impact all aspects of daily life, with countless examples in engineering and science.
- What about the future? Just, imagine. . . !

SOLVED PROBLEMS

P1.1 Ocean tides are caused by changes in the gravitational field due to the rotation of the Earth and its relative position with respect to the Moon and the Sun. A typical data set is presented in Figure 1.1 (For more information and

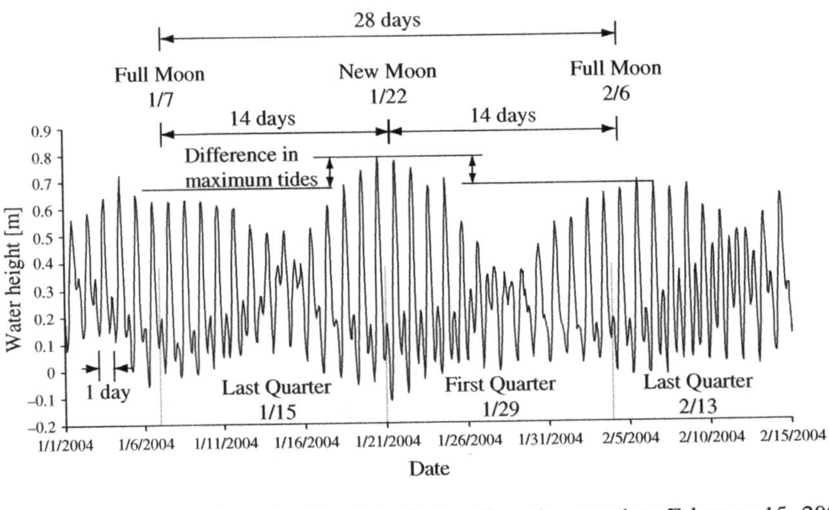

Figure P1.1 Tide levels at the Honolulu Harbor from January 1 to February 15, 2004. The sampling interval is one hour

data, visit NOAA's Center for Operational Oceanographic Products and Services on the Internet.) Determine the main periodicities in the record and identify the underlying physical phenomena that cause them.
Solution: The beat function observed in Figure P1.1 is the result of concurrent events with three different periods. The one-day period is caused by the daily rotation of the Earth and the gravitational pull of the Moon on ocean waters. The 14-day period is related to the alignment of the Sun and the Moon, causing maximum high tides and minimum low tides for the New Moon and the opposite for the Full Moon. The 28-day period is caused by the completion of the Moon cycle. The different periods are shown in Figure P1.1.

P1.2 Many have attempted to identify trends in the stock market in order to improve trading decisions. Consider extrapolating simple polynomial fittings to the New York Stock Exchange (NYSE) weekly closing values. Fit polynomials order 5 and 10 to data from January 1990 until June 2003 (Figure 1.2). Then, extrapolate to predict stock market trends until June 2004. Compare predictions against observed values. Conclude about the potential use of this technique to become a successful stockbroker.
Solution: The polynomial trends are fitted by minimizing the square error and are superimposed on Figure P1.2. While polynomials fit past data well, the prediction of future trading is poor. Regression methods are analyzed in Chapters 8 and 9.

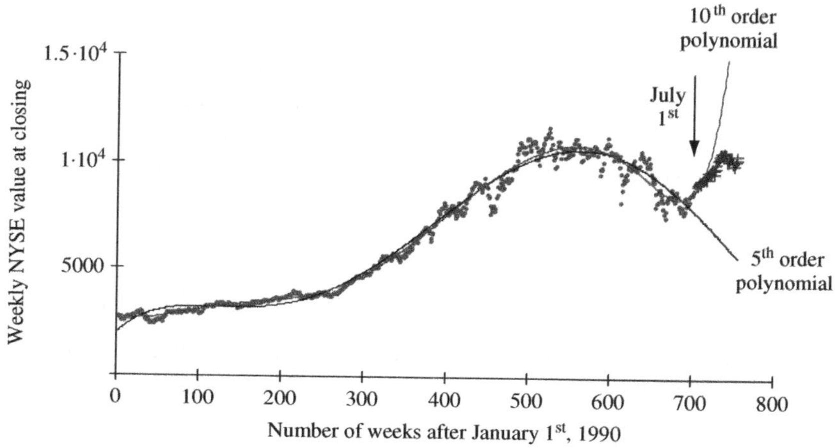

Figure P1.2 Evolution of the NYSE weekly closing values (data downloaded from URL: http://yahoo.com/finance)

ADDITIONAL PROBLEMS

P1.3 Identify important signals in your field of interest. Briefly describe their characteristics.

P1.4 Identify and describe inverse problems in your field of interest.

P1.5 *Digital image processing.* Identify an application of digital image processing in your area of interest, list the information to be extracted from the image, required resolution and image size. Then, visit a video camera shop and a computer store to learn about the *hardware*. Verify system compatibility. Study specifications to determine the speed of digitization, which is critical for some real-time applications. Recognize the trade-off between object size and resolution; as a general guideline, the smallest object must be at least $\sim 3 \times 3$ pixels in size. Then, download public domain digital image processing *software* available at multiple sites on the Internet, test their capabilities with simulated images, and study the underlying mathematical procedures.

P1.6 *Nondestructive testing: acoustic source.* A versatile source for wave propagation studies in the sound range (20 Hz to 20 kHz frequency range) can be built connecting the sound output from your computer through an audio power amplifier into an old speaker cone. Visit your local computer and electronic stores and review specifications. Then design the system and estimate its cost.

P1.7 *Analog-to-digital conversion: storage.* Consider a sensing transducer (photosensor, accelerometer, thermocouple, linear variable differential transducer, or piezocrystal) that provides an analog output. Design a system that digitizes and stores the signal. Search for available components, read catalogs of electronic suppliers, and carefully review specifications. Describe the meaning of each of the following terms: sampling frequency per channel, memory per channel, stacking capabilities, internal noise, preamplification capabilities, and input impedance. Note: A digital storage oscilloscope is the most versatile device to prototype a monitoring system; most units include a computer interface to download the discrete time series for postprocessing.

P1.8 *Step response: thermal diffusion.* Make a cylindrical specimen out of gelatin (length-to-diameter ratio ~ 2). Insert one thermometer at the center of the cylinder and place a second thermometer adjacent to the cylinder. Place the setup inside a refrigerator and keep overnight to homogenize the specimen at a low temperature. The following morning, remove the setup and expose to room temperature. Take temperature readings every five minutes until the temperature in both thermometers equals the room temperature. Use the signals gathered with the two thermometers to determine the "thermal properties" of gelatin given the imposed step-like thermal change.

P1.9 *Music*. Design a musical instrument to produce a 2 kHz frequency sound (e.g. wind, percussion, string). Understand the underlying physical processes and develop an analytical model to predict the resonant frequency of the instrument. Use the audio capabilities in your computer to digitize the signal and corroborate the frequency content. What is the shape of the signal? How can you alter the frequency? Whistle to match the frequency of sound emitted by the instrument; verify the frequency match using the same monitoring system.

2

Mathematical Concepts

The discrete mathematical representation of signals and transformations lends itself to transparent storage and processing in the form of matrices and arrays. Additional mathematical tools required for the efficient analysis of discrete signals and inverse problems include complex numbers and exponentials. A convenient review of definitions and salient properties invoked in subsequent chapters is presented next.

2.1 COMPLEX NUMBERS AND EXPONENTIAL FUNCTIONS

Sinusoidal signals are among the most frequently used functions in signal processing, system analysis, and transformations. Although the manipulation of sinusoidals is often cumbersome, operations can be efficiently implemented with complex numbers and exponential functions.

2.1.1 Complex Numbers

The amplitude of the response is not sufficient to characterize a system. For example, if you shake a car with a sinusoidal varying force $x(t) = \cos(\omega \cdot t)$, the car vibration $y(t)$ will be a sinusoidal, with the same frequency ω, and some amplitude "A". But the peaks of the input and the output time histories will not occur at the same time. In other words, there will be a phase angle φ and the response will be $y(t) = A \cdot \cos(\omega \cdot t - \varphi)$.

Discrete Signals and Inverse Problems J. C. Santamarina and D. Fratta

The shifted sinusoid y(t) is equivalent to the sum of a cosine (in-phase) and a sine (90° out-of-phase). The amplitude of each of these two components is determined using trigonometric identities:

$$\begin{aligned} y &= A \cdot \cos(\omega \cdot t - \varphi) \\ &= \underbrace{[A \cdot \cos(\varphi)]}_{a} \cdot \cos(\omega \cdot t) + \underbrace{[A \cdot \sin(\varphi)]}_{b} \cdot \sin(\omega \cdot t) \\ &= a \cdot \cos(\omega \cdot t) + b \cdot \sin(\omega \cdot t) \end{aligned} \tag{2.1}$$

Therefore, the amplitudes of the cosine and sine components are (Figure 2.1)

$$a = A \cdot \cos(\varphi) \tag{2.2}$$

$$b = A \cdot \sin(\varphi) \tag{2.3}$$

Complex numbers facilitate the mathematical representation and solution of this type of problem. In complex number notation, the signal y(t) is represented as a construct that captures the two values, a and b:

$$Y = a + j \cdot b \qquad \textit{corresponds to frequency}\ \omega \tag{2.4}$$

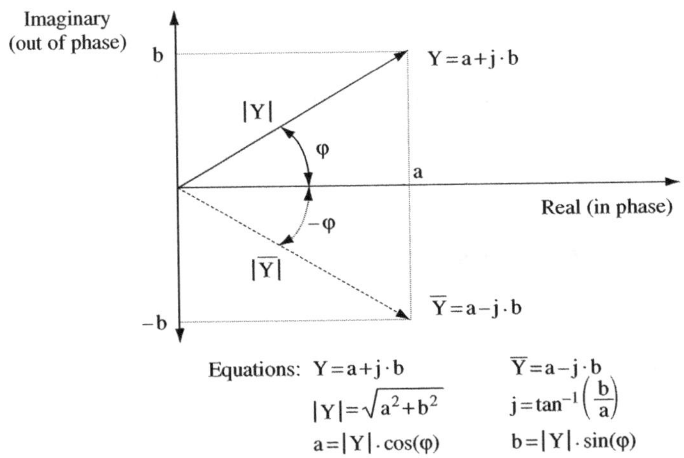

Figure 2.1 Complex numbers. The graphical representation of a complex number is a vector in a complex plane with a real in-phase component and an imaginary out-of-phase component. A complex number and its conjugate have the same magnitude but opposite phase

where the imaginary unit is $j^2 = -1$. The numbers a and b are known as the *real* and *imaginary* parts of the complex number (yet, both are very real numbers!). The amplitude A and the phase φ of the original sinusoid y(t) are recovered as

$$A = |Y| = \sqrt{a^2 + b^2} = \sqrt{\mathrm{Re}(Y)^2 + \mathrm{Im}(Y)^2} \tag{2.5}$$

$$\varphi = \tan^{-1}\left(\frac{b}{a}\right) = \tan^{-1}\left[\frac{\mathrm{Im}(Y)}{\mathrm{Re}(Y)}\right] \tag{2.6}$$

This graphical representation of a complex number is shown in Figure 2.1, where both rectangular $(a + jb)$ and polar coordinates (A, φ) are indicated.

The *complex conjugate* $\overline{Y}$ of the complex number Y is defined as follows:

$$\overline{Y} = a - j \cdot b \tag{2.7}$$

Figure 2.1 also shows the representation of a complex conjugate in the complex plane. The amplitude of the complex conjugate is the same as the amplitude of the original complex number, but the phase angle φ has opposite sign.

Mathematical operations with complex numbers are implemented by treating them as binomials:

addition $$(a + j \cdot b) + (c + j \cdot d) = (a + c) + j \cdot (b + d) \tag{2.8}$$

multiplication $$(a + j \cdot b) \cdot (c + j \cdot d) = (a \cdot c - b \cdot d) + j \cdot (a \cdot d + b \cdot c) \tag{2.9}$$

The trick required to compute the division of two complex numbers is to leave a real quantity in the denominator. This is achieved by multiplying the numerator and the denominator by the complex conjugate of the denominator:

division $$\frac{(a + j \cdot b)}{(c + j \cdot d)} = \frac{(a + j \cdot b) \cdot (c - j \cdot d)}{(c + j \cdot d) \cdot (c - j \cdot d)} = \frac{(a \cdot c + b \cdot d) + j \cdot (-a \cdot d + b \cdot c)}{c^2 + d^2} \tag{2.10}$$

Operations with complex numbers satisfy the commutative, associative, and distributive rules.

2.1.2 Exponential Functions

The exponential function is defined by

$$y = a^x \tag{2.11}$$

where "a" is a constant. A special exponential function is the Napierian exponential where $a = e = 2.718....$ The exponent x may be complex. Common operations with exponential functions include

$$\textit{multiplication} \quad e^x \cdot e^y = e^{x+y} \tag{2.12}$$

$$\textit{division} \quad \frac{e^x}{e^y} = e^{x-y} \tag{2.13}$$

$$\textit{power} \quad (e^x)^y = e^{x \cdot y} \tag{2.14}$$

$$\textit{derivative} \quad \frac{d(e^u)}{dx} = \frac{du}{dx} \cdot e^u \tag{2.15}$$

$$\textit{integral} \quad \int \left(\frac{du}{dx}\right) e^u \cdot du = e^u + cte \tag{2.16}$$

The importance of exponential functions is partially alluded to in these expressions. First, they convert multiplication into addition (Equations 2.12 and 2.13). Second, the derivative of an exponential function is the function itself times a factor (Equation 2.15); therefore, exponential functions are solutions of differential equations of the form $dy/dx = y$, such as the motion of harmonic oscillators.

In addition, complex exponentials are linked to trigonometric functions, as captured in *Euler's identities*,

$$e^{j \cdot \varphi} = \cos(\varphi) + j \cdot \sin(\varphi) \tag{2.17}$$

$$e^{-j \cdot \varphi} = \cos(\varphi) - j \cdot \sin(\varphi) \tag{2.18}$$

Thus, the following equalities hold (Equations 2.1–2.6):

$$\begin{aligned} Y &= a + j \cdot b \\ &= |Y| \cdot [\cos(\varphi) + j \cdot \sin(\varphi)] \\ &= |Y| \cdot e^{j \cdot \varphi} \end{aligned} \tag{2.19}$$

where $a = |Y| \cdot \cos(\varphi)$ and $b = |Y| \cdot \cos(\varphi)$. From Euler's identity, and for any integer k,

$$e^{j \cdot (2\pi \cdot k)} = \left[e^{j \cdot (2\pi)}\right]^k = [\cos(2\pi) + j \cdot \sin(2\pi)]^k = 1 \tag{2.20}$$

and trigonometric periodicity in exponential form becomes

$$e^{j \cdot (\theta + 2\pi \cdot k)} = e^{j \cdot \theta} \tag{2.21}$$

Given: the complex number X and operator H,

$$X = a + j \cdot b = |X| \cdot [\cos(\alpha) + j \cdot \sin(\alpha)] = |X| \cdot e^{j \cdot \alpha}$$

$$H = c + j \cdot d = |H| \cdot [\cos(\beta) + j \cdot \sin(\beta)] = |H| \cdot e^{j \cdot \beta}$$

Compute: $Y = X \cdot H$

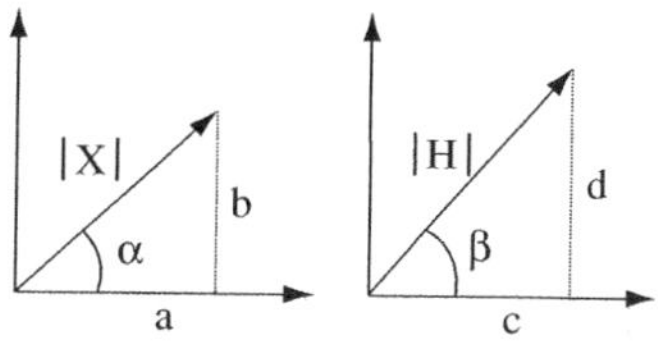

1) Procedure with complex numbers:

$$Y = X \cdot H = (a + j \cdot b) \cdot (c + j \cdot d) = (a \cdot c - b \cdot d) + j \cdot (a \cdot d + b \cdot c)$$

$$|X \cdot H| = \sqrt{a^2 \cdot c^2 + b^2 \cdot d^2 + a^2 \cdot d^2 + b^2 \cdot c^2} \qquad \text{magnitude}$$

$$\delta = \tan^{-1}\left(\frac{a \cdot d + b \cdot c}{a \cdot c - b \cdot d}\right) \qquad \text{phase}$$

2) Procedure in polar notation with magnitude and phase:

$$X \cdot H = |X| \cdot [\cos(\alpha) + j \cdot \sin(\alpha)] \cdot |H| \cdot [\cos(\beta) + j \cdot \sin(\beta)]$$

$$X \cdot H = |X| \cdot |H| \cdot [\cos(\alpha) + j \cdot \sin(\alpha)] \cdot [\cos(\beta) + j \cdot \sin(\beta)]$$

$$X \cdot H = |X| \cdot |H| \cdot [\cos(\alpha + \beta)] + j \cdot \sin(\alpha + \beta)]$$

3) Procedure with exponential functions:

$$X \cdot H = |X| \cdot |H| \cdot e^{j \cdot (\alpha + \beta)}$$

Figure 2.2 Multiplication of two complex quantities

2.1.3 Example

The addition and multiplication of two quantities, each with its own magnitude and phase, are common operations in signal processing and system analysis. The rectangular representation is more convenient for addition (Equation 2.8) whereas the exponential notation facilitates multiplication (Equation 2.12). The multiplication of two complex quantities is demonstrated in Figure 2.2 using complex, polar, and exponential forms. Note the efficient implementation using exponentials.

2.2 MATRIX ALGEBRA

A matrix is an arrangement of numbers in columns and rows. A review of fundamental matrix operations follows.

2.2.1 Definitions and Fundamental Operations

The following notation is used to designate the matrix $\underline{\underline{a}}$ by its elements:

$$\underline{\underline{a}} = \begin{bmatrix} a_{1,1} & \cdots & a_{1,N} \\ \cdots & a_{i,k} & \cdots \\ a_{M,1} & \cdots & a_{M,N} \end{bmatrix} \qquad (2.22)$$

where the index i refers to the row number and varies from 1 to M, and k indicates column number and varies from 1 to N, where M and N are integers. Sometimes it is more convenient to vary i from 0 to $M-1$, and k from 0 to $N-1$. For example, this is the case when the first entry refers to zero time or zero frequency. A matrix $\underline{\underline{a}}$ is

- *square* if $M=N$
- *real* when all its elements are real numbers
- *complex* if one or more of its elements are complex numbers
- *nonnegative* if all $a_{i,k} \geq 0$
- *positive* if all $a_{i,k} > 0$

Negative and nonpositive matrices are similarly defined.

The *trace* of a square matrix is the sum of the elements in the main diagonal, $a_{i,i}$. The *identity* matrix $\underline{\underline{I}}$ is a square matrix where all its elements are zeros, except for the elements in the main diagonal, which are ones: $I_{i,k} = 1.0$ if $i = k$, else $I_{i,k} = 0$. Typical operations with matrices include the following:

- addition : $\underline{\underline{c}} = \underline{\underline{a}} + \underline{\underline{b}}$ $\quad c_{i,k} = a_{i,k} + b_{i,k}$ (2.23)
- subtraction : $\underline{\underline{d}} = \underline{\underline{a}} - \underline{\underline{b}}$ $\quad d_{i,k} = a_{i,k} - b_{i,k}$ (2.24)
- scalar multiplication : $\underline{\underline{e}} = \alpha \cdot \underline{\underline{a}}$ $\quad e_{i,k} = \alpha \cdot a_{i,k}$ (2.25)
- matrix multiplication : $\underline{\underline{f}} = \underline{\underline{a}} \cdot \underline{\underline{b}}$ $\quad f_{i,k} = \sum_p a_{i,p} \cdot b_{p,k}$ (2.26)

Note that *matrix multiplication is a summation of binary products*; this type of expression is frequently encountered in signal processing (Chapter 4).

The *transpose* $\underline{\underline{a}}^T$ of the matrix $\underline{\underline{a}}$ is obtained by switching columns and rows:

$$a_{k,i} \text{ in } \underline{\underline{a}} \text{ is equal to } a_{i,k} \text{ in } a^T \tag{2.27}$$

A square matrix $\underline{\underline{a}}$ is *symmetric* if it is identical to its transpose ($\underline{\underline{a}}^T \equiv \underline{\underline{a}}$ or $a_{i,k} = a_{k,i}$). The matrices $(\underline{\underline{a}}^T \cdot \underline{\underline{a}})$ and $(\underline{\underline{a}} \cdot \underline{\underline{a}}^T)$ are square and symmetric for any matrix $\underline{\underline{a}}$.

The *Hermitian adjoint* $\underline{\underline{a}}^H$ of a matrix is the transpose of the complex conjugates of the individual elements. For example, if an element in $\underline{\underline{a}}$ is $a_{i,k} = b + j \cdot c$, the corresponding element in the Hermitian adjoint is $a_{k,i} = b - j \cdot c$. A square matrix is *Hermitian* if it is identical to its Hermitian adjoint; the real symmetric matrix is a special case.

The matrix $\underline{\underline{a}}^{-1}$ is the *inverse* of the square matrix $\underline{\underline{a}}$ if and only if

$$\underline{\underline{a}}^{-1} \cdot \underline{\underline{a}} = \underline{\underline{a}} \cdot \underline{\underline{a}}^{-1} = \underline{\underline{I}} \tag{2.28}$$

A matrix is said to be *orthogonal* if $\underline{\underline{a}}^{T} \equiv \underline{\underline{a}}^{-1}$; then

$$\underline{\underline{a}}^{T} \cdot \underline{\underline{a}} = \underline{\underline{I}} \tag{2.29}$$

Finally, a matrix is called *unitary* if the Hermitian adjoint is equal to the inverse, $\underline{\underline{a}}^{H} \equiv \underline{\underline{a}}^{-1}$.

The *determinant* of the square matrix $\underline{\underline{a}}$ denoted as $|\underline{\underline{a}}|$ is the number whose computation can be defined in recursive form as

$$\left|\underline{\underline{a}}\right| = \sum_{k} (-1)^{i+k} \cdot a_{i,k} \cdot \left|\underline{\underline{\mathrm{minor}}}\right| \tag{2.30}$$

where the $\underline{\underline{\mathrm{minor}}}$ is the submatrix obtained by suppressing row i and column k. The determinant of a single element is the value of the element itself. If the determinant of the matrix is zero, the matrix is *singular* and noninvertible. Conversely, if $|\underline{\underline{a}}| \neq 0$ the matrix is invertible.

The following relations hold:

$$\left(\underline{\underline{a}} \cdot \underline{\underline{b}}\right)^{-1} = \underline{\underline{b}}^{-1} \cdot \underline{\underline{a}}^{-1} \tag{2.31}$$

$$\left(\underline{\underline{a}} \cdot \underline{\underline{b}}\right)^{T} = \underline{\underline{b}}^{T} \cdot \underline{\underline{a}}^{T} \tag{2.32}$$

$$\left(\underline{\underline{a}} \cdot \underline{\underline{b}}\right)^{H} = \underline{\underline{b}}^{H} \cdot \underline{\underline{a}}^{H} \tag{2.33}$$

$$\left(\underline{\underline{a}}^{-1}\right)^{T} = \left(\underline{\underline{a}}^{T}\right)^{-1} \tag{2.34}$$

$$\left|\underline{\underline{a}}\right| = \left|\underline{\underline{a}}^{T}\right| \tag{2.35}$$

$$\left|\underline{\underline{a}} \cdot \underline{\underline{b}}\right| = \left|\underline{\underline{a}}\right| \cdot \left|\underline{\underline{b}}\right| \tag{2.36}$$

$$\left|\underline{\underline{a}}^{-1}\right| = \frac{1}{\left|\underline{\underline{a}}\right|} \tag{2.37}$$

A row in a matrix is linearly independent if it cannot be computed as a linear combination of the other rows. The same applies to the columns in the matrix. The *rank* of a matrix $r[\underline{\underline{a}}]$ is the number of linearly independent rows or columns in $\underline{\underline{a}}$. If $r[\underline{\underline{a}}] = S$, then there is a square submatrix size $S \times S$ whose determinant is nonzero.

2.2.2 Matrices as Transformations

So far, matrices have been described as isolated rectangular arrays of real or complex numbers. Consider now the matrix $\underline{\underline{a}}$ as an operator that transforms an "input" vector $\underline{x}$ into an "output" vector $\underline{y}$:

$$\underline{y} = \underline{\underline{a}} \cdot \underline{x} \tag{2.38}$$

Computationally, the transformation $\underline{y} = \underline{\underline{a}} \cdot \underline{x}$ is a linear combination of the columns of $\underline{\underline{a}}$ according to the entries in $\underline{x}$.

If matrix $\underline{\underline{a}}$ N×N is noninvertible, there will be vectors $\underline{x}$ that are normal to the columns of $\underline{\underline{a}}$ and map to $\underline{y} = \underline{\underline{a}} \cdot \underline{x} = \underline{0}$; those vectors are the subspace of $\underline{x}$ called the *null space* (Figure 2.3). On the other hand, not all the space of $\underline{y}$ is reachable from $\underline{x}$; the *range* of $\underline{\underline{a}}$ is the subset of the space of $\underline{y}$ reachable by the transformation (Equation 2.38). The fact that $\underline{\underline{a}}$ is noninvertible indicates that some of the columns in $\underline{\underline{a}}$ are linearly dependent, and they will not contribute to the dimensionality of the range. Hence, the dimension of the range is the rank $r[\underline{\underline{a}}]$:

$$\dim(\text{range}) = r[\underline{\underline{a}}] \tag{2.39}$$

It follows from these definitions that the sum of the dimensions of the null space and the range is N:

$$\dim(\text{null space}) + \dim(\text{range}) = N \tag{2.40}$$

If $\underline{\underline{a}}$ is invertible, $\underline{y} = \underline{\underline{a}} \cdot \underline{x} = \underline{0}$ only if $\underline{x} = \underline{0}$, and the dimension of the null space is zero. For a simple visualization of these concepts, consider the transformation matrix $\underline{\underline{a}} = [(1,0,0),(0,1,0),(0,0,0)]$, with rank $r[\underline{\underline{a}}] = 2$. All vectors $\underline{x} = (0,0,x_3)$ map to $\underline{y} = \underline{\underline{a}} \cdot \underline{x} = (0,0,0)$; therefore, they are in the null space of the transformation. On the other hand, only vectors $\underline{y} = (y_1, y_2, 0)$ are reachable by the transformation; this is the range.

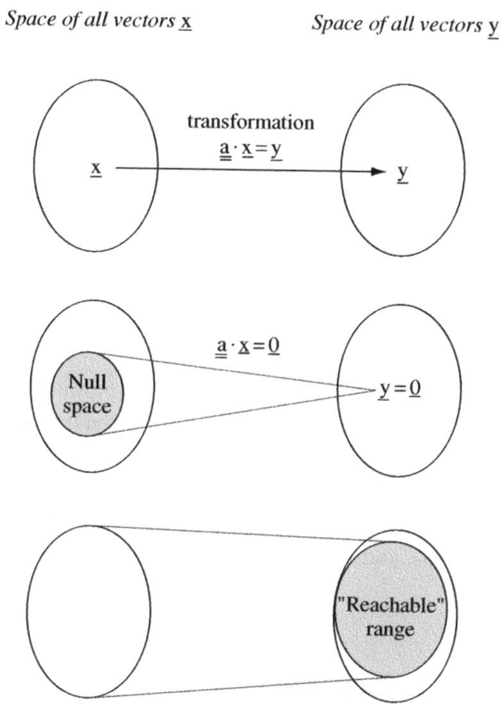

Figure 2.3 Definition of null space and range of a transformation $\underline{\underline{a}}$ [N × N]. The dimension of the range is equal to the rank of $\underline{\underline{a}}$. The dimension of the range plus the dimension of the null space is equal to N

The transformation matrix $\underline{\underline{a}}$ is *positive definite* if

$$\underline{x}^T \cdot \underline{\underline{a}} \cdot \underline{x} > 0 \tag{2.41}$$

for all nonzero vectors $\underline{x}$. If $\underline{x}^T \cdot \underline{\underline{a}} \cdot \underline{x} \geq 0$ then $\underline{\underline{a}}$ is positive semidefinite. Typically, the matrix $\underline{\underline{a}}$ is positive definite when the elements along the main diagonal of $\underline{\underline{a}}$ are positive and when they are also the largest elements in the matrix.

2.2.3 Eigenvalues and Eigenvectors

If $\underline{\underline{a}}$ is a square matrix, and $\underline{y}$ is obtained either as matrix multiplication $\underline{\underline{a}} \cdot \underline{x}$ or as scalar multiplication $\lambda\underline{x}$,

$$\underline{y} = \underline{\underline{a}} \cdot \underline{x} = \lambda \cdot \underline{x} \tag{2.42}$$

then $\underline{x}$ is an *eigenvector* of $\underline{\underline{a}}$ and λ is its corresponding *eigenvalue*. The eigenvalues of $\underline{\underline{a}}$ are obtained by solving the polynomial

$$\left|\underline{\underline{a}} - \lambda \cdot \underline{\underline{I}}\right| = 0 \tag{2.43}$$

where $\underline{\underline{I}}$ is the identity matrix. For each eigenvalue λ_p, the corresponding eigenvector $\underline{x}_p$ is computed by replacing λ_p in Equation 2.42,

$$\left(\underline{\underline{a}} - \lambda_p \cdot \underline{\underline{I}}\right) \cdot \underline{x}_p = \underline{0} \tag{2.44}$$

where $\underline{0}$ is an array of zeros. The eigenvectors corresponding to distinct eigenvalues of a Hermitian or symmetric matrix are orthogonal vectors; that is, the dot product is equal to zero. The eigenvalues of a Hermitian or symmetric matrix are real. The eigenvalues of a symmetric, positive-definite matrix are real and positive $\lambda_i > 0$, and the matrix is invertible. Last, for any given matrix $\underline{\underline{a}}$ $[M \times N]$, the eigenvalues of $(\underline{\underline{a}}^T \cdot \underline{\underline{a}})$ and $(\underline{\underline{a}} \cdot \underline{\underline{a}}^T)$ are nonnegative and their nonzero values are equal.

2.2.4 Matrix Decomposition

The solution of systems of equations, including matrix inversion, can be more effectively implemented by decomposing the matrix into factors.

Eigen Decomposition

A invertible square matrix $\underline{\underline{a}}$ with distinct eigenvalues can be expressed as the multiplication of three matrices

$$\underline{\underline{a}} = \underline{\underline{X}} \cdot \underline{\underline{\Lambda}} \cdot \underline{\underline{X}}^{-1} \tag{2.45}$$

where the columns of $\underline{\underline{X}}$ are the eigenvectors of $\underline{\underline{a}}$, and $\underline{\underline{\Lambda}}$ is a diagonal matrix that contains the eigenvalues of $\underline{\underline{a}}$. The order of the eigenvectors in matrix $\underline{\underline{X}}$ must be the same as the order of the eigenvalues in matrix $\underline{\underline{\Lambda}}$. The inverse of $\underline{\underline{a}}$ is (Equation 2.31)

$$\underline{\underline{a}}^{-1} = \left(\underline{\underline{X}} \cdot \underline{\underline{\Lambda}} \cdot \underline{\underline{X}}^{-1}\right)^{-1} = \left(\underline{\underline{\Lambda}} \cdot \underline{\underline{X}}^{-1}\right)^{-1} \underline{\underline{X}}^{-1} = \underline{\underline{X}} \cdot \underline{\underline{\Lambda}}^{-1} \cdot \underline{\underline{X}}^{-1} \tag{2.46}$$

The elements in the diagonal matrix $\underline{\underline{\Lambda}}^{-1}$ are the inverse of the eigenvalues $1/\lambda_i$. If $\underline{\underline{a}}$ is symmetric, then $\underline{\underline{X}}^{-1} = \underline{\underline{X}}^T$.

Singular Value Decomposition (SVD)

Any real matrix $\underline{\underline{a}}$ [M×N] with $M \geq N$, and rank $r \leq N$, can be expressed as

$$\underline{\underline{a}} = \underline{\underline{U}} \cdot \underline{\underline{\Lambda}} \cdot \underline{\underline{V}}^T \tag{2.47}$$

where

$\underline{\underline{U}}$ [M × M]	Orthogonal matrix Its columns are eigenvectors of $\underline{\underline{a}} \cdot \underline{\underline{a}}^T$ (in order as in $\underline{\underline{\Lambda}}$) Vectors $u_1 \ldots u_r$ span the range of $\underline{\underline{a}}$
$\underline{\underline{\Lambda}}$ [M × N]	Diagonal matrix $\Lambda_{i,k} = 0$ for $i \neq k$ Values $\Lambda_{i,i} = \lambda_i$ are the singular values in descending order λ_i are the nonnegative square root of eigenvalues of $\underline{\underline{a}} \cdot \underline{\underline{a}}^T$ or $\underline{\underline{a}}^T \cdot \underline{\underline{a}}$ Singular values $\lambda_1 > \ldots \lambda_r > 0$ and singular values $\lambda_{r+1} = \ldots = \lambda_N = 0$
$\underline{\underline{V}}$ [N × N]	Orthogonal matrix Its columns are eigenvectors of $\underline{\underline{a}}^T \cdot \underline{\underline{a}}$ (in same order as λ in $\underline{\underline{\Lambda}}$) The null space of $\underline{\underline{a}}$ is spanned by vectors $v_{r+1} \ldots v_N$

For $\underline{\underline{a}}$ real, the resulting three matrices are also real. The SVD is generalized to complex matrices using the Hermitian instead of the transpose. The method is equally applicable when the size of the matrix is $M < N$, with proper changes in indexes.

Other Decompositions

Two efficient algorithms are used to solve systems of equations that involve square matrices $\underline{\underline{a}}$ [N × N]. The LU decomposition converts $\underline{\underline{a}}$ into the multiplication of a lower triangular matrix $\underline{\underline{L}}$ ($L_{i,j} = 0$ if $i < j$) and an upper triangular matrix $\underline{\underline{U}}$ ($U_{i,j} = 0$ if $i > j$), such that $\underline{\underline{a}} = \underline{\underline{L}} \cdot \underline{\underline{U}}$. Furthermore, if the matrix $\underline{\underline{a}}$ is symmetric and positive definite, the Cholesky decomposition results in $\underline{\underline{a}} = \underline{\underline{U}}^T \cdot \underline{\underline{U}}$, where $\underline{\underline{U}}$ is upper triangular.

2.3 DERIVATIVES – CONSTRAINED OPTIMIZATION

A linear function of multiple variables $f = a_1 \cdot x_1 + a_2 \cdot x_2 + \ldots$ can be expressed in matrix form as

$$f = a_1 \cdot x_1 + a_2 \cdot x_2 + \ldots = \begin{bmatrix} a_1 & a_2 & \ldots \end{bmatrix} \cdot \begin{bmatrix} x_1 \\ x_2 \\ \ldots \end{bmatrix} = \underline{a}^T \cdot \underline{x} \qquad (2.48)$$

(Note that for a family of functions $f_1 \ldots f_M$, the array $\underline{a}$ becomes matrix $\underline{\underline{a}}$.) Likewise, the partial derivatives $\partial f / \partial x_i$ are organized into an array

$$\frac{\partial f}{\partial \underline{x}} = \begin{bmatrix} \dfrac{\partial f}{\partial x_1} \\ \dfrac{\partial f}{\partial x_2} \\ \ldots \end{bmatrix} = \begin{bmatrix} a_1 \\ a_2 \\ \ldots \end{bmatrix} = \underline{a} \qquad (2.49)$$

The following set of equations facilitates the derivation of explicit solutions in optimization problems that are captured in matrix form (Chapter 9):

$$f = \underline{a}^T \cdot \underline{x} \qquad \frac{\partial f}{\partial \underline{x}} = \underline{a} \qquad \text{(as shown above)} \qquad (2.50)$$

$$f = \underline{x}^T \cdot \underline{a} \qquad \frac{\partial f}{\partial \underline{x}} = \underline{a} \qquad (2.51)$$

$$f = \underline{x}^T \cdot \underline{x} \qquad \frac{\partial f}{\partial \underline{x}} = 2 \cdot \underline{x} \qquad (2.52)$$

$$f = \underline{x}^T \cdot \underline{\underline{a}} \cdot \underline{x} \qquad \frac{\partial f}{\partial \underline{x}} = 2 \cdot \underline{\underline{a}} \cdot \underline{x} \qquad \text{for } \underline{\underline{a}} \text{ symmetric} \qquad (2.53)$$

In each case, the function is written in explicit form, partial derivatives are computed, and the result is once again expressed in matrix form.

Given M-measurements y_i that depend on N-parameters x_k, the partial derivative $\partial y_i / \partial x_k$ indicates the sensitivity of the i-th measurement to the k-th parameter. The *Jacobian matrix* is the arrangement of the $M \times N$ partial derivatives in matrix form. The Jacobian matrix is useful to identify extrema and to guide optimization algorithms.

The extremum of a function is tested for minimum or maximum with the *Hessian matrix* $\underline{\underline{Hes}}$ formed with the second derivatives of $f(\underline{x})$:

$$Hes_{i,k} = \frac{\partial^2 f}{\partial x_i \cdot \partial x_k} \qquad (2.54)$$

The function has a minimum if $\underline{\underline{Hes}}$ is positive definite.

The extremum of a function of N-variables $f(x_1, \ldots, x_N) = 0$ subject to V-constraints $\phi\ (x_1, \ldots, x_N) = 0$ can be obtained using Lagrange multipliers λ. First, a new objective function Γ is formed,

$$\Gamma(x_1, \ldots, x_N) = f(x_1, \ldots, x_N) + \lambda_1 \cdot \phi_1(x_1, \ldots, x_N) + \cdots + \lambda_V \cdot \phi_V(x_1, \ldots, x_N) \tag{2.55}$$

that involves $N + V$ unknowns $(x_1, \ldots, x_N, \lambda_1, \ldots \lambda_V)$. These unknowns are found by solving the following system of $N + V$ simultaneous equations:

$$\text{N-equations} \quad \frac{\partial \Gamma}{\partial x_i} = 0 = \frac{\partial f}{\partial x_i} + \lambda_1 \cdot \frac{\partial \phi_1}{\partial x_i} + \cdots + \lambda_v \cdot \frac{\partial \phi_v}{\partial x_i}$$

$$\text{V-equations} \quad 0 = \phi_1(x_1, \cdots, x_N) \tag{2.56}$$

2.4 SUMMARY

The analysis of signals and systems makes extensive use of sinusoidal functions. The mathematical manipulation of sinusoids is effectively implemented with complex numbers and exponential functions.

The representation of discrete signals and transformations involves arrays and matrices. The inversion of a transformation implies matrix inversion. A symmetric, positive-definite matrix is invertible.

FURTHER READING

Golub, G. H., and Van Loan, C. F. (1989). Matrix Computations. Johns Hopkins University Press, Baltimore. 642 pages.

Goult, R. J., Hoskins, R. F., Milner, J. A., and Pratt, M. J. (1974). Computational Methods in Linear Algebra. John Wiley & Sons, New York. 204 pages.

Horn, R. A., and Johnson, C. R. (1985). Matrix Analysis. Cambridge University Press, London. 561 pages.

Press, W. H., Teukolsky, S. A., Betterling, W. T., and Flannery, B. P. (1992). Numerical Recipes in FORTRAN. The Art of Scientific Computing. Cambridge University Press, New York. 963 pages.

Strang, G. (1980). Linear Algebra and Its Applications. Academic Press, New York. 414 pages.

Trefethen, L. N. and Bau, III, D. (1997). Numerical Linear Algebra. Society for Industrial and Applied Mathematics, Philadelphia. 361 pages.

SOLVED PROBLEMS

P2.1 Given two complex numbers: $A = 3 - 4j$ and $B = 3 + 3j$, compute:

(a) Magnitude and phase of A:

$$|A| = \sqrt{[\mathrm{Re}\,(A)]^2 + [\mathrm{Im}\,(A)]^2} = \sqrt{3^2 + (-4)^2} = \sqrt{25} = 5$$

$$\varphi = \mathrm{a\,tan}\left[\frac{\mathrm{Im}\,(A)}{\mathrm{Re}\,(A)}\right] = \mathrm{a\,tan}\left[\frac{-4}{3}\right] = -0.928\,\mathrm{rad} = -53.1^\circ$$

(b) Magnitude and phase of $C = A + B$:

$$C = A + B = (3 - 4j) + (3 + 3j) = 6 - 1j$$

$$|C| = \sqrt{[\mathrm{Re}\,(C)]^2 + [\mathrm{Re}\,(C)]^2} + \sqrt{6^2 + (-1)^2} = \sqrt{37} = 6.08$$

$$\varphi = \mathrm{a\,tan}\left[\frac{\mathrm{Im}\,(C)}{\mathrm{Re}\,(C)}\right] = \mathrm{a\,tan}\left[\frac{-1}{6}\right] = -0.165\,\mathrm{rad} = -9.46^\circ$$

(c) $D = A \cdot B$ and $E = A/B$:

$$\begin{aligned} D = A \cdot B &= (3 - 4j) \cdot (3 + 3j) \\ &= 3 \cdot 3 + 3 \cdot 3j - 4j \cdot 3 - 4j \cdot 3j = 9 + 9j - 12j + 12 = 21 - 3j \end{aligned}$$

$$\begin{aligned} E = \frac{A}{B} &= \frac{3 - 4j}{3 + 3j} = \frac{3 - 4j}{3 + 3j} \cdot \frac{3 - 3j}{3 - 3j} \\ &= \frac{9 - 9j - 12j - 12}{9 + 9} = \frac{-3 - 21j}{18} = -\frac{1}{6} - \frac{7}{6}j \end{aligned}$$

(d) $D = A \cdot B$ and $E = A/B$ using the exponentials:

$$A = 3 - 4j = |3 - 4j|\,e^{1j \cdot \mathrm{a\,tan}\left(\frac{-4}{3}\right)} = 5e^{-0.928j}$$

$$B = 3 + 3j = |3 + 3j|\,e^{1j \cdot \mathrm{a\,tan}\left(\frac{3}{3}\right)} = 4.24e^{0.785j}$$

$$D = A \cdot B = \left(5e^{-0.928j}\right) \cdot \left(4.24e^{0.785j}\right) = 21.2e^{-0.142j}$$

$$E = \frac{A}{B} = \frac{5e^{-0.928j}}{4.24e^{0.785j}} = 1.18e^{-1.713j}$$

P2.2 Given the square matrix $\underline{\underline{a}}$, calculate $\underline{\underline{b}} = \underline{\underline{a}} \cdot \underline{\underline{a}}^T$, the determinant of $\underline{\underline{a}}$, the inverse matrix $\underline{\underline{a}}^{-1}$, $\underline{\underline{a}} \cdot \underline{\underline{a}}^{-1}$, and the determinant of $\underline{\underline{a}}^{-1}$.

$$\underline{\underline{a}} = \begin{bmatrix} 2 & 3 \\ 1 & 1 \end{bmatrix}$$

$$\underline{\underline{b}} = \underline{\underline{a}} \cdot \underline{\underline{a}}^T = \begin{bmatrix} 2 & 3 \\ 1 & 1 \end{bmatrix} \cdot \begin{bmatrix} 2 & 1 \\ 3 & 1 \end{bmatrix} = \begin{bmatrix} 2\cdot 2+3\cdot 3 & 2\cdot 1+3\cdot 1 \\ 1\cdot 2+1\cdot 3 & 1\cdot 1+1\cdot 1 \end{bmatrix} = \begin{bmatrix} 13 & 5 \\ 5 & 2 \end{bmatrix}$$

$$\left|\underline{\underline{a}}\right| = \left|\begin{bmatrix} 2 & 3 \\ 1 & 1 \end{bmatrix}\right| = 2\cdot 1 - 1\cdot 3 = -1$$

Inverse: the matrix $\underline{\underline{a}}$ is reduced by rows and the same operation is performed on the identity matrix $\underline{\underline{I}}$:

$\underline{\underline{a}} = \begin{bmatrix} 2 & 3 \\ 1 & 1 \end{bmatrix}$	$\begin{bmatrix} 1 & 0 \\ 0 & 1 \end{bmatrix} = \underline{\underline{I}}$
$\begin{bmatrix} -1 & 0 \\ 1 & 1 \end{bmatrix}$	$\begin{bmatrix} 1 & -3 \\ 0 & 1 \end{bmatrix}$
$\begin{bmatrix} 1 & 0 \\ 1 & 1 \end{bmatrix}$	$\begin{bmatrix} -1 & 3 \\ 0 & 1 \end{bmatrix}$
$\begin{bmatrix} 1 & 0 \\ 0 & 1 \end{bmatrix}$	$\begin{bmatrix} -1 & 3 \\ 1 & -2 \end{bmatrix} = \underline{\underline{a}}^{-1}$

$$\underline{\underline{a}} \cdot \underline{\underline{a}}^{-1} = \begin{bmatrix} 2 & 3 \\ 1 & 1 \end{bmatrix} \cdot \begin{bmatrix} -1 & 3 \\ 1 & -2 \end{bmatrix}$$

$$= \begin{bmatrix} 2\cdot(-1)+3\cdot 1 & 2\cdot 3+3\cdot(-2) \\ 1\cdot(-1)+1\cdot 1 & 1\cdot 3-1\cdot(-1) \end{bmatrix} = \begin{bmatrix} 1 & 0 \\ 0 & 1 \end{bmatrix} = \underline{\underline{I}}$$

P2.3 Given matrix $\underline{\underline{b}}$, show that $1/|\underline{\underline{b}}|$ is equal to $|\underline{\underline{b}}^{-1}|$ (Equation 2.37).

$$\underline{\underline{b}} = \begin{bmatrix} 2 & 3 \\ 2 & 2 \end{bmatrix} \text{ by row reduction } \underline{\underline{b}}^{-1} = \begin{bmatrix} -1 & 1.5 \\ 1 & -1 \end{bmatrix}$$

$$\left|\underline{\underline{b}}\right| = \begin{vmatrix} 2 & 3 \\ 2 & 2 \end{vmatrix} = 2\cdot 2 - 2\cdot 3 = -2$$

$$\left|\underline{\underline{b}}^{-1}\right| = \begin{vmatrix} -1 & 1.5 \\ 1 & -1 \end{vmatrix} = (-1)\cdot(-1) - 1\cdot 1.5 = 1 - 1.5 = -0.5$$

Indeed, $\dfrac{1}{\left|\underline{\underline{b}}\right|} = \dfrac{1}{-2} = \left|\underline{\underline{b}}^{-1}\right|$

P2.4 Determine the eigenvalues and eigenvectors of matrix

$$\underline{\underline{a}} = \begin{bmatrix} 2 & 3 \\ 1 & 1 \end{bmatrix}$$

Evaluation of eigenvalues λ:

$$\left|\begin{bmatrix} 2-\lambda & 3 \\ 1 & 1-\lambda \end{bmatrix}\right| = 0$$

$$(2-\lambda)\cdot(1-\lambda) - 1\cdot 3 = 2 - 2\lambda - \lambda + \lambda^2 - 3 = \lambda^2 - 3\lambda - 1 = 0$$

Solving for the roots : $\lambda = \dfrac{-(-3) \pm \sqrt{(-3)^2 - 4\cdot 1\cdot(-1)}}{2\cdot 1} = \dfrac{3 \pm \sqrt{13}}{2}$

$$\lambda_1 = 3.303 \text{ and } \lambda_2 = -0.303$$

Eigenvectors associated with eigenvalue λ_1:

$$\left(\underline{\underline{a}} - \lambda_1 \cdot \underline{\underline{I}}\right)\cdot \underline{x}_1 = 0$$

$$\left(\begin{bmatrix} 2 & 3 \\ 1 & 1 \end{bmatrix} - \lambda_1 \cdot \begin{bmatrix} 1 & 0 \\ 0 & 1 \end{bmatrix}\right)\cdot \underline{x}_1 = 0$$

$$\begin{bmatrix} 2-\lambda_1 & 3 \\ 1 & 1-\lambda_1 \end{bmatrix}\cdot \underline{x}_1 = \begin{bmatrix} 2-\lambda_1 & 3 \\ 1 & 1-\lambda_1 \end{bmatrix}\cdot \begin{bmatrix} x_{1,1} \\ x_{2,1} \end{bmatrix} = 0$$

$$\begin{bmatrix} (2-3.303)\cdot x_{1,1} + 3\cdot x_{2,1} \\ x_{1,1} + (1-3.303)\cdot x_{2,1} \end{bmatrix} = \begin{bmatrix} -1.303\cdot x_{1,1} + 3\cdot x_{2,1} \\ x_{1,1} - 2.303\cdot x_{2,1} \end{bmatrix} = 0$$

Assuming that $x_{2,1} = 1$ the eigenvector is

$$\begin{bmatrix} x_{1,1} \\ x_{2,1} \end{bmatrix} = \begin{bmatrix} 2.303 \\ 1 \end{bmatrix}$$

Eigenvectors associated with eigenvalue λ_2:

$$\left(\underline{\underline{a}} - \lambda_2 \cdot \underline{\underline{I}}\right) \cdot \underline{x}_2 = 0$$

$$\left(\begin{bmatrix} 2 & 3 \\ 1 & 1 \end{bmatrix} - \lambda_2 \cdot \begin{bmatrix} 1 & 0 \\ 0 & 1 \end{bmatrix}\right) \cdot \underline{x}_2 = 0$$

$$\begin{bmatrix} 2-\lambda_2 & 3 \\ 1 & 1-\lambda_2 \end{bmatrix} \cdot \underline{x}_1 = \begin{bmatrix} 2-\lambda_2 & 3 \\ 1 & 1-\lambda_2 \end{bmatrix} \cdot \begin{bmatrix} x_{1,2} \\ x_{2,2} \end{bmatrix} = 0$$

$$\begin{bmatrix} (2+0.303) \cdot x_{1,2} + 3 \cdot x_{2,2} \\ x_{1,2} + (1+0.303) \cdot x_{2,2} \end{bmatrix} = \begin{bmatrix} 2.303 \cdot x_{1,2} + 3 \cdot x_{2,2} \\ x_{1,2} + 1.303 \cdot x_{2,2} \end{bmatrix} = 0$$

Assuming that $x_{2,2} = 1$ the eigenvector is

$$\begin{bmatrix} x_{1,2} \\ x_{2,2} \end{bmatrix} = \begin{bmatrix} -1.303 \\ 1 \end{bmatrix}$$

ADDITIONAL PROBLEMS

P2.5 Write the following matrix operations in the subindex format:

$$\underline{\underline{c}} = \underline{\underline{a}}^T \cdot \underline{\underline{b}}$$

$$\underline{\underline{d}} = k \cdot \left(\underline{\underline{a}} \cdot \underline{\underline{b}}\right) + \underline{\underline{c}}$$

P2.6 Demonstrate:

(a) Operations with complex numbers satisfy commutative, associative, and distributive rules.

(b) Equality $\left(\underline{\underline{a}} \cdot \underline{\underline{b}}\right)^{-1} = \underline{\underline{b}}^{-1} \cdot \underline{\underline{a}}^{-1}$

(c) Equality $\left(\underline{\underline{a}} \cdot \underline{\underline{b}}\right)^{T} = \underline{\underline{b}}^{T} \cdot \underline{\underline{a}}^{T}$

(d) If $f = \underline{x}^T \cdot \underline{x}$ then $\dfrac{\partial f}{\partial \underline{x}} = 2 \cdot \underline{x}$

(e) If $f = \underline{x}^T \cdot \underline{\underline{a}} \cdot \underline{x}$ then $\dfrac{\partial f}{\partial \underline{x}} = 2 \cdot \underline{\underline{a}} \cdot \underline{x}$ for $\underline{\underline{a}}$ symmetric

P2.7 Compute:

(a) Given complex numbers $X = 1 + 3j$ and $H = 2 - 7j$, compute $Y = H \cdot X$ using complex numbers, polar notation and exponential notation. Verify that the three results are identical.

(b) The determinant, eigenvectors, and eigenvalues of matrices A and B:

$$A = \begin{bmatrix} 4 & 2 & 7 \\ 3 & 1 & 3 \\ 5 & 4 & 4 \end{bmatrix} \quad \text{and} \quad B = \begin{bmatrix} 6 & 2 & 1 \\ 3 & 2 & 7 \\ 1 & 4 & 5 \end{bmatrix}$$

(c) The determinant of $A \cdot B$ given

$$A = \begin{bmatrix} 4 & 2 & 7 \\ 3 & 1 & 3 \\ 5 & 4 & 4 \end{bmatrix} \quad \text{and} \quad B = \begin{bmatrix} 6 & 2 & 1 \\ 3 & 2 & 7 \\ 1 & 4 & 5 \end{bmatrix}$$

(d) The value of $\underline{x} = \left[\left(\underline{\underline{a}}^T \cdot \underline{\underline{a}} + \lambda \cdot \underline{\underline{R}}^T \cdot \underline{\underline{R}} \right)^{-1} \right] \cdot \underline{\underline{a}}^T \cdot \underline{y}$ given

$$\underline{\underline{a}} = \begin{bmatrix} 2 & 1 & 3 \\ 3 & 2 & 1 \\ 1 & 5 & 4 \end{bmatrix} \quad \underline{\underline{R}} = \begin{bmatrix} 1 & 0 & 0 \\ 0 & 1 & 0 \\ 0 & 0 & 1 \end{bmatrix} \quad \underline{y} = \begin{bmatrix} 4 \\ 1 \\ 7 \end{bmatrix} \quad \lambda = 10^{-3}$$

Note: This operation is similar to the regularized least square solution (RLSS) of inverse problems (Chapter 9).

P2.8 What is the determinant of a triangular matrix (lower or upper triangular)? What are its eigenvalues? Try 2×2, 3×3, and 4×4 matrices. Conclude.

P2.9 *Singular value decomposition and image compression.* Generate a 128×128 pixel image showing a block amplitude 1.0 at the center of the image amplitude 0.0. This is done by creating a 128×128 matrix $\underline{\underline{a}}$, assigning values $a_{i,k} = 0$ for the background and $a_{i,k} = 1.0$ wherever the block is. Determine the singular value decomposition of the image $\underline{\underline{a}}$ and regenerate the image using the largest 8, 16, 32, 64, and 128 singular values. Repeat the exercise adding random noise to the image $\underline{\underline{a}}$. Draw conclusions.

3

Signals and Systems

Signal processing and inverse problem solving are common tasks in engineering and science applications (Chapter 1). This chapter focuses on the essential characteristics of signals and systems, highlights important implications of analog-to-digital conversion, describes elemental signals that are used to analyze all other signals, and redefines the superposition principle in the context of linear time-invariant systems.

3.1 SIGNALS: TYPES AND CHARACTERISTICS

A signal is information encoded as the variation of a parameter with respect to one or more independent variables (Section 1.1). Time or spatial coordinates are the most frequently used independent variables. Consider, for example, the spatial variation of annual precipitation in a region, or the daily fluctuations of the Dow Jones index in one year.

The independent variable that represents either the temporal or spatial coordinate is herein called "time" and denoted by the letter t. Furthermore, the period of any event is denoted by the letter T, in spite of the fact that the event may take place in space with "spatial period" or wavelength λ. In the same spirit, the Greek letter ω is generically used to refer to angular frequency ($\omega = 2\pi/T$) in either time or space domains, even though the spatial frequency is the wavenumber $\kappa = 2\pi/\lambda$.

3.1.1 Continuous and Discrete Signals

A continuous signal is the *ceaseless* and *uninterrupted* observation of a parameter in time or space. A discrete signal, on the other hand, is the *intermittent*

Discrete Signals and Inverse Problems J. C. Santamarina and D. Fratta

observation of the parameter, that is, a sequence of values separated in time (or space). A mercury thermometer senses temperature continuously, yet the recording of temperature every five minutes produces a discrete signal that corresponds to the continuous variation of temperature in time. Likewise, the evaluation of brightness at different locations on a wall results in a matrix of values, leading to a digital image that corresponds to the true continuous image. Figure 3.1 shows a continuous signal and its discrete counterpart.

A continuous sinusoid exists for all values of the continuous independent variable t

$$x(t) = \sin(\omega \cdot t + \varphi) \qquad \textit{continous signal} \qquad (3.1)$$

Conversely, discrete signals are defined at discrete values t_i. The *sampling interval* Δt is the separation between two contiguous discrete times. Then, the i-th time is

$$t_i = i \cdot \Delta t \qquad \textit{discrete time} \qquad (3.2)$$

Each entry x_i is a discrete value of the parameter x being monitored. The index i indicates the order or location of x_i in the array of values. For example, the discrete signal obtained by sampling the continuous signal defined in Equation 3.1 becomes

$$x_i = \sin(\omega \cdot i \cdot \Delta t + \varphi) \qquad \textit{discrete signal} \qquad (3.3)$$

where the subindex denotes the sequence of discrete data points in the array. Consider the case of water flowing through an irrigation channel. The flow rate is sampled every 18 hours, that is $\Delta t = 18$ h. The recorded discrete signal is:

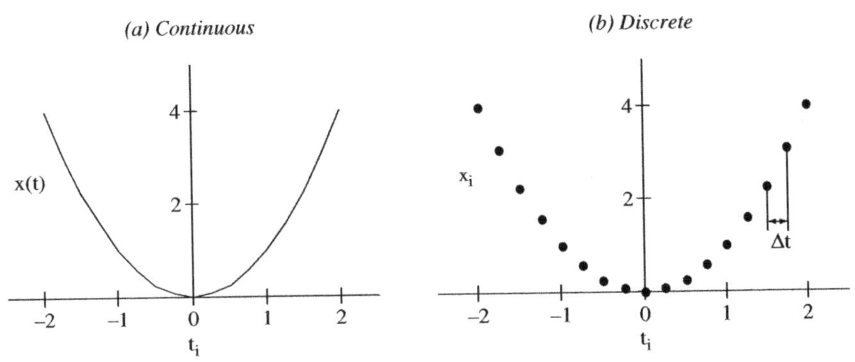

Figure 3.1 (a) A continuous signal; (b) a digital version of the signal obtained with a sampling interval Δt

Index i	0	1	2	3	4	5	6	7	8	9	10	11
Time $t_i = i \cdot \Delta t$ [h]	0	18	36	54	72	90	108	126	144	162	180	198
Signal x_i [m^3/min]	2.2	2.8	3.4	5.0	4.7	3.5	3.7	3.2	3.4	4.3	8.3	3.5

Because the sampling interval is 18 h, daily peaks and valleys in demand may go undetected. And if detected by coincidence, they will bias the interpretation of measurements.

Digital technology facilitates capturing, storing, and postprocessing signals in discrete form. Digital storage oscilloscopes store signals as arrays of individual voltage values that are equally spaced by a constant sampling interval Δt. Optical disks use a laser to "burn" digital information onto a flat substrate; the disk geometry permits fast access without having to wind long tapes.

3.1.2 One-dimensional (1D) and Multidimensional Signals

The dimension of a signal is the number of independent variables used to define it. When a stone falls on a quiet pond, the ripples define a three-dimensional (3D) signal where the surface displacement varies in the two dimensions of space and in time. Figure 3.2a shows the instantaneous position of the surface at a given time. This is a two-dimensional (2D) signal where displacement varies in space; it is stored as a 2D array or matrix. A slice of this instantaneous signal along the radial line A–A is the 1D signal shown in Figure 3.2b. In general, the time series produced by a single transducer is a 1D signal, e.g. accelerometers, strain gages, or photosensors.

3.1.3 Even and Odd Signals

The symmetry of signals with respect to the origin of the independent variable determines whether the signal is even or odd. An even signal satisfies (Figure 3.3a)

$$x_i = x_{-i} \qquad \textit{even signal} \tag{3.4}$$

whereas in an odd signal (Figure 3.3b)

$$x_i = -x_{-i} \qquad \textit{odd signal} \tag{3.5}$$

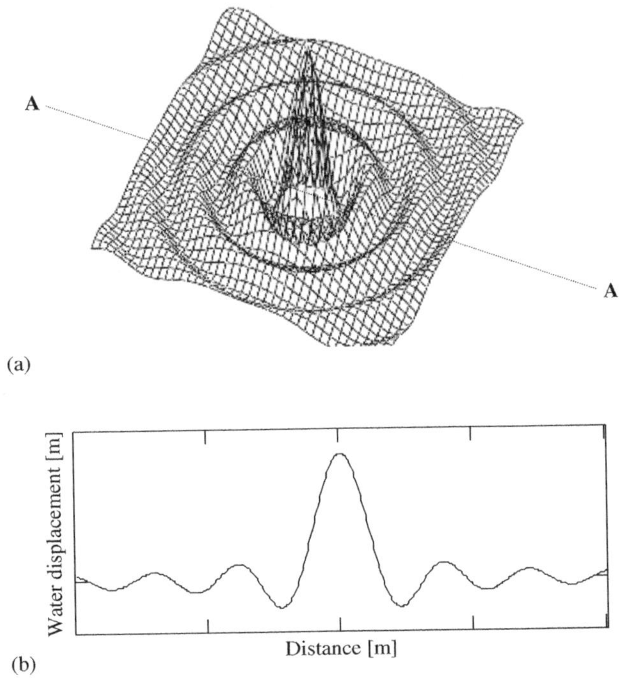

Figure 3.2 Ripples in a pond – 2D and 1D signals: (a) instantaneous surface displacement; (b) water surface displacement along the plane A–A

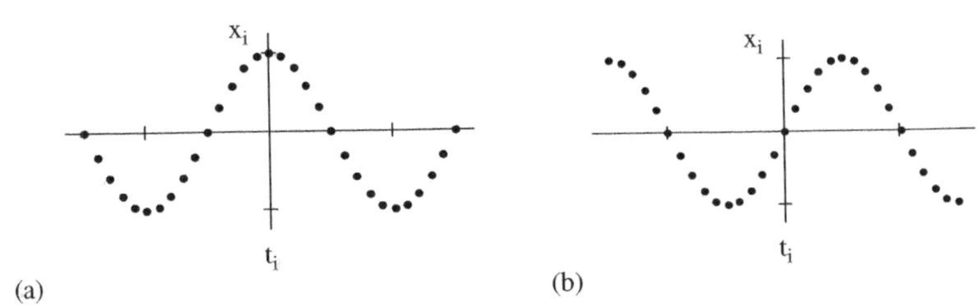

Figure 3.3 Signal symmetry: (a) even signal; (b) odd signal

3.1.4 Periodic and Aperiodic Signals (and Transformations)

A periodic signal is a repetitive sequence of values with a well-defined timescale or period $T = p \cdot \Delta t$, so that

$$x_i = x_{i+p} \tag{3.6}$$

This general definition of periodicity applies as well to periodicity in space, in which the characteristic scale would be the wavelength λ. However, as indicated earlier, "time" is the generic term used to name the independent variable. A periodic signal is shown in Figure 3.4a. An aperiodic signal is a one-of-a-kind variation of a parameter that does not repeat itself at least within the duration of the observation D (see Figure 3.4b).

It is often convenient to consider an aperiodic signal as a periodic signal that repeats itself with periodicity D. In other words, even though there are no observations outside the interval 0 to D, it is assumed that the same signal repeats before $t = 0$ and after $t = D$, with a periodicity $T = D$ (see Figure 3.4c).

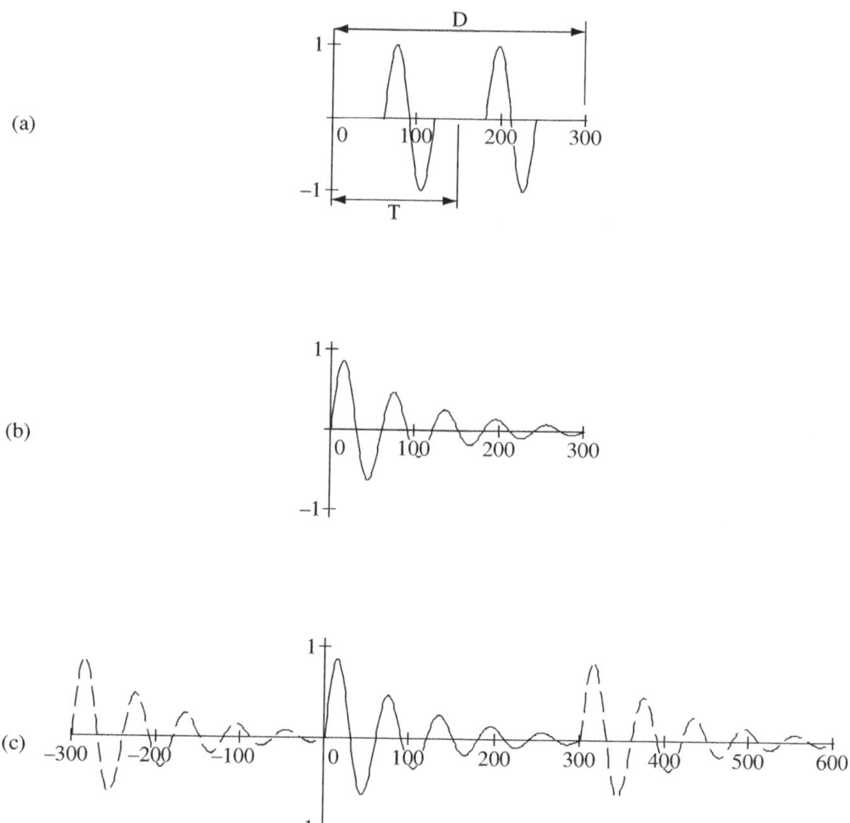

Figure 3.4 Signal periodicity – transformation: (a) periodic signal; (b) aperiodic signal; (c) periodicity assumption in common transformations

The presumption of periodicity for aperiodic signals is tacitly made in many analyses (Chapter 5). Its implications are important and often misleading.

All signals can be decomposed into a sum of aperiodic or periodic components, or into a sum of even and odd components (Chapters 4 and 5).

3.1.5 Stationary and Ergodic Signals

A physical event can be captured in multiple time segments that form an ensemble of signals (Figure 3.5). Statistical parameters such as mean and variance can be computed for each record. In addition, ensemble statistics can be determined for the set of values formed by the k-th element in each signal. Signals are stationary if the ensemble statistics at two different times are the same (for example at times t_1 and t_2 in Figure 3.5). The signal is ergodic if ensemble statistics are the same as the statistics for any record. Ergodic signals are stationary, but stationary signals need not be ergodic.

3.2 IMPLICATIONS OF DIGITIZATION – ALIASING

Sampling a signal at discrete time intervals may cause profound effects that must be either avoided or accounted for. These implications are discussed in this section using numerical examples.

Consider the periodic signal shown in Figure 3.6a. Figure 3.6b shows the signal digitized with a sampling interval $T_0/\Delta t = 25$ (integer). Figure 3.6c shows the same signal digitized with $T_0/\Delta t = 8.33$ (noninteger). In the latter case, the original periodicity of the signal is lost, because there is no value of p for which $x_i = x_{i+p}$ for all i (Equation 3.6).

Time shift δt and phase shift $\delta\varphi$ are related when a periodic continuous signal of period T is considered:

$$\frac{\delta t}{T} = \frac{\delta\varphi}{2\pi} \quad \text{then} \quad \delta\varphi = 2\pi\frac{\delta t}{T} \tag{3.7}$$

However, a time shift in the sampled signal results in another signal, still periodic, but with different entries in the array (see Figure 3.7 and notice the numerical values in the arrays). In discrete signals, the correspondence $\delta t \leftrightarrow \delta\varphi$ is only satisfied when $\delta t = k \cdot \Delta t$, where k is an integer and Δt is the sampling interval.

The most often discussed consequence of digitization is frequency aliasing by undersampling. (The semantic meaning refers to "alias" or pseudo.) The continuous

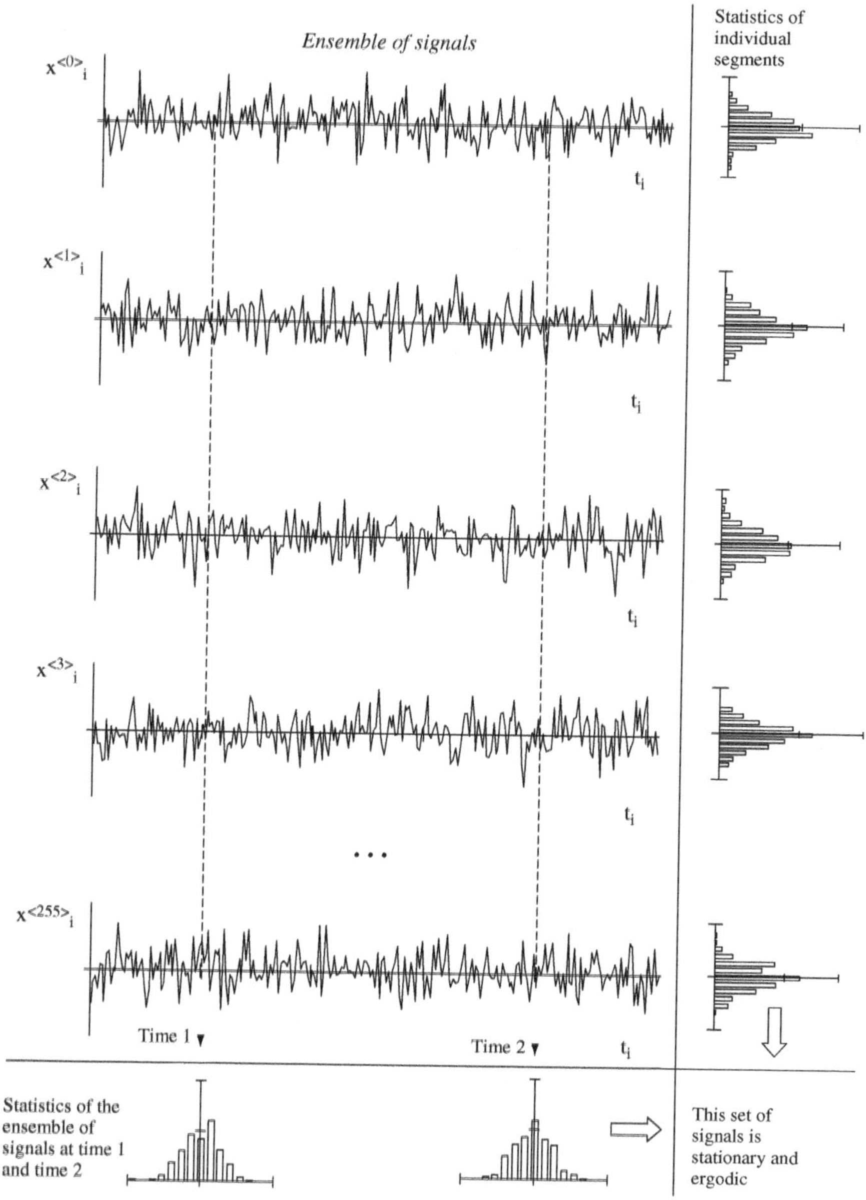

Figure 3.5 Ensemble of signals or "segments". A signal is stationary if the ensemble statistics at times t_1 and t_2 are equal. A signal is ergodic if the ensemble statistics at a given time are the same as the statistics of any segment

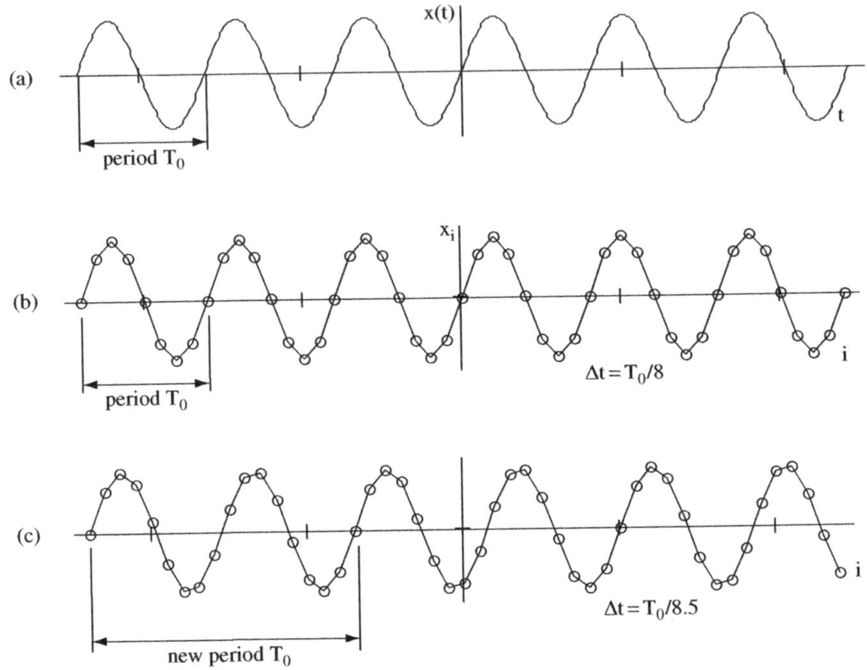

Figure 3.6 Implications of digitization – effect of periodicity: (a) continuous signal; (b) signal digitized with sampling interval $T_0/\Delta t =$ integer; (c) signal digitized with sampling interval $T_0/\Delta t \neq$ integer. Notice the position of points

sinusoid of period T_0 shown in Figure 3.8a is digitized in Figures 3.8b–d with different sampling intervals. When the sampling interval is greater or equal to half the period, $\Delta t \geq T_0/2$ (Figure 3.8d), the signal is undersampled and its periodicity appears "aliased" into a signal of lower frequency content.

Consider the following mental experiment. A white disk with a black radial line is turned clockwise at 600 rpm. A stroboscopic light is aimed at the disk and used to "sample" the position of the line:

- If the frequency of the light is 600 times per minute (it flashes at 10 Hz), the line appears still. For this reason, fluorescent lights must not be used when operating turning machinery.
- If it flashes slightly faster than 10 Hz, the next flash will illuminate the line slightly before the still position. Therefore, the disk will appear as if it were spinning counterclockwise, with some "negative frequency"; this accounts for what looks like wheels turning backwards in movies.

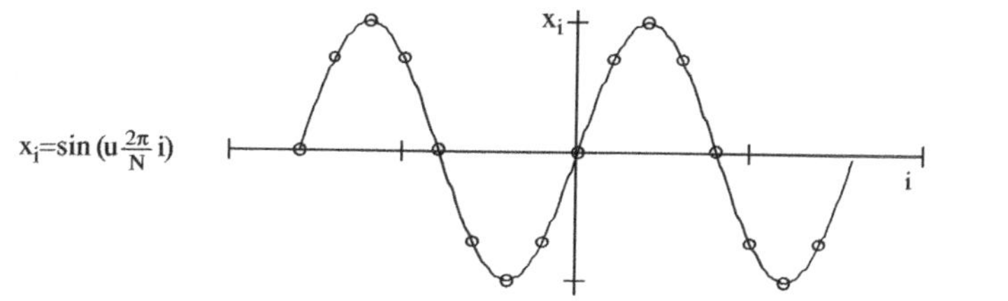

x^T =[0 0.71 1 0.71 0 −0.71 −1 −0.71 0 0.71 1 0.71 0 −0.71 −1 −0.71]

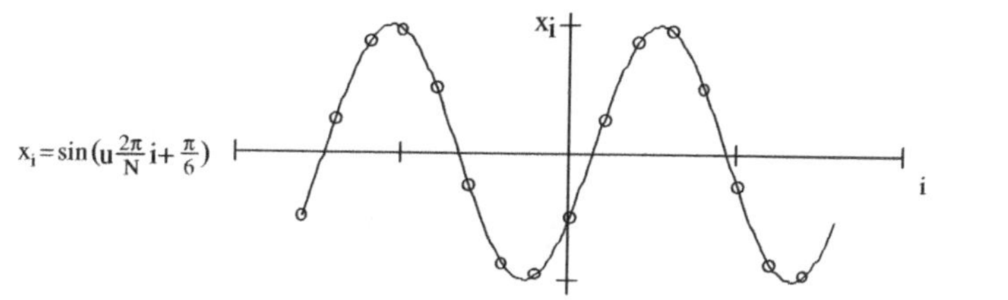

x^T =[−0.5 0.26 0.87 0.97 0.5 −0.26 −0.87 −0.97 −0.5 0.26 0.87 0.97 0.5 −0.26 −0.87 −0.97]

Figure 3.7 Implications of digitization – time shift. A time shift $\delta t = T/12$ (owing to $\delta\varphi = \pi/6$) in a discrete periodic signal sampled with $\Delta t = T/8$ leads to another discrete periodic signal. The periods of both signals are the same, but the elements in the arrays are different

- If the frequency is 20 Hz, the line will be seen twice in each cycle, on opposite sides, and both lines will appear still.

It can be shown that the frequency of the continuous periodic signal is properly identified from the discrete signal if the sampling frequency f_{samp} exceeds the *Nyquist frequency* f_{ny}

$$f_{samp} = \frac{1}{\Delta t} > f_{ny} = \frac{2}{T_0} \tag{3.8}$$

In practice, a minimum of ~10 points per cycle is recommended. The highest expected frequency should be considered when selecting the sampling rate. Analog antialiasing filters must be placed in series before digitization to remove frequency components higher than $1/(2 \cdot \Delta t)$. Engineered devices such as oscilloscopes and signal analyzers typically include antialiasing filters built-in; however, this may not be the case with simple A/D boards.

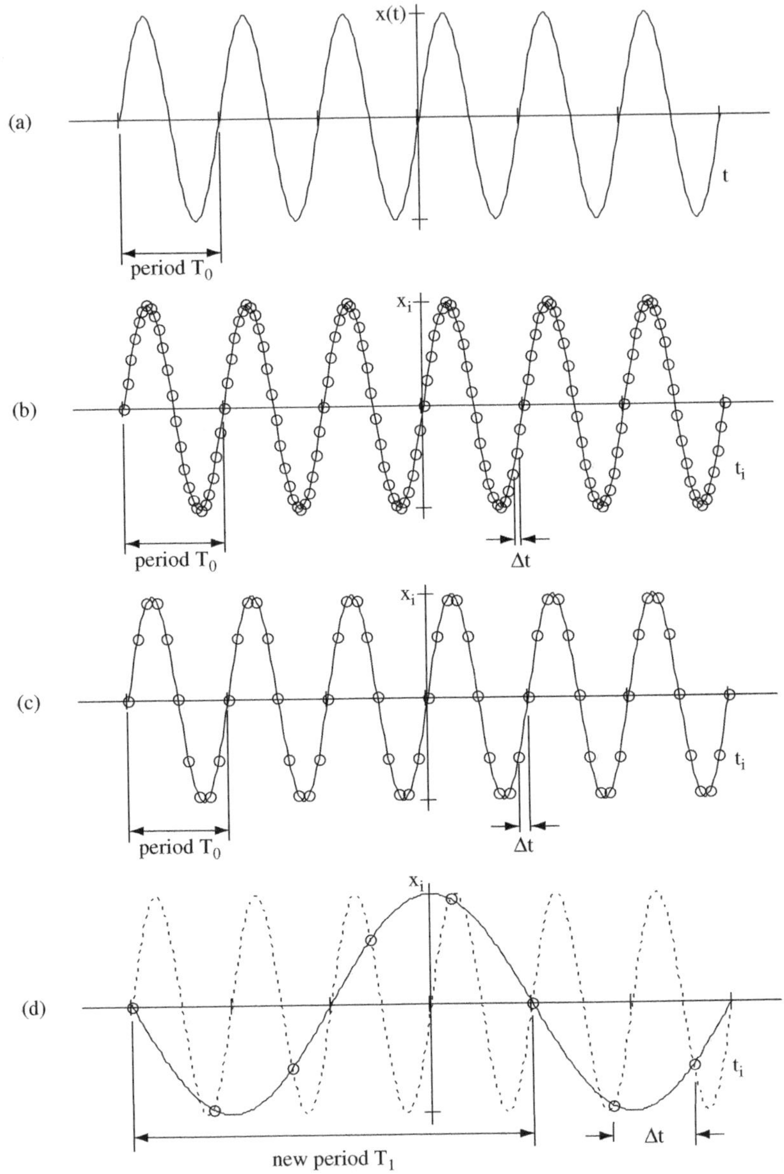

Figure 3.8 Sampling interval and aliasing – numerical example: (a) continuous signal; (b) sampling interval $\Delta t = T_0/25$; (c) sampling interval $\Delta t = T_0/10$; (d) sampling interval $\Delta t = T_0/1.25$. The original periodicity T_0 is lost as the sampling interval exceeds the Nyquist criterion and the signal is aliased into a lower frequency sinusoid

It is not necessary to associate the concept of Nyquist frequency with periodic signals. In more general terms, the sampling theorem states that *the sampling interval must be significantly smaller than the scale of interest.* Consider the determination of the stress–strain curve for a steel specimen. If the sampling interval in strain $\Delta\varepsilon$ is too large, the initial yielding of the material is undersampled, information is lost, and the wrong conclusion about the behavior of the material could be drawn. Fractal systems – such as surface roughness – lack a characteristic scale and the digital signal will continue gathering new information as the sampling interval decreases.

The undersampled signal in Figure 3.8d is not random: it reflects the information in the continuous sinusoid in Figure 3.8a, but folded onto a new predictable frequency that is determined by the frequency of the sinusoid and the sampling frequency. This observation suggests that undersampling is an effective approach to capture, process, and store signals as long as the continuous signal does not contain information in the folded frequency. Hence, narrow bandwidth signals can be undersampled; for example, a 100 MHz center frequency communications signal may have a 5 MHz bandwidth (see problems at the end of this Chapter).

3.3 ELEMENTAL SIGNALS AND OTHER IMPORTANT SIGNALS

Several "elemental" signals play an essential role in the analysis of signals and systems in engineering and science applications. Their importance results from their simplicity, information content, or physical interaction with the systems under study. The definition of these elemental signals in discrete time follows.

3.3.1 Impulse

The impulse signal δ_i is defined at the origin of time $i = 0$ and it is the sudden change in the value of the signal from $x_i = 0$ everywhere else to $x_0 = 1$ at $i = 0$:

$$\begin{aligned} \delta_i &= 1 \quad \text{if } i = 0 \\ \delta_i &= 0 \quad \text{if } i \neq 0 \end{aligned} \tag{3.9}$$

The graphical representation of an impulse in discrete form is shown in Figure 3.9a. The impulse can be shifted to any other location. However, in order to fulfill the mathematical expression that defines it, the shifted impulse must be denoted by the amount of shift. For example, an impulse at location $i = 10$ is defined as δ_{i-10}. When $i = 10$, the subindex becomes $10 - 10 = 0$ and $\delta_0 = 1$ in agreement with the definition in Equation 3.9.

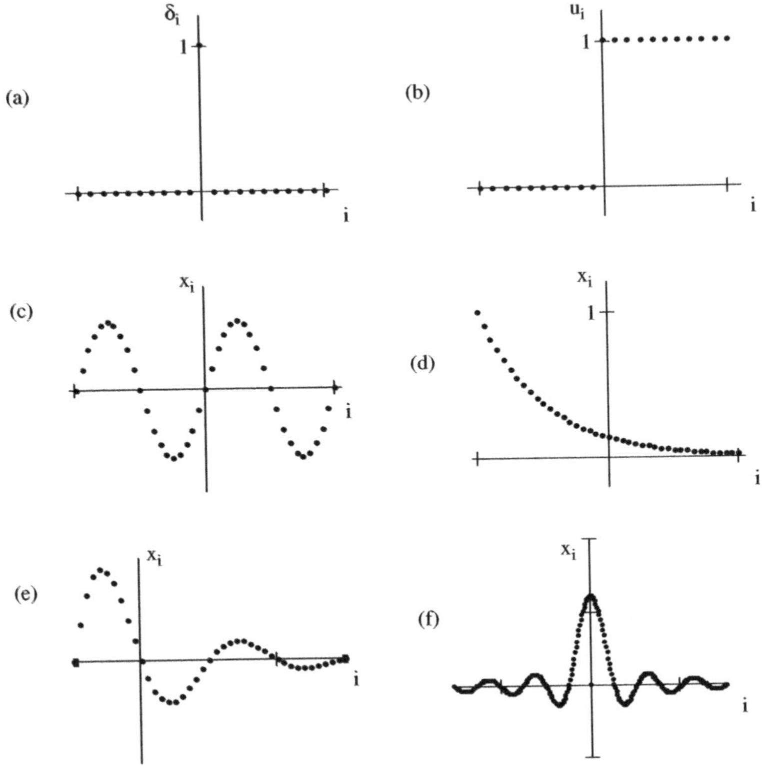

Figure 3.9 Elemental signals: (a) impulse; (b) step; (c) sinusoid; (d and e) exponential; (f) wavelet

3.3.2 Step

A step signal u_i is the sudden change in value of the signal from a constant value of 0 to a constant value of 1. It is defined at time zero; therefore,

$$\begin{aligned} u_i &= 0 \quad \text{if } i < 0 \\ u_i &= 1 \quad \text{if } i \geq 0 \end{aligned} \tag{3.10}$$

The step signal in discrete time is shown in Figure 3.9b. Note that the step can also be obtained by accumulating the impulse signal from left to right:

$$u_k = \sum_{i=-\infty}^{k} \delta_i \tag{3.11}$$

Conversely, the impulse is obtained by differentiating the step in time, $\delta_i = u_i - u_{i-1}$. The step signal can also be shifted in time, following the same guidelines described above for the shifting of the impulse signal.

3.3.3 Sinusoid

A sinusoidal signal is defined by the trigonometric functions sine and cosine, as indicated in Equation 3.3, $x_i = A \cdot \sin(\omega \cdot t_i + \varphi)$. For a signal with period $T = N \cdot \Delta t$, the frequency is

$$f = \frac{1}{T} = \frac{1}{N \cdot \Delta t} \quad \text{and} \quad \omega = 2\pi \cdot f = \frac{2\pi}{N \cdot \Delta t} \tag{3.12}$$

Then the argument of the sinusoid when samples are determined at discrete times $t_i = i \cdot \Delta t$ becomes

$$\omega \cdot t_i = \left(\frac{2\pi}{N \cdot \Delta t}\right) \cdot (i \cdot \Delta t) = \frac{2\pi}{N} i \tag{3.13}$$

Finally, the expression of a sinusoid in discrete time is

$$x_i = A \cdot \sin\left(\frac{2\pi}{N} i + \varphi\right) \tag{3.14}$$

where φ is the phase angle. Whereas the step and impulse signals are nonperiodic, sinusoids are inherently periodic signals. A discrete time sinusoid is shown in Figure 3.9c. This could be the response of an undamped harmonic oscillator.

3.3.4 Exponential

The exponential function is one of the most important functions in mathematics and science. It is described as:

$$x_i = A \cdot e^{b \cdot i \cdot \Delta t} \tag{3.15}$$

There are several important cases of exponential functions:

- If the parameter b is a real number, the resulting signal either increases $b > 0$, or decreases $b < 0$ with the independent variable (growth and decay processes).
- If the parameter b is imaginary, $b = j \cdot \omega = j \cdot 2\pi/(N \cdot \Delta t)$, the exponential signal represents a sinusoid (from Euler's identities – Chapter 2):

$$x_i = A \cdot e^{j \frac{2\pi}{N} i} = A \cdot \left[\cos\left(\frac{2\pi}{N} i\right) + j \cdot \sin\left(\frac{2\pi}{N} i\right)\right] \tag{3.16}$$

where "i" identifies the index or counter of the discrete signal, and "j" denotes the imaginary component of a complex number ($j^2 = -1$).

- And, if the parameter b is complex, $b = \alpha + j \cdot \omega$, the resulting signal is a sinusoid with either increasing or decreasing amplitude, depending on the sign of the real component α:

$$x_i = A \cdot e^{(\alpha + j \cdot \omega) \cdot i \cdot \Delta t} \tag{3.17}$$

For example, Equation 3.17 is used to represent the response of a damped single degree of freedom oscillator. In the most general case A is also complex and permits changing the phase of sinusoids. Exponential signals are sketched in Figures 3.9d and e.

3.3.5 Wavelets

Wavelets are signals with relatively short duration. The "sinc" signal is defined as

$$x_i = \frac{\sin\left(\frac{2\pi}{M} i\right)}{\frac{2\pi}{M} i} \tag{3.18}$$

where the frequency content is determined by the parameter M. Another example is the Morlet wavelet defined as

$$x_i = e^{j \cdot (v \cdot i)} e^{-4 \cdot \ln(2) \cdot \left(\frac{i}{M}\right)^2} \tag{3.19}$$

where the central frequency is $\omega = v/\Delta t$, the width of the wavelet is $M \cdot \Delta t$, and $v < \pi$. This wavelet is sketched in Figure 3.9f.

3.3.6 Random Noise

Random noise or white noise is *not* an "elemental signal" in the sense that it is not used to analyze or decompose other signals. Yet, it is a convenient signal in testing and simulation. Random noise is characterized by a sequence of values that are uncorrelated in any scale of the independent time variable:

$$x_i = \text{random}(a) \tag{3.20}$$

where a is the amplitude of the noise. There are different types of random noise. The amplitude distribution can be uniform or Gaussian, typically with zero mean. The energy distribution in frequency determines the "color" of noise: *white noise* carries equal energy in all frequency bins, while *pink noise* has equal energy in bins defined in "log-frequency". Pink noise is preferred for perception-related studies.

3.4 SIGNAL ANALYSIS WITH ELEMENTAL SIGNALS

Complex signals may be decomposed or *"analyzed"* into elemental signals. Conversely, the signal is *synthesized* by summing across an ensemble of scaled elemental signals.

3.4.1 Signal Analysis with Impulses

The most evident decomposition of a discrete signal is in terms of "scaled and shifted" impulses. For example, the step function u_i defined in Equation 3.10 is obtained as (see Figure 3.10)

$$u_i = \sum_{k=0}^{\infty} \delta_{i-k} \tag{3.21}$$

where the i-th value of the step u_i at discrete time t_i is obtained as a sum across the ensemble of shifted impulses δ_{i-k}, as sketched in Figure 3.10; this is a summation in "k". Note that there is a subtle yet important difference between this equation

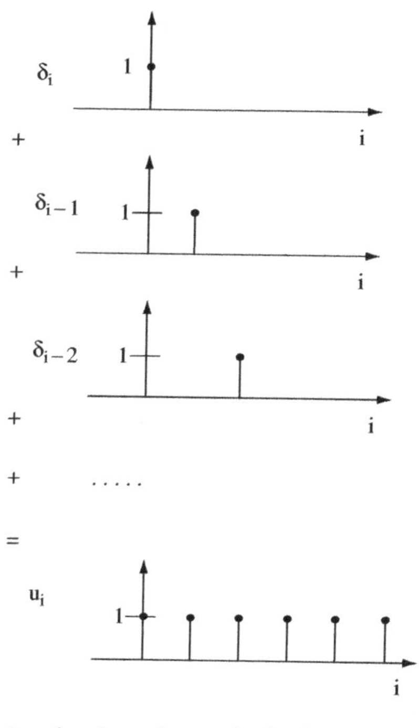

Figure 3.10 The step signal can be synthesized as a sum of shifted impulses

and Equation 3.11, where the step was obtained as a time accumulation along a single impulse; that is a summation in "i".

Any discrete signal $\underline{x}$ can be represented in terms of *scaled and shifted impulses*. The amplitude of $\underline{x}$ at position $i = k$ is x_k. Then x_k is used to scale the shifted impulse δ_{i-k}. For a discrete signal $\underline{x}$ with N entries, the summation involves N scaled and shifted impulses from $i = 0$ to $N-1$:

$$x_i = \sum_{k=0}^{N-1} x_k \cdot \delta_{i-k} \tag{3.22}$$

This is the *synthesis* equation. The summation of binary products implied in Equation 3.22 is equivalent to matrix multiplication. Each shifted impulse is an array of 0's, except for an entry of 1 at the time of the impulse. If these arrays are assembled into a matrix, each column represents a shifted impulse, and Equation 3.22 is written as

$$\begin{bmatrix} x_0 \\ x_1 \\ x_2 \\ \cdots \\ x_N \end{bmatrix} = \left[\begin{pmatrix} 1 \\ 0 \\ 0 \\ \cdots \\ 0 \end{pmatrix} \begin{pmatrix} 0 \\ 1 \\ 0 \\ \cdots \\ 0 \end{pmatrix} \begin{pmatrix} 0 \\ 0 \\ 1 \\ \cdots \\ 0 \end{pmatrix} \begin{pmatrix} \cdots \\ \cdots \\ \cdots \\ \cdots \\ \cdots \end{pmatrix} \begin{pmatrix} 0 \\ 0 \\ 0 \\ \cdots \\ 1 \end{pmatrix} \right] \cdot \begin{bmatrix} x_0 \\ x_1 \\ x_2 \\ \cdots \\ x_N \end{bmatrix} \tag{3.23}$$

Indeed, $\underline{x} = \underline{\underline{I}} \cdot \underline{x}$, where $\underline{\underline{I}}$ [N×N] is the identity matrix! While these expressions are self-evident, expressing a discrete signal in the form of Equations 3.22 or 3.23 facilitates understanding the convolution operation in Chapter 4.

The signal $\underline{x}$ could also be analyzed in terms of step functions placed at each discrete time t_i. The amplitude of each step is equal to the change in the amplitude of the signal at that discrete time interval $x_i - x_{i-1}$. In this case, the synthesis equation is a summation of scaled and shifted steps, similar to Equation 3.22.

3.4.2 Signal Analysis with Sinusoids

Consider the square wave shown in Figure 3.11. It is readily synthesized as the sum of scaled and shifted impulses, as shown previously. But, it can also be decomposed into sinusoids, whereby the signal $\underline{x}$ is expressed as a sum of scaled sines and cosines:

$$\begin{aligned} x_i &= \sum_{u=0}^{N-1} [a_u \cdot \cos(\omega_u \cdot t_i) + b_u \cdot \sin(\omega_u \cdot t_i)] \\ &= \sum_{u=0}^{N-1} \left[a_u \cdot \cos\left(u \frac{2\pi}{N} i\right) + b_u \cdot \sin\left(u \frac{2\pi}{N} i\right) \right] \end{aligned} \tag{3.24}$$

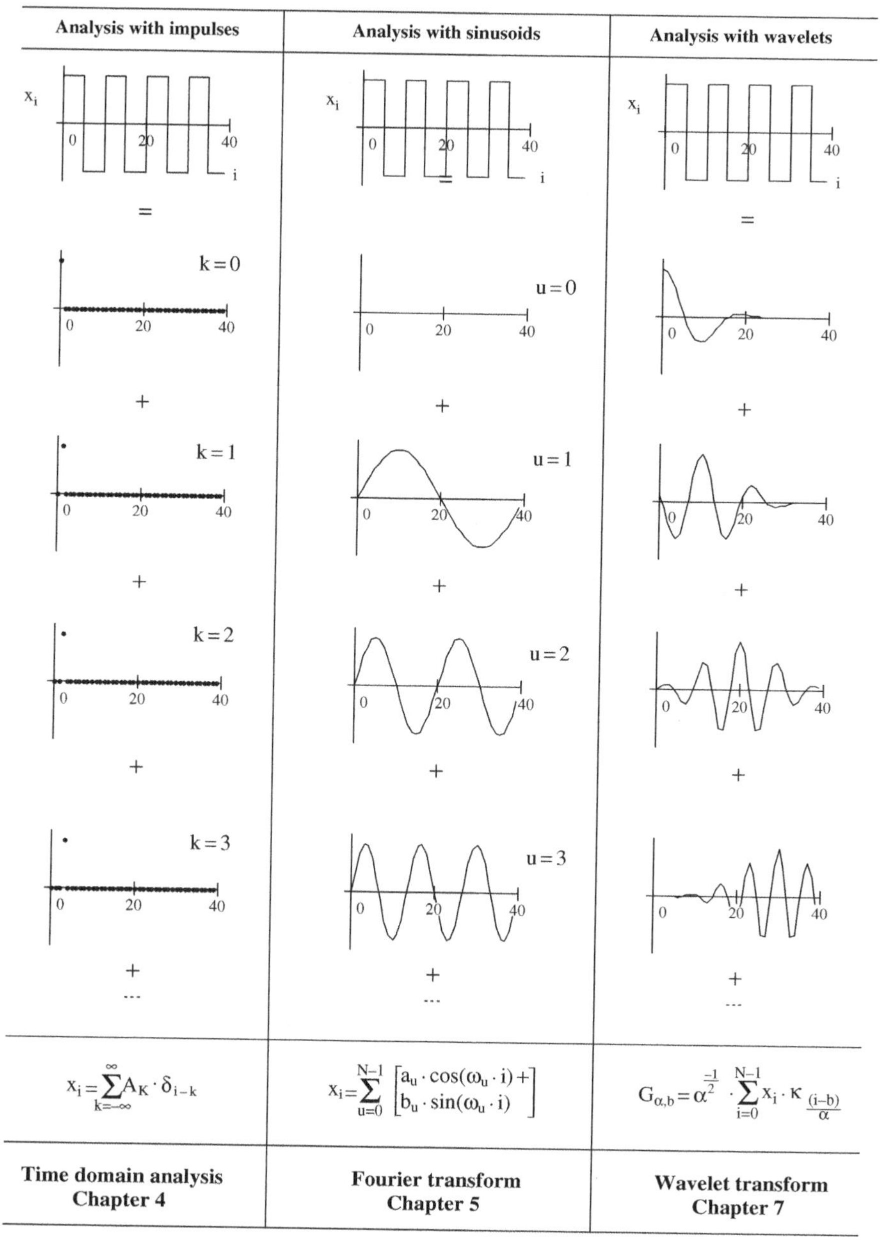

Figure 3.11 Analysis and synthesis of discrete signals. Any discrete signal can be represented as a linear combination of elemental signals

The scaling factors a_u and b_u indicate the participation of frequency ω_u in the signal $\underline{x}$. Once again, this synthesis equation is a summation of products and is rewritten as matrix multiplication, but in this case, the columns are the discrete values of sinusoids at times t_i. Each column corresponds to a different angular frequency ω_u. The assembled matrix multiplies the vector that contains the corresponding scaling coefficients a_u and b_u,

$$\underline{x} = \left[\begin{pmatrix} \cos \\ u=0 \\ \cdots \\ \cdots \end{pmatrix} \begin{pmatrix} \sin \\ u=0 \\ \cdots \\ \cdots \end{pmatrix} \begin{pmatrix} \cdots \\ \cdots \\ \cdots \\ \cdots \end{pmatrix} \begin{pmatrix} \cos \\ u=n \\ \cdots \\ \cdots \end{pmatrix} \begin{pmatrix} \sin \\ u=n \\ \cdots \\ \cdots \end{pmatrix} \right] \cdot \begin{bmatrix} a_0 \\ b_0 \\ \cdots \\ \cdots \\ a_N \\ b_N \end{bmatrix} \tag{3.25}$$

Coefficients a_u and b_u can be determined following standard least squares curve-fitting procedures (Chapters 5 and 9).

3.4.3 Summary of Decomposition Methods - Domain of Analysis

The decomposition of signals into elemental signals is the starting point for signal processing and system characterization. The choice of the elemental signal defines the type of "transformation" and affects subsequent operations (Figure 3.11):

- If the signal $\underline{x}$ is decomposed into scaled and shifted impulses (or steps), the analysis will take place in the *time domain* (Chapter 4).
- If the signal $\underline{x}$ is decomposed into scaled sinusoids of different frequency, subsequent operations will be conducted in the *frequency domain*. This is the Fourier transform of the signal (Chapter 5).

Signals could also be decomposed in terms of other elemental signals:

- The wavelet transform consists of expressing signals as a summation of wavelets (Figure 3.11, Chapter 7).
- The Laplace transform consists of decomposing signals in terms of growing or decaying sinusoids (complex exponentials).
- The Walsh transform consists of analyzing signals in terms of elemental square signals made of 1 and -1 values (see problems in Chapter 5).

Analysis and synthesis operations are linear combinations. Therefore, the same signal-processing operation can be implemented with any of these transformations as long as the operation preserves linearity. In particular, time domain operations are implemented in the frequency domain, and vice versa. Then, what domain should be preferred? Computation efficiency and ease of data interpretation will affect this decision.

3.5 SYSTEMS: CHARACTERISTICS AND PROPERTIES

A system transforms an input signal $\underline{x}$ into the output signal $\underline{y}$ (Figure 3.12). Consider the following examples (see also Table 1.1):

- A rubber band stretches when a load is applied, a metal rod contracts when cooled, and the electrical current in a conductor increases when the applied voltage difference increases.
- A lamp swings after a house is shaken by a tremor.
- A sharp sound is reflected from various objects and the multiple reflections arrive at a microphone like distinct echoes, each with its own time delay, amplitude, and frequency content.

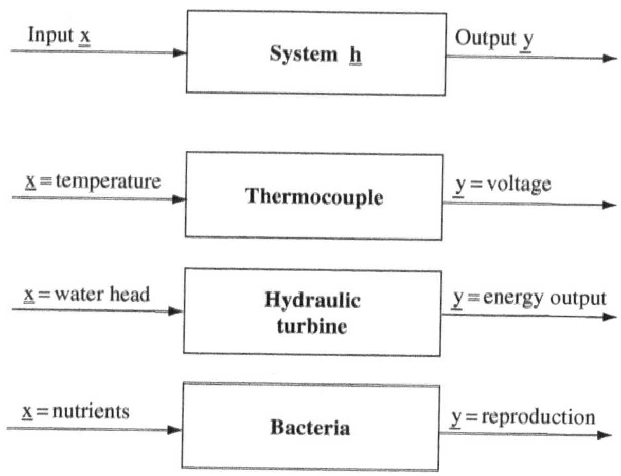

Figure 3.12 Definition of a system. Examples of systems with different degrees of complexity

- Water flows into a reservoir with a certain chemical composition and leaves the reservoir with a different chemistry after a delay time.
- Voice is encoded by a cellular phone to be transmitted through the airwaves.

These systems are characterized according to some salient aspects of the transformations they impose on the input signal, as described next.

3.5.1 Causality

A system satisfies causality if the response at time $i = k$ is only because of input at time $i \leq k$. Causality is the fundamental hypothesis of science: the search for an explanation to a given event presumes the existence of a cause. A system that appears to violate causality must be reassessed to identify undetected inputs or improper system definition, or incorrect analysis.

3.5.2 Linearity

A system is linear when the output is proportional to the input. Consider two springs: one is the standard cylindrical spring with linear elastic force–deformation response, and the other is a conical spring with nonlinear elastic response (Figure 3.13). Loads F_1 and F_2 cause deformations δ_1 and δ_2 in each spring. If the linear spring is loaded with a force $F_3 = F_1 + F_2$, then the measured deformation is $\delta_3 = \delta_1 + \delta_2$. This is not the case in the conical spring as seen in the figure. Likewise, a k-fold load produces a k-fold deformation only in the linear spring.

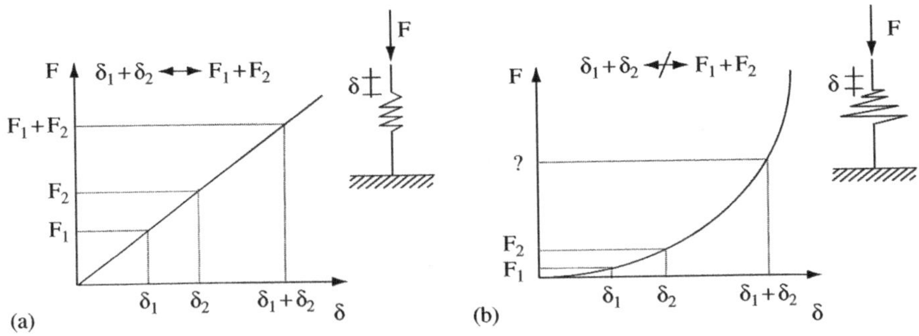

Figure 3.13 Linearity: (a) linear system; (b) nonlinear system

These two observations combine in the superposition principle: "*sum of causes → sum of effects*". Given two signals $\underline{x}^{<1>}$ and $\underline{x}^{<2>}$

$$\begin{array}{llll} \text{if} & \underline{x}^{<1>} & \text{causes} & \underline{y}^{<1>} \\ \text{and} & \underline{x}^{<2>} & \text{causes} & \underline{y}^{<2>} \\ \text{then} & a \cdot \underline{x}^{<1>} + b \cdot \underline{x}^{<2>} & \text{causes} & a \cdot \underline{y}^{<1>} + b \cdot \underline{y}^{<2>} \end{array} \tag{3.26}$$

Therefore, the linearity of a system is tested by verifying the superposition principle. "True linearity" is not a property of real systems, yet it is a valid hypothesis for small amplitude input signals or perturbations. Furthermore, it is often possible to identify an equivalent linear system that resembles the response of the nonlinear system for a certain input level.

3.5.3 Time Invariance

A system is time-invariant if its response to a given input does not vary with time, but only experiences a time shift equal to the input time shift. All systems evolve in time: electronic devices change their response while warming up, the response of a building varies as damage accumulates during an earthquake, materials age throughout the years, and the properties of the atmosphere experience daily and seasonal fluctuations. These examples suggest that systems encountered in engineering and science are inherently time-variant or "dynamic". However, the systems in these examples may be considered time-invariant for a very short, one-millisecond-long input. In other words, *time invariance must be assessed in reference to the duration of signals and events of interest.* Then, it is correct to assume that the atmosphere is "time-invariant" during the passage of a short laser signal that is used to remotely explore changes in chemical composition throughout the day. In general, phenomena with very different timescales are independently studied.

3.5.4 Stability

System stability implies "*bounded input → bounded output*". System stability is also apparent in the magnification of input uncertainty. The uncertainty in the initial location of the ball in Figure 3.14a is not relevant for the final location after it is freed; this is a stable system. By contrast, any uncertainty in the initial location will be magnified in the unstable system sketched in Figure 3.14b. Systems that manifest chaotic behavior are unstable, as in the case of a thin ruler that suddenly buckles when subjected to compressive loading at the ends.

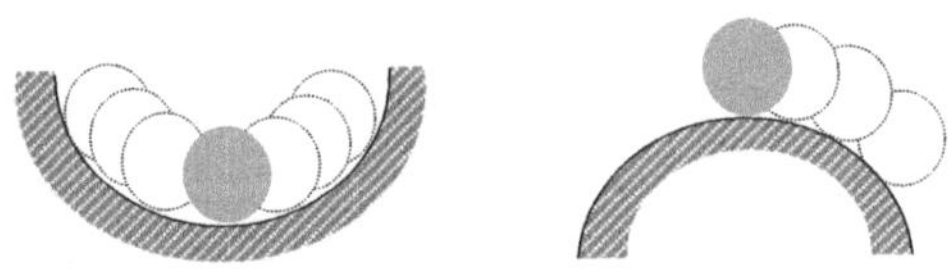

Figure 3.14 Stability. (a) The output of a stable system diminishes the uncertainty of the input: the final position of the ball is not sensitive to the initial position. (b) An unstable system is characterized by the magnification of initial uncertainties: a small uncertainty in the initial position of the ball has an important effect on its final position

3.5.5 *Invertibility*

A system is invertible if there is an inverse transformation that renders the input from the output (inverse problem)

$$\text{if} \quad \underline{x} \xrightarrow{\text{system}} \underline{y} \quad \text{then} \quad \underline{y} \xrightarrow[]{\substack{\text{inverse}\\ \text{system}}} \underline{x} \tag{3.27}$$

An analog telephone system consists of a microphone, a transmission line, and the earpiece on the other end. The voice spoken into the microphone is encoded into the electromagnetic wave that is transmitted, and later converted back into sound at the other end. In this case, the speaker inverts the transformation imposed at the microphone, and although the inversion is not perfect, it is acceptable for communication purposes. Invertibility is the central theme in Chapters 8–11.

3.5.6 *Linear Time-invariant (LTI) Systems*

The analysis of a system is considerably simpler when it is linear and time-invariant. A salient characteristic of this type of system is that it preserves the statistics of the input signal onto the output signal. For example, if the input signal has Gaussian statistics, the output signal will also have Gaussian statistics. This important observation leads to a possible procedure to test whether a system is LTI:

- Input a signal with known statistics. The signal duration must be relevant to signals and events of interest.
- Measure the output signal and compute the statistics.
- Compare input and output statistics.

The superposition principle applicable to linear systems is now extended to LTI systems: "*sum of time-shifted causes* → *sum of time-shifted effects*". Deviations from linear time invariance and implications for signal processing and system analysis are discussed in Chapter 7.

3.6 COMBINATION OF SYSTEMS

Engineering tasks and scientific studies often involve systems with multiple components. These systems are analyzed into a sequence of interconnected subsystems (Figure 3.15). Consider the following two systems used to measure material properties:

- *Sound velocity*. The measurement system involves several subsystems in *series*: signal generator → cable → source transducer → coupler → specimen → coupler → receiving transducer → signal conditioner → cable → A/D converter and storage (a simpler system is sketched in Figure 3.16).

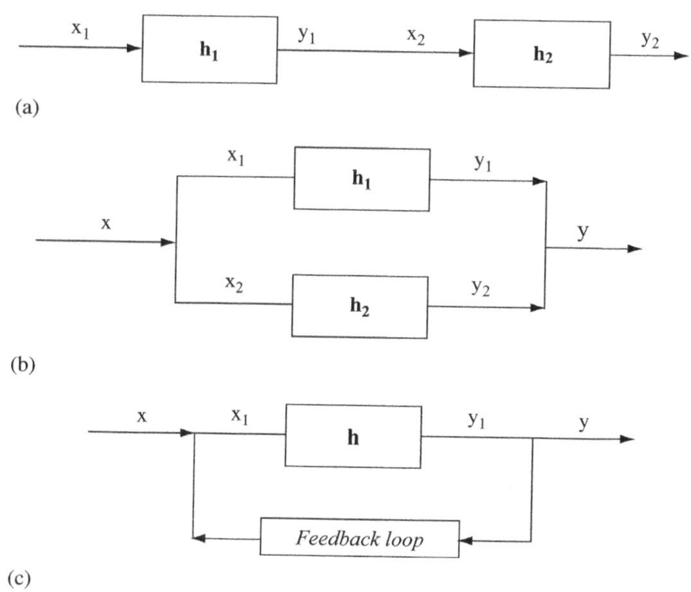

Figure 3.15 Combination of systems: (a) systems in series; (b) systems in parallel; (c) system with feedback loop

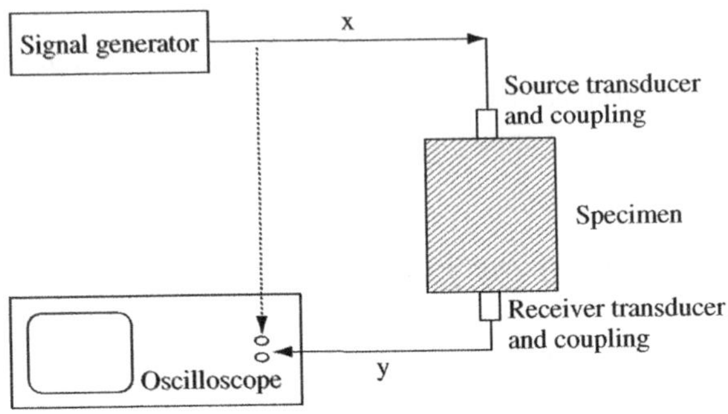

Figure 3.16 A measurement system is the combination of several subsystems. The response of peripheral must be removed from the total response to obtain the specimen properties

- *Complex permittivity*. The material is placed in a capacitor-type cell and the complex impedance is measured with an impedance analyzer. Stray capacitance and induction develop both in *series* as well as in *parallel* with the material, in addition to the resistance in *series* contributed by the connecting cables.

In both cases, the effects of all peripheral subsystems must be removed from the measured signal to recover the sought material response.

Many mechanical and electronic devices may also include feedback loops. Feedback in electromechanical systems is facilitated by the high speed of electrical signals and computer processors compared to the speed of mechanical events. For example, audio amplifiers include feedback to enhance fidelity, airplanes are equipped with computer-controlled stabilizers, and feedback is used to damp ringing effects in accelerometers and large amplitude oscillations in buildings.

The global system response is obtained by combining the individual subsystems' response according to their interconnectivity. Let us consider the simplest case of linear springs, with transformation $F = k \cdot \delta$. The equivalent stiffness of M springs is

$$k_{equiv} = k_1 + k_2 + \ldots. + k_M \qquad \textit{connected in parallel} \qquad (3.28)$$

$$\text{and} \quad k_{equiv} = \left(\frac{1}{k_1} + \frac{1}{k_2} + \ldots. + \frac{1}{k_M} \right)^{-1} \qquad \textit{connected in series} \qquad (3.29)$$

The sequential order of the components does not affect the equivalent global response in each case. This is generalized to all LTI subsystems (Chapter 5).

3.7 SUMMARY

- Signals may be periodic or aperiodic, even or odd, one-dimensional or multi-dimensional, stationary-ergodic, or nonstationary.
- Signal digitization may alter the periodicity of the signal and cause information loss and aliasing of undersampled frequencies. The Nyquist criterion must be fulfilled during digitization of baseband signals with energy from DC to the maximum signal frequency.
- There are several elemental signals including steps, impulses, sinusoids, and exponentials. Other important signals include wavelets and random noise.
- Any discrete signal can be decomposed into a linear combination of elemental signals.
- The selection of the elemental signal determines the type of analysis and the domain of operation. The analysis of signals into scaled and shifted impulses leads to "time domain" operations, whereas the decomposition of signals into scaled sinusoids conduces to the "frequency domain". Equivalent linear signal processing operations can be defined in either domain.
- A system enforces a transformation on the input signal. Linear time-invariant (LTI) systems are the most tractable. The generalized superposition principle "sum of time-shifted causes → sum of time-shifted effects" applies to LTI systems. LTI systems preserve the statistics of the input signal in the output signal.
- Any real system consists of several subsystems connected in series or in parallel. The sequential order of LTI subsystems does not affect the global output. Subsystems may include a feedback loop to help control the response.
- The measurement of a system characteristics always involves other peripheral subsystems; their response must be removed from the measured values.

FURTHER READING

Cruz, J. B. and Can Valkenburg, M. E. (1974). Signals in Linear Circuits. Houghton Miffling Company, Boston. 594 pages.

Oppenhein, A. V., Willsky, A. S., and Young, I. T. (1983). Signals and Systems. Prentice-Hall, Inc., Englewood Cliffs, NJ. 796 pp.

Taylor, F. J. (1994). Principles of Signals and Systems. McGraw-Hill, Inc., New York. 562 pages.

SOLVED PROBLEMS

P3.1 *Linear system.* Assume a Gaussian distributed input signal x_i = random. Show that the distribution of the input is preserved in the output when the system is linear but that it is not preserved when the system is nonlinear. *Solution*: Let us generate a vector $\underline{x}$ of N = 512 normal distributed random numbers and compute the responses $\underline{y}$ and $\underline{z}$ for a linear and a nonlinear transformation. In each case we verify the histogram and compute skewness s.

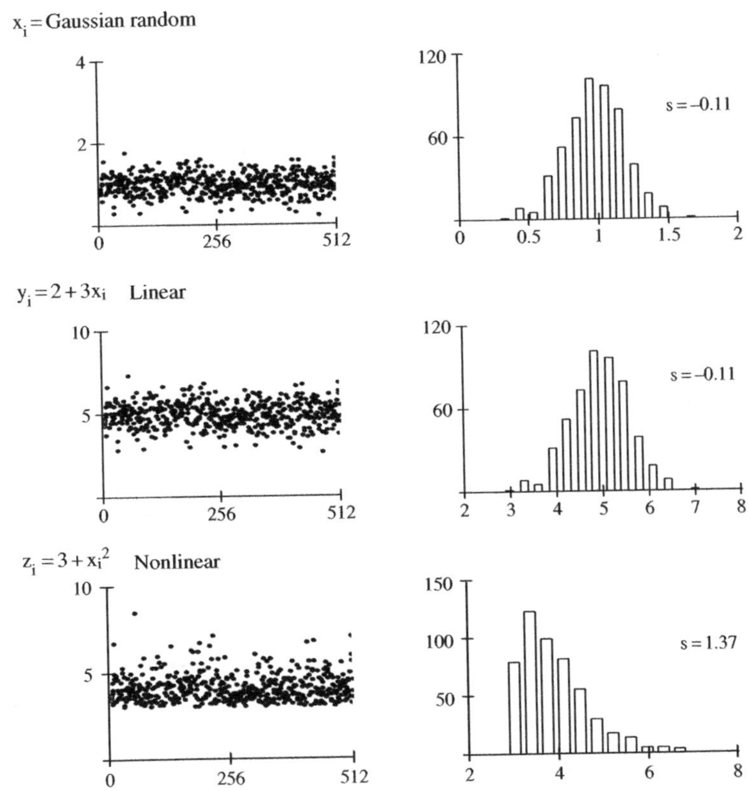

The histograms corresponding to $\underline{x}$ and $\underline{y}$ approach Gaussian distributions; however, the histogram for $\underline{z}$ shows that the nonlinear transformation alters the Gaussian nature of the input.

P3.2 *Elemental signals.* Define an impulse at location $i = -15$ and a step function at location $i = -15$. Implement these definitions numerically and plot the signals.

Solution:

Impulse at location $i = -15$:

$x_i = \delta_{i+15}$

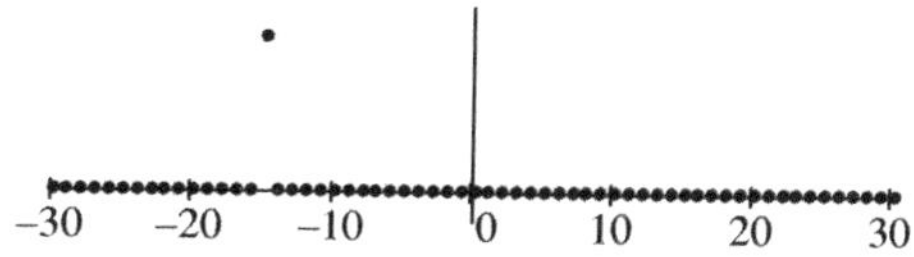

Step at location $i = -15$:

$x_i = u_{i+15}$

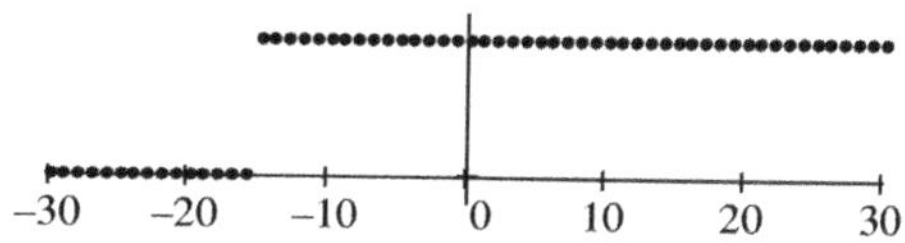

P3.3 *Signal analysis and synthesis.* Consider a periodic triangular time history $\underline{x} = (3, 2, 1, 0, -1, -2, -3, -2, -1, 0, 1, 2, \cdots)$. Approximate this array as a sum of discrete time cosine signals $y_i = \Sigma a_u \cdot \cos[u \cdot (2\pi/N) \cdot i]$, where N is the number of points $N = 12$, and u is an integer $u \geq 0$. The goal is to determine the coefficients $a_0, a_1, a_2, \cdots$ that minimize the total square error E between the array $\underline{x}$ and the approximation $\underline{y}$, where E is computed as $E = \Sigma(x_i - y_i)^2$. What is the residual E when only the $u = 1$ cosine function is included, and when the first four cosine functions are included?
Solution: The single frequency cosine $y_i = a_1 \cdot \cos[(2\pi/N) \cdot i]$ that fits the triangular signal closest is determined by iteratively fixing a_1 and computing the total error E. The value of a_1 that renders E minimum is the sought value:

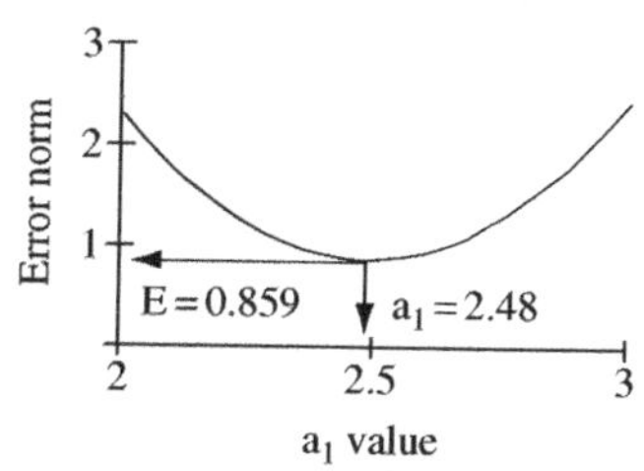

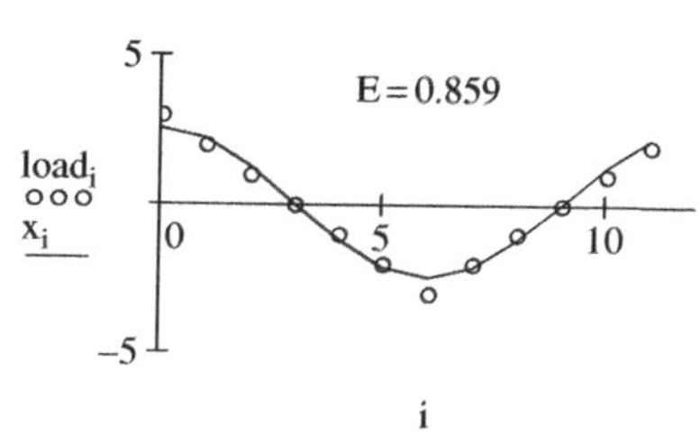

The coefficient a_2 can now be determined using the same approach, and the procedure is repeated to obtain all other higher terms. The first four coefficients are $a_0 = 0$, $a_1 = 2.48$, $a_2 = 0$, and $a_3 = 0.34$. The triangular signal and the signal synthesized with these first four terms are:

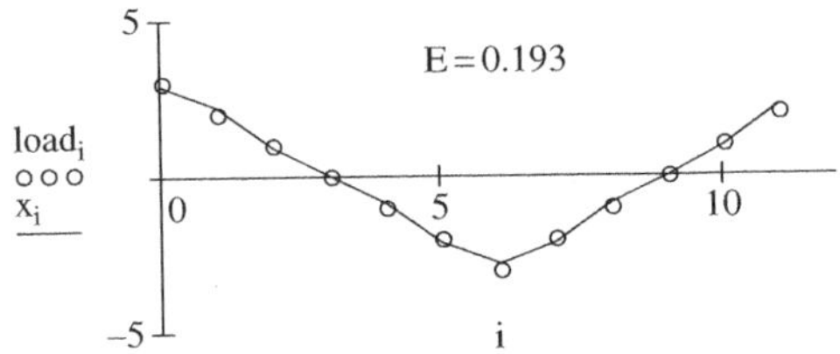

How many terms are needed in the summation to obtain E = 0?

P3.4 *Stationary and ergodic signals.* Form an ensemble of sinusoidals (four cycles in each segment) with additive zero-mean Gaussian noise. Verify stationary and ergodic criteria.
Solution: The ensemble is formed with 512 random signals or "segments". Each signal is 512 points long.

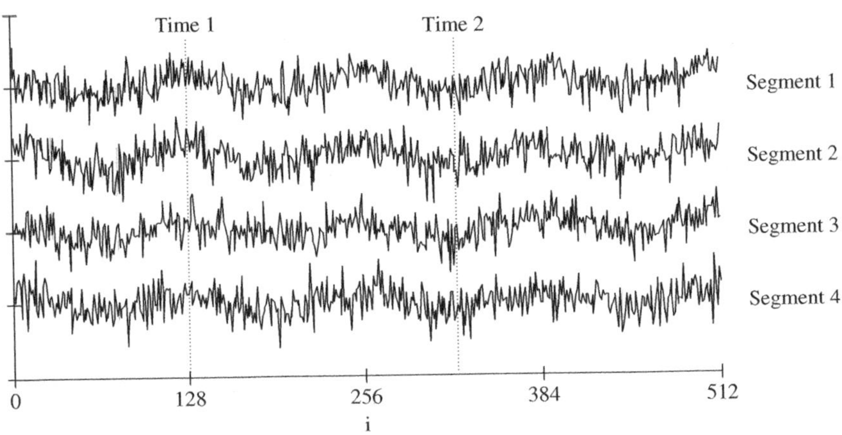

Histograms for selected ensemble values at selected times and segments:

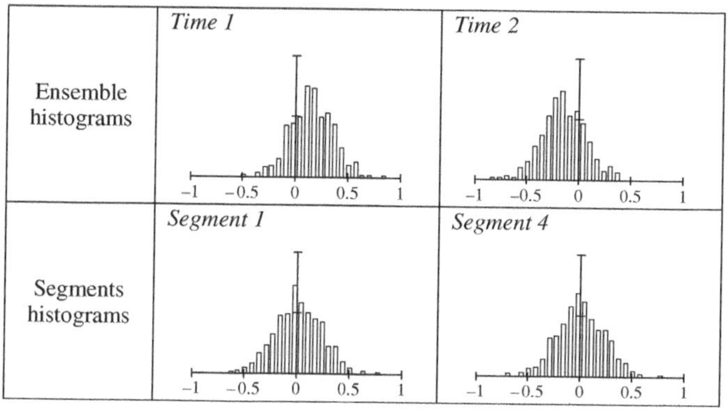

Ensemble statistics vary in time and are not the same as segment statistics: the signal is nonstationary and nonergodic.

ADDITIONAL PROBLEMS

P3.5 *Signals and systems.* Identify stationary and nonstationary signals, linear and nonlinear systems, and time-varying and time-invariant systems in your area of interest.

P3.6 *Amplitude modulation.* The multiplication of a sinusoidal signal with an exponential decaying signal yields a sinusoidal signal that decays with time (Figure 3.9e). What type of signal results from the multiplication of two sinusoidals of different frequencies? Plot the two signals and the product.

P3.7 *Signal digitization – undersampling.* A sinusoid frequency f is undersampled with a sampling frequency $f_{sam} < f_{Ny}$. Derive an expression for the frequency it folds into, as a function of the original frequency ω and the sampling frequency f_{sam}. Use numerical simulation to verify the equation for a single-frequency sinusoid. Then extend the study to explore undersampling effects for a beat function composed of two signals. Vary the frequency gap between the two signals relative to the sampling frequency.

P3.8 *Photography.* Explore the application of photography in your field of interest (engineering, science, sports, etc.):

(a) Explore commercially available cameras and flashes.

(b) What is the highest shutter speed and rewind rate?

(c) What is the shortest flash duration and the highest repetition rate?

(d) What type of events can you study with these "sampling rates"?

(e) Can you design your experiment to undersample?

(f) Explore ways to use a stroboscopic light with your photographic system.

P3.9 *Signal analysis and synthesis*. Repeat problem P3.3 by fitting a polynomial function instead of sinusoids. What order polynomial is needed? What is the total error when only the first four terms are included?

P3.10 *Stationary and ergodic signals*. Use the audio system in your computer to gather multiple records (segments) of background noise in your working environment. Analyze the ensemble to determine whether stationary and ergodic conditions are satisfied.

P3.11 *Combination of systems*. The electrical impedance $Z = V/I$ of the three fundamental circuit elements R, C, and L are:

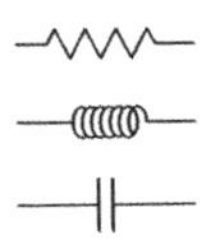

Resistor	R	$Z = R + j0$
Inductor	L	$Z = 0 + j\omega L$
Capacitor	C	$Z = 0 - j\frac{1}{\omega C}$

Complex Z values indicate a phase shift between current I and voltage V (Chapter 2). The inverse of the impedance is called the admittance Y. According to Equations 3.28 and 3.29, the equivalent impedance Z_{eq} of elements in series in the sum of the impedances, and the equivalent admittance Y_{eq} of elements in parallel is the sum of their admittances. Given three elements $R = 10^6$ ohm, $C = 2.5 \cdot 10^{-10}$ farad, and $L = 10^3$ henry, compute the equivalent impedance and plot admittance (amplitude and phase) versus frequency for (a) series and (b) parallel connection of the three elements.

P3.12 *Application: birds singing*. Knowing that a single tune lasts about 2 s and that you can whistle at the same frequency as the birds (about 2 kHz), select the sampling frequency and buffer memory for a portable A/D system.

4

Time Domain Analyses of Signals and Systems

Signal processing and system analysis operations are frequently encountered in most engineering and science applications. The fundamental signal processing operations are related to noise control to improve signal interpretation, and cross-correlation to identify similarities between signals. When a system is involved, data processing operations are developed to assess the system characteristics and to "convolve" a given input signal with the characteristic response of the system to compute the output signal.

4.1 SIGNALS AND NOISE

The presence of noise is one of the most pervasive difficulties in measurements. Given a signal amplitude V_S and noise amplitude V_N (same units as V_S), the signal-to-noise ratio is $SNR = V_S/V_N$. In applications where SNR varies in a wide range, decibel notation is used:

$$SNR = \frac{V_S}{V_N} \quad \text{or} \quad SNR[dB] = 20 \cdot \log_{10}\left(\frac{V_S}{V_N}\right) \tag{4.1}$$

A value of $SNR = 1 = 0\,dB$ means that the amplitude of the signal V_S is the same as the amplitude of noise V_N and the signal is almost indistinguishable.

The process of digitizing an analog signal adds noise. The analog-to-digital converter can resolve a limited number of discrete values related to the number of bits "n". For example, an $n = 8$ bit A/D board can map an analog value to one

Discrete Signals and Inverse Problems J. C. Santamarina and D. Fratta

of $2^8 = 256$ discrete values; hence, the potential noise level is one step. In general, if the signal amplitude occupies the n-bits, then, $V_S = 2^n$ steps, $V_N = 1$ step, and the signal-to-noise ratio associated with digitization is $SNR = 2^n = 6.02 \cdot n\,dB$.

The first and most important step to increase SNR is to design a proper experiment to minimize noise *before* signals are recorded. Consider the careful selection of transducers, peripheral electronics and A/D converter; the proper control of boundary conditions, including grounding and shielding; vibration isolation and thermal noise reduction, which may require cooling circuitry to near absolute zero. Once recorded, versatile signal processing algorithms can be used to enhance the signal-to-noise ratio. Time domain signal processing algorithms are discussed next.

4.1.1 Signal Detrending and Spike Removal

There are some known and undesired signal components that can be removed prior to processing. Low-frequency noise can be removed by signal detrending in the time domain. This operation consists of least squares fitting a low-frequency function $\underline{tr}$ to the noisy signal $\underline{x}$, and subtracting the trend from the measurements to obtain the detrended signal $y_i = x_i - tr_i$. Selected functions typically include a constant value, a straight line, or a long period sinusoid. Guidelines and procedures to fit a trend to a signal by least squares are presented in Chapter 9. Figure 4.1 shows examples of detrended signals.

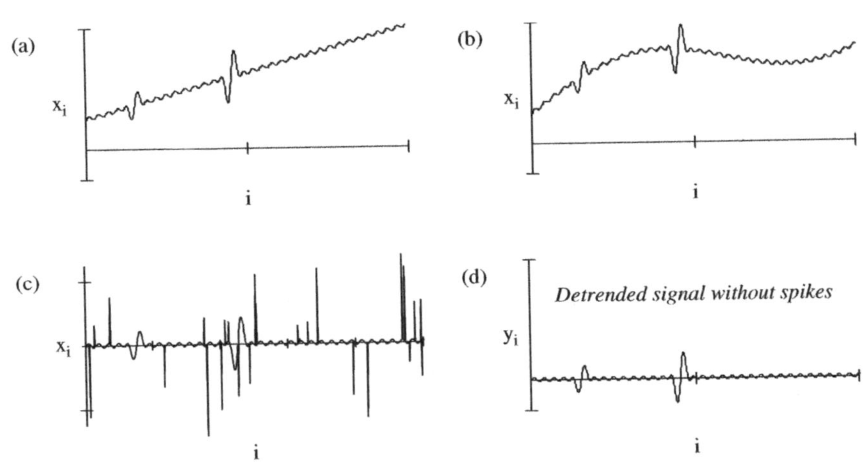

Figure 4.1 Detrending and spike removal: (a and b) signal riding on a low-frequency trend; (c) signal with spikes; (d) detrended signal without spikes

The signal mean value is known as the DC component or the static zero-frequency offset. The detrended signal without the DC offset is

$$y_i = x_i - \frac{1}{N} \sum_{i=0}^{i=N-1} x_i \tag{4.2}$$

Spikes are impulses randomly distributed along the signal (Figure 4.1c). Spikes can be "clipped", or removed and replaced by locally compatible signal values. For example, given a signal $\underline{x}$ with spikes, the signal $\underline{y}$ without spikes can be obtained with the following algorithm that compares the current value of x_i with the previous despiked value y_{i-1}: if $|x_i - y_{i-1}| <$ 'threshold' then $y_i = x_i$, otherwise $y_i = (x_{i-1} + x_{i+1})/2$, where the "threshold" value is selected to remove the spikes with minimal effect on the signal. Spike removal is demonstrated in Figure 4.1d.

4.1.2 Stacking: Improving SNR and Resolution

Signal stacking is an effective alternative to gather clear signals above the level of background noise. The operation consists of measuring the signal multiple times and averaging across the ensemble: the i-th element in the mean signal is the average of all the i-th elements in the measured signals.

The underlying assumption is that noise has zero mean, so that averaging reduces the noise level in the mean signal and increases the SNR of the correlated component. Figure 4.2 shows a noisy signal that is simulated by adding random noise to a decaying periodic sinusoid. The fluctuation of the background noise is the same as the amplitude of the periodic signal (SNR = 1). The different frames show the effect of stacking for an increasing number of signals.

The following two theorems from statistics help analyze the effects of stacking on signal-to-noise ratio for zero-mean noise:

1. The i-th value of the mean signal $x_i^{<mean>}$ is a good predictor of the true value $x_i^{<true>}$.
2. The mean value of averaged noise has Gaussian statistics, fluctuates around zero, and the standard deviation is proportional to the standard deviation of noise $\sigma^{<noise>}$ and decreases with the square root of the number of stacked signals M, $\sigma^{<noise>}/\sqrt{M}$.

Therefore, the signal-to-noise ratio SNR increases with $\sqrt{M}$.

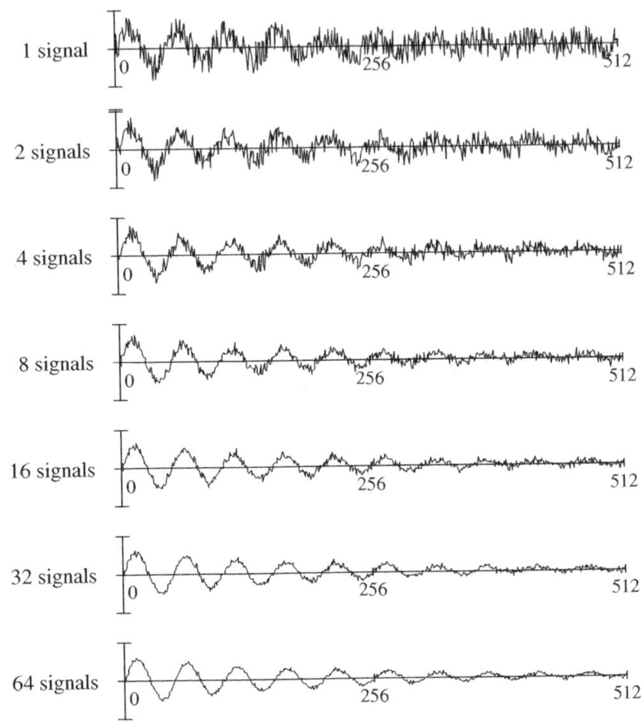

Figure 4.2 Noise control by signal stacking in the time domain. The SNR increases as the number of stacked signals increases

Number of Required Signals

One can expect with a certain probability p that the average value $x_i^{<mean>}$ does not deviate from the true value more than a prefixed quantity "E", which is related to the standard deviation of mean noise $|E| \leq \alpha\sigma^{<noise>}/\sqrt{M}$. Furthermore, the error E should be a small part β of the mean signal amplitude $|E| \leq \beta \cdot x^{<max>}$. Combining these two expressions, the required number of signals M to be stacked can be estimated as

$$M = \left(\frac{\alpha \cdot \sigma^{<noise>}}{\beta \cdot x^{<max>}} \right)^2 \tag{4.3}$$

The value of α is a function of the probability p:

probability p	$p = 80\%$	$p = 90\%$	$p = 95\%$	$p = 99\%$
coefficient α	$\alpha = 1.28$	$\alpha = 1.65$	$\alpha = 1.96$	$\alpha = 2.58$

For example, consider a signal with an estimated mean peak amplitude $x^{<max>} = 10$ and a measured background noise standard deviation $\sigma^{<noise>} = 2$. If one expects with a 90% probability ($\alpha = 1.65$) that the mean peak value of the stacked signal will deviate from the true value within 5% ($\beta = 0.05$), then the required number of signals in the ensemble is $M = 44$. (This analysis is revisited in the frequency domain, Chapter 6.)

Improved Resolution and Dynamic Range

The best resolution an n-bit A/D converter can attain is when the input signal is preamplified to the maximum input value in the converter without saturating it, so that the available 2^n discrete values are utilized. Signal stacking enhances resolution and the dynamic range between the largest and smallest recorded value when noisy signals are recorded. This is readily demonstrated with the following A/D conversion simulation:

- analog values $x(t) < 0.5$ are digitized into discrete values $x_i = 0$, and
- analog values $x(t) = 0.5$ are digitized into discrete values $x_i = 1$.

Then, an incoming noiseless analog signal value $x(t) = 0.6$ is stored as $x_i = 1$ in all the individual signals in the ensemble. Therefore, the stacked mean signal value will be $x_i^{<mean>} = 1$, and there is no advantage on resolution. However, when the incoming analog signal value is noisy, $x(t) = 0.6 \pm \text{noise}$, the digitized value will be $x_i = 1$ in some cases and $x_i = 0$ in others: 1 0 0 1 0 1 1 1 1 0 0 0 1 1 1... The mean value in the stacked signal approaches $x_i^{<mean>} \approx 0.6$ if the noise level is *at least* one digitizing step and a sufficient number of signals is stacked.

Likewise, noiseless signal values smaller than one digitization step remain undetected and signal stacking does not enhance the dynamic range of the A/D converter. Yet, noise adds to small signal amplitudes so that their values are registered with some probability, and the average value in the stacked signal asymptotically converges to the true value given adequate noise level and sufficient number of stacked signals.

It follows from the previous discussion that there is some "most favorable noise level" for which one can attain optimal detectability and maximum dynamic range. The effect of noise on A/D conversion resembles the physical phenomenon of "stochastic resonance".

Restrictions

Signal stacking presumes that the signal can be repeated. This implies that the source must be identical, that the system must remain time-invariant from one signal to the other, and that the triggering of the recording device can be synchronized with the signal to avoid random time shifting of successive signals. These are *not* readily attainable conditions in many situations. Consider, for example, a source of seismic signals for subsurface characterization consisting of a hammer and an aluminum plate resting on the ground. Successive hammer blows gradually sink the plate into the ground, change the stiffness of the soil beneath the plate, cause differences in triggering times (inertial switch response), and progressively change the frequency content in each signal.

4.1.3 Moving Kernels

Moving kernels are used to transform a signal $\underline{x}$ into a signal $\underline{y}$. For example, high-frequency noise can be reduced by running a moving average: the i-th value in the smoothed signal $\underline{y}$ is computed as an average of neighboring values around the i-th entry in the original noisy signal $\underline{x}$. The m-coefficients used in computing the weighted average are stored in the "kernel" $\underline{\kappa} = (\kappa_1, \kappa_2, \ldots, \kappa_m)$. Typically, the kernel is symmetric, m is an odd number, and the sum of all weights equals $\Sigma\kappa_p = 1$. Mathematically, the smoothed signal $\underline{y}$ is obtained as

$$y_i = \sum_{p=1}^{p=m} \kappa_p \cdot x_{\left(i-\frac{m-1}{2}\right)+p} \tag{4.4}$$

The noisy signal in Figure 4.3a is smoothed using the kernels in Figures 4.3b and c (m = 11 elements). Smoothing permits enhancing signals obtained from one-of-a-kind events or that contain coherent high-frequency noise where stacking cannot be applied.

Kernel length m and the weights κ_p determine the effect of moving kernels. The study of frequency domain operations in Chapter 5 facilitates kernel designing (see also related discussion in Chapter 9 in the context of regularization). In the meantime, a few guiding criteria follow .

Kernel Length

Very short kernels remove only very high-frequency noise. On the other hand, very long kernels may remove frequency components that are relevant to the signal. Thus, the effective kernel time span, $m \cdot \Delta t$, should be shorter than the shortest relevant period in the signal T; as a practical guideline, keep $m \le T/(10 \cdot \Delta t)$. Noise components in the same frequency as the signal of interest cannot be filtered with moving kernels, yet the signal-to-noise ratio can still be improved by stacking.

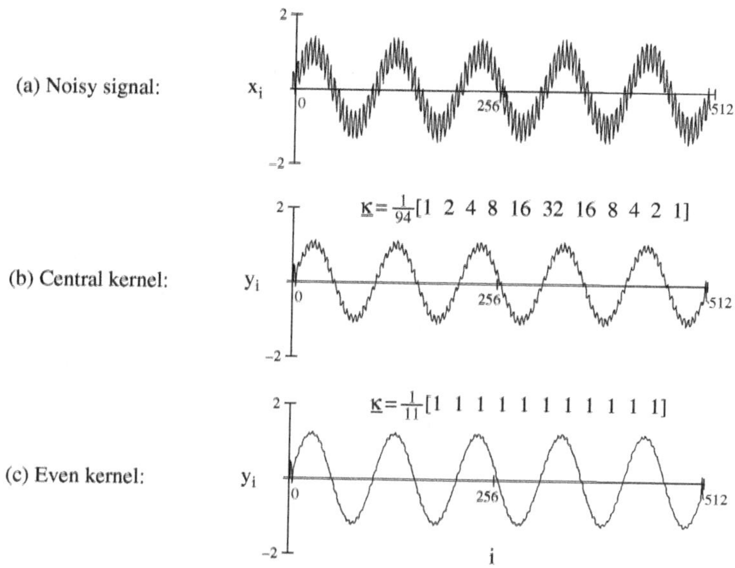

Figure 4.3 Noise control by moving average in the time domain: (a) noisy signal; (b and c) signal obtained after running the moving average kernels shown

Kernel Values

Physical criteria may guide kernel selection. For example, a kernel could be designed to set the second derivative equal to zero if the signal corresponds to a physical process modeled by a zero Laplacian, such as steady-state conduction phenomena. Expressing the second derivative in finite differences, a zero Laplacian becomes

$$\frac{\Delta^2 x}{\Delta t^2} = \frac{1}{\Delta t^2} \cdot (x_{i-1} - 2x_i + x_{i+1}) = 0 \tag{4.5}$$

Then, the smoothed value of the signal at location i is computed as

$$y_i = \frac{x_{i-1} + x_{i+1}}{2} \tag{4.6}$$

and the corresponding kernel for one-dimensional signals is

1/2	0	1/2

Moving kernels can be used to perform other discrete operations. For example, it follows from Equation 4.5 that the kernel $\kappa = (1, -2, 1)$ is a "second-order differentiator". Then, when signal $\underline{x}$ is processed with this moving kernel, the resulting signal $\underline{y}$ is the discrete second derivative of $\underline{x}$.

Adaptive filters

If the noisy signal $\underline{x}$ is nonstationary, the kernel length and weights may be adapted to the time-varying characteristics of the signal, whether the signal has been stored or it is streaming in real time, such as in adaptive feedback control. In this case, the kernel to be applied to the current entry x_i depends on x_i and the prior m-values of the signal: $x_i, x_{i-1}, \ldots x_{i-m}$. A simple adaptation strategy consists of selecting the kernel length $m^{<i>}$ at location i as a function of the signal variance around x_i. Other strategies assume zero-mean Gaussian noise and locally fit a presumed smooth signal behavior to the measured signal by minimizing the square error; this is a form of inverse problem (see also ARMA models – Chapters 7 and 8). In general, adaptive filtering is a nonlinear operation.

Kernels for Two-dimensional Signals

The concept of moving kernels is readily extended to two-dimensional signals, such as digital images. For example, the Laplacian in finite differences is expressed in terms of the values corresponding to the pixels above $x_{i,k+1}$, below $x_{i,k-1}$, to the left $x_{i-1,k}$ and to the right $x_{i+1,k}$ of the current pixel $x_{i,k}$:

$$\left(x_{i+1,k} - 2x_{i,k} + x_{i-1,k}\right) + \left(x_{i,k+1} - 2x_{i,k} + x_{i,k-1}\right) = 0 \tag{4.7}$$

where i and k are the position indices in the two normal directions, and the sampling interval or pixel size is equal in both directions. Then, the Laplacian-smoothing kernel becomes

- Laplacian smoothing $\frac{1}{4} \cdot \begin{array}{|c|c|c|} \hline 0 & 1 & 0 \\ \hline 1 & 0 & 1 \\ \hline 0 & 1 & 0 \\ \hline \end{array}$

Other kernels for 2D signals are summarized in Figure 4.4. The smoothed pixel value $y_{i,k}$ is obtained as a weighted average of pixel values in the original noisy

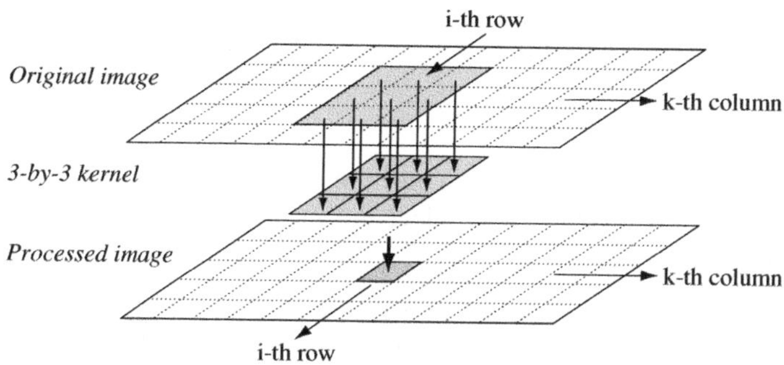

Smoothing — $\frac{1}{16}\cdot$

1	2	1
2	4	2
1	2	1

Vertical smoothing — $\frac{1}{15}\cdot$

1	3	1
1	3	1
1	3	1

Vertical edge detection

1	0	−1
1	0	−1
1	0	−1

Laplacian smoothing — $\frac{1}{4}\cdot$

0	1	0
1	0	1
0	1	0

Horizontal smoothing — $\frac{1}{15}\cdot$

1	1	1
3	3	3
1	1	1

Horizontal edge detection

1	1	1
0	0	0
−1	−1	−1

Figure 4.4 Filtering 2D signals in the time domain. Conceptual algorithm for convolutional 2D filters and typical 2D filters and typical 2D kernels

image around $x_{i,k}$ according to the coefficients in the 2D kernel $\underline{\underline{K}}$, and the operation is repeated for all i and k positions.

Values on Boundaries?

When the moving kernel approaches the signal boundaries, it requires values that are outside the signal. There are several alternatives to overcome this difficulty: external values are disregarded in the weighted average, special boundary kernels are defined, or imaginary signal values are assumed outside the boundary following symmetric or antisymmetric criteria, depending on physical considerations (see related discussion in Chapter 9 under regularization).

4.1.4 Nonlinear Signal Enhancement

A weighted average is a linear operation; therefore, signal processing with moving kernels is a linear operation, except when the kernel varies, such as in adaptive filtering. There are other nonlinear operators frequently used in noise control. The following common procedures are reviewed in the context of digital image processing:

- *Median smoothing*. For each (i, k) position of the analysis window, sort pixel values within the window, and select the median as the (i, k) value in the filtered image.
- *Selective smoothing*. For each (i, k) position of the analysis window, consider "neighbors" those pixels that have similar value to the central pixel "c", that is when $|(x_i - x_c)|/x_c$ is less than some threshold "t". Then, compute the weighted average, taking into consideration only the accepted neighbors. Selective smoothing is capable of removing noise without blurring contrast.
- *Thresholding and recoloring*. Compute and display the histogram of pixel values to guide the selection of a threshold value. Then, repaint the image by assigning the same color to pixel values above the threshold and another color to pixels with values below the threshold. Thresholding is a powerful trick to enhance the visual display of a homogeneous parameter with an anomalous region. The underlying assumption is that cells with similar pixel values correspond either to the background or to the anomaly.

These operations are nonlinear. When nonlinear procedures are part of a sequence of signal processing operations, the final result will depend on the order in which they are implemented. Linear and nonlinear filter effects are demonstrated in Figure 4.5.

4.1.5 Recommendations on Data Gathering

The best approach to noise control is to *improve the data at the lowest possible level*. Start with a proper experimental design:

- *Carefully design the experimental procedure to attain the best raw data.* Explore various testing methodologies and select the most robust procedure you can implement within the available facilities. Whenever possible and relevant, select variable ranges where the phenomenon has a clear response distinguishable from random response. Explore different excitations and boundary conditions. For example, a material with low-strain dynamic stiffness and damping can be characterized using pulse propagation, resonance, logarithmic

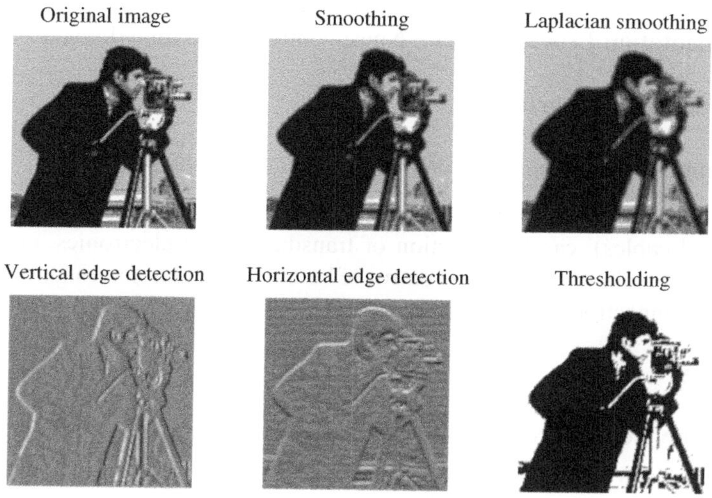

Figure 4.5 Examples of digital image processing © MIT. Printed with permission

decrement, and quasi-static hysteric behavior; each of these techniques presents advantages and limitations.

- *Select the transducers that are best fitted to sense the parameter under study*. For example: if the intent is to monitor low-frequency oscillations, avoid accelerometers because their response is proportional to the displacement multiplied by ω^2; therefore, high-frequency noise will be magnified in the signal.
- *Match the impedance between the transducer and the medium*. Improper impedance matching reduces the detected signal amplitude and aggravates poor signal-to-noise situations.
- *Increase signal level* whenever possible, but avoid amplitude-related effects such as unwanted nonlinearities.
- *Noise level*. Reduce the noise level by isolating the system. Consider noise in all forms of energy: electromagnetic (shield and ground), mechanical (vibration isolation), thermal and chemical (environmental chamber), and biological (prior decontamination).
- *Use quality peripheral electronics* and match electrical impedances.

The Implementation Procedure 4.1 summarizes the techniques for noise control in the time domain.

Implementation Procedure 4.1 Noise control in the time domain

First and most important

- Attempt to reduce noise before measurements are obtained.
- Consider proper grounding and shielding (including the use of coaxial and twisted cables), careful selection of transducers and electronics, enhanced vibration isolation, adequate control of boundary conditions, enhanced quality of connections.

Stacking

- Measure the background noise and determine its statistics.
- Estimate the number M of signals to be stacked (Equation 4.3). The error in the stacked measurement decreases with the square root of the number of stacked signals M.
- Detrend individual signals and remove spikes.
- Arrange the M stored signals in matrix form $x_{i,k}$ where the index i relates to the discrete time and k is the label of each record.
- Compute the average signal $x_i^{<avr>}$

$$x_i^{<avr>} = \frac{1}{M} \sum_k x_{i,k}$$

Moving average

- Select a kernel. Define its length m (and odd number) and weights κ_p. Recall that if T is the shortest relevant period in the signal, then $m \le T/(10 \cdot \Delta t)$.
- Convolve the kernel κ_p with the signal x_i:

$$y_i = \sum_{p=1}^{p=m} \kappa_p \cdot x_{\left(i-\frac{m-1}{2}\right)+p}$$

- When using this equation, the sum of all weights in a smoothing kernel must equal $\Sigma\kappa_p=1$.
- Kernels must be redefined at the boundaries of the arrays, where $i-p<0$ or $i+p>N-1$ (where the array $\underline{x}$ contains N elements $0 \ldots N-1$). Physical principles must be taken into consideration (see Chapter 9).

Example

Signal enhancement by noise control underlies all measurement and signal processing tasks. The effectiveness of stacking and moving average is demonstrated in Figures 4.2 and 4.3.

Note: Noise control with frequency domain operations is presented in Chapter 6. That discussion will facilitate the design of filtering kernels.

4.2 CROSS- AND AUTOCORRELATION: IDENTIFYING SIMILARITIES

Cross-correlation is a very robust signal processing operation that permits identifying similarities between signals even in the presence of noise. How can a computer be trained to identify similarities between two arrays of discrete values? Consider the two similar but time-shifted signals $\underline{x}$ and $\underline{z}$ in Figure 4.6. The cross-correlation operation gradually time-shifts the second signal $\underline{z}$ to the left. For each time shift $k \cdot \Delta t$, the pair of values facing each other in the two arrays are multiplied $x_i . z_{i+k}$ and summed for all i-entries. This result is the cross-correlation between $\underline{x}$ and $\underline{z}$ for the k-shift:

$$cc_k^{<x,z>} = \sum_i x_i \cdot z_{i+k} \tag{4.8}$$

The process is repeated for different k-shifts, and cross-correlation values $cc_k^{<x,z>}$ are assembled in the array $\underline{cc}^{<x,z>}$.[1]

The cross-correlation between signals $\underline{x}$ and $\underline{z}$ in Figure 4.6 when the time shift is zero, i.e. $k = 0$, leads to the multiplication of nonzero $\underline{x}$ amplitudes with zero $\underline{z}$ amplitudes at low values of t_i; the opposite happens at high values of t_i. Therefore, the cross-correlation of $\underline{x}$ with $\underline{z}$ when $k = 0$ is equal to zero. As the signal $\underline{z}$ is shifted relative to $\underline{x}$, $k > 0$, the cross-correlation sum begins to show nonzero values. The best match is obtained when the signal $\underline{x}$ is sufficiently shifted to superimpose with signal $\underline{z}$, and the cross-correlation reaches its maximum value.

---

[1] The cross-correlation in continuous time is a function of the time shift τ:

$$cc^{<x,z>}(\tau) = \int_{-\infty}^{\infty} x(t) \cdot z(t+\tau) \cdot dt.$$

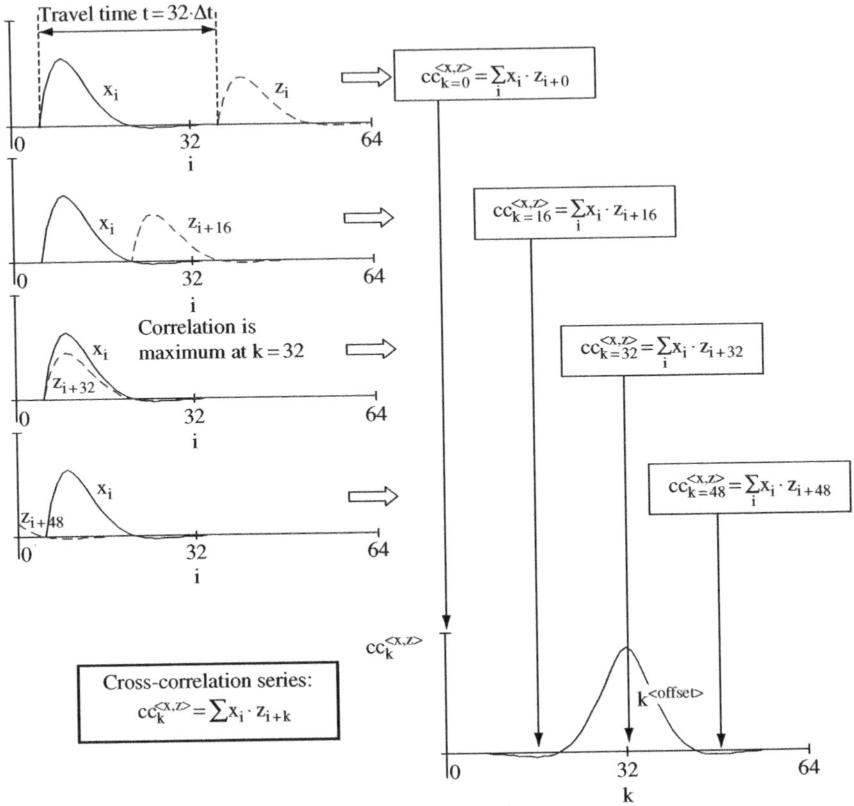

Figure 4.6 Cross-correlation. The signal $\underline{z}$ shifts to the left as k increases. The peak in the cross-correlation sum takes place when $k = 32$ and the two signals are superimposed

As the signal $\underline{z}$ is shifted past signal $\underline{x}$, values on the right end of $\underline{x}$ face empty sites. These sites are filled with "imaginary entries". If the two arrays have N entries each $\underline{x} = (x_o, x_1, ..x_{N-1})$ and $\underline{z} = (z_o, z_1, ..z_{N-1})$, and cross-correlation is explored for the full length of the signals, the signal $\underline{z}$ will be tail-padded from $i = N$ to $i = 2N - 1$ so that it can be shifted past $\underline{x}$ from $k = 0$ to $k = N - 1$. The imaginary entries can be either zeros (when signals have been detrended), values compatible with the signal trend, or the same "circular" signal $\underline{z}$ wrapped around so that $z_N = z_0$, $z_{N+1} = z_1$, etc. (This requires a detrended signal). The selected padding alternative must be compatible with the physical reality under study.

For clarity, the computation of cross-correlation in the form of a spreadsheet is shown in Figure 4.7. Each column in the central block shows the shifted signal

	k = 0	k = 1	k = 2	k = 3	k = 4	k
i	$\mathbf{x_i \cdot z_i}$	$\mathbf{x_i \cdot z_{i+1}}$	$\mathbf{x_i \cdot z_{i+2}}$	$\mathbf{x_i \cdot {}_{i+3}}$	$\mathbf{x_i \cdot z_{i+4}}$	$\mathbf{x_i \cdot z_{i+k}}$
0	$x_0 \cdot z_0$	$x_0 \cdot z_1$	$x_0 \cdot z_2$	$x_0 \cdot z_3$	$x_0 \cdot z_4$	...
1	$x_1 \cdot z_1$	$x_1 \cdot z_2$	$x_1 \cdot z_3$	$x_1 \cdot z_4$	$x_1 \cdot z_5$	...
2	$x_2 \cdot z_2$	$x_2 \cdot z_3$	$x_2 \cdot z_4$	$x_2 \cdot z_5$	$x_2 \cdot z_6$	...
3	$x_3 \cdot z_3$	$x_3 \cdot z_4$	$x_3 \cdot z_5$	$x_3 \cdot z_6$	$x_3 \cdot z_7$	...
4	$x_4 \cdot z_4$	$x_4 \cdot z_5$	$x_4 \cdot z_6$	$x_4 \cdot z_7$	$x_4 \cdot z_8$	...
5	$x_5 \cdot z_5$	$x_5 \cdot z_6$	$x_5 \cdot z_7$	$x_5 \cdot z_8$	$x_5 \cdot z_9$	...
6	$x_6 \cdot z_6$	$x_6 \cdot z_7$	$x_6 \cdot z_8$	$x_6 \cdot z_9$	$x_6 \cdot z_{10}$	...
7	$x_7 \cdot z_7$	$x_7 \cdot z_8$	$x_7 \cdot z_9$	$x_7 \cdot z_{10}$	$x_7 \cdot z_{11}$	...
Σ	$\Downarrow$	$\Downarrow$	$\Downarrow$	$\Downarrow$	$\Downarrow$	$\Downarrow$
cc_k	$\sum_i x_i \cdot z_i$	$\sum_i x_i \cdot z_{i+1}$	$\sum_i x_i \cdot z_{i+2}$	$\sum_i x_i \cdot z_{i+3}$	$\sum_i x_i \cdot z_{i+4}$	$\sum_i x_i \cdot z_{i+k}$

Figure 4.7 Spreadsheet for the computation of cross-correlation

z_{i+k} for increasing values of k. The signal $\underline{x}$ remains unshifted in all columns. The sum of each column is equal to the cross-correlation of $\underline{x}$ and $\underline{z}$ for each shift k: the first column corresponds to zero shift, $k = 0$, the second column for a shift of one time interval, $k = 1$, etc. Implementation Procedure 4.2 presents the step-by-step computation of the cross-correlation between two signals.

Implementation Procedure 4.2 Cross-correlation sum

1. Arrange signals $\underline{x}$ and $\underline{z}$ in vector form. The length of array $\underline{x}$ is N.
2. Tail-pack signal $\underline{z}$ so that it can be shifted along the N entries in signal $\underline{x}$.
3. For a given k-shift in $\underline{z}$, the k-th element of the cross-correlation sum is equal to

$$\sum_i x_i \cdot z_{i+k}$$

4. Continue with next k until $k = N - 1$.
5. The resulting array of N entries is the cross-correlation sum between signals $\underline{x}$ and $\underline{z}$.

If the signal $\underline{z}$ has reversed polarity, the peak of the cross-correlation is negative. A plot of the absolute value of the cross-correlation $|\underline{cc}^{<x,z>}|$ often facilitates comparing the magnitude of positive and negative peaks.

Examples

Figures 4.6, 4.8, and 4.9 show several numerical examples of cross-correlation.

Note: The cross-correlation can be efficiently computed in the frequency domain (see Chapter 6).

4.2.1 Examples and Observations

Identifying Similarities

The first signal $\underline{x}$ in Figure 4.8a is a single-frequency sinusoid whereas the second signal $\underline{z}$ consists of $\underline{x}$ plus a high-frequency sinusoid. The cross-correlation is plotted on the right-hand side of Figure 4.8a. It was obtained by tail-duplicating signal $\underline{z}$, given the periodicity of these signals. The cross-correlation of $\underline{x}$ and $\underline{z}$ depicts the lower-frequency component, which is common to both signals $\underline{x}$ and $\underline{z}$. The effects of positive and negative high-frequency fluctuations in signal $\underline{z}$ cancel each other in the cross-correlation sum.

Determining Travel Time

Consider the nondestructive evaluation of some material of interest. In this particular case, the intent is to measure the sound wave velocity to characterize the low-strain stiffness of the material. Sent and received noiseless signals are shown in Figure 4.8b. Visual observation indicates that the received signal is an attenuated version of the input signal but shifted 64 time intervals; hence, the travel time across the specimen is $64 \cdot \Delta t$. When the cross-correlation is computed, the peak in the cross-correlation takes place at $k = 64$. (Note: if the received signal $\underline{z}$ had opposite amplitude, the cross-correlation peak would be a negative value – Figure 4.8c.)

Identifying Replicas in Noisy Signals

Cross-correlation is very robust with respect to noise. Figure 4.8d shows the same received signal $\underline{z}$ as Figure 4.8b but masked in noise. The cross-correlation of signal $\underline{x}$ with the noisy signal $\underline{z}$ is shown in Figure 4.8d. Once again, the peak

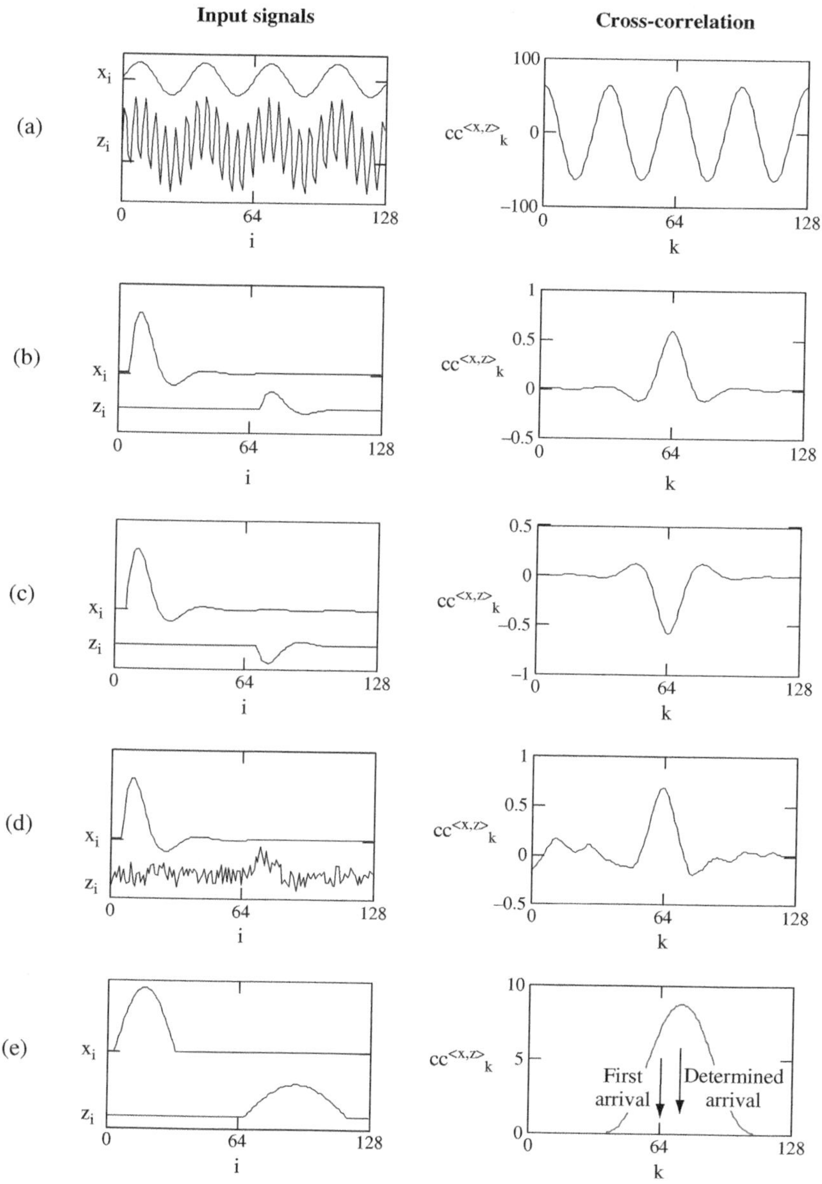

Figure 4.8 Examples of cross-correlation: (a) identifying similarities; (b) determining travel time; (c) effect of reverse polarization; (d) identifying replicas in noisy signals; (e) lossy and dispersive media

of the cross-correlation takes place at $k = 64$. The clarity of the cross-correlation peak in the noisy signal is surprising and most relevant to laboratory and field testing. Based on observations made in relation to Figure 4.8a, the effect of random fluctuations between the two signals tends to cancel each other in the cross-correlation sum.

Biases: Lossy Media, Dispersive Media and Multiple Paths

A wave experiences fairly complex transformations as it traverses a material. For example, different frequency components travel with different phase velocities and are subjected to different levels of attenuation. Consider the two signals shown in Figure 4.8e. The difference in the first arrival corresponds to $k = 64$. However, the cross-correlation reaches its maximum when the energy peaks in the two signals are aligned, $k = 74$ (group velocity). If the medium is not homogeneous, there will be more than one path for energy propagation (due to reflections and refractions). The maximum value of cross-correlation will correspond to the path that conducted most energy, not necessarily the shortest time path.

Observations

These examples allows us to make the following observations:

- The cross-correlation procedure is able to identify replicas of a signal in other signals, even in the presence of significant background noise.
- The value of the shift k for the peak cross-correlation indicates the time delay between replicas, $t^{<delay>} = k^{<peak>} \cdot \Delta t$. Changes in frequency content between input $\underline{x}$ and output $\underline{z}$ require the reinterpretation of cross-correlation when determining time shift between signals.
- If the signal and its replica were of opposite polarity, for example $z_i = -x_i$, the largest value of $\underline{cc}^{<x,z>}$ would be negative. A positive peak value indicates that both signals have the same polarity. Polarity reversal can be equipment related such as the wiring of transducers, or it can be of physical nature and provide important information about the system, such as the reflection of sound from a free boundary versus a fixed boundary.
- Periodic components common to both signals $\underline{x}$ and $\underline{z}$ manifest in the cross-correlation $\underline{cc}^{<x,z>}$. Therefore, cross-correlation can be used to assess the presence of selected frequency components.
- The cross-correlation of two infinitely long sinusoids of different frequency is null, $cc_k^{<x,z>} = 0$ for all k. This is also the case for the cross-correlation of any

signal with zero-mean random noise. Cancellation is not complete when the number of points is small because there are not enough terms in the summation to attain statistical equilibrium.

4.2.2 Autocorrelation

Autocorrelation is the cross-correlation of a signal with itself.[2]

$$ac_k^{<x>} = cc_k^{<x,x>} = \sum_i x_i \cdot x_{i+k} \tag{4.9}$$

Autocorrelation permits identifying internal timescales (or length scales) within a signal such as the repetitive appearance of a pattern or feature. These internal scales manifest as peaks in the autocorrelation array $\underline{ac}^{<x>}$.

The highest autocorrelation value occurs for zero shift $k = 0$ and it is equal to $ac_0^{<x>} = \Sigma x_i^2$. The autocorrelation of a finite-length array of zero-mean Gaussian noise tends to zero everywhere else but at the origin, that is $ac_k^{<x>} = 0$ for $k>0$.

Example

Consider a long steel rod excited with a short signal. An accelerometer is mounted at one end of the rod (Figure 4.9a). The excitation travels in the rod back and forth with velocity V_{rod}. The signal detected with the accelerometer shows successive passes of the excitation signal, changing polarity each time, with a well-defined interval that corresponds to the travel time for twice the length L of the rod, $t = 2 \cdot L/V_{rod}$ (Figure 4.9b). Successive repetitions of the signal have lower amplitude due to attenuation and become gradually masked in the background noise. The autocorrelation of this signal depicts the multiple reflections and the characteristic time of the process, in this case: $k \cdot \Delta t = 180\Delta t$ (Figure 4.9c). The result is clear even in the presence of significant background noise.

4.2.3 Digital Images – 2D Signals

Correlation studies with digital images help identify the location(s) of a selected pattern $\underline{\underline{z}}$ in an image $\underline{\underline{x}}$ (cross-correlation) or discover repetitive internal scales in

---

[2] The definition of autocorrelation in continuous time is

$$ac^{<x>}(\tau) = \int_{-\infty}^{\infty} x(t) \cdot x(t+\tau) \cdot dt.$$

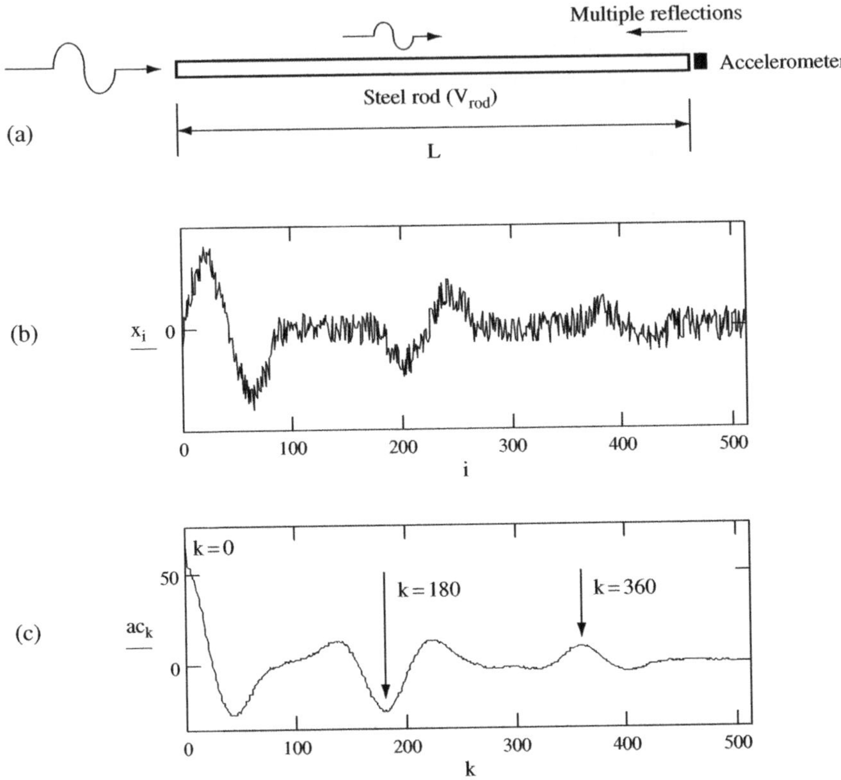

Figure 4.9 Autocorrelation: (a) experimental setup to detect multiple reflections in a steel rod; (b) a noisy signal with several reflections; (c) autocorrelation. The autocorrelation sum has the largest peak at k = 0, then a negative peak at k = 180, and a smaller positive peak at k = 360. Peaks point to the time when the signal "finds" itself

a single image $\underline{\underline{x}}$ (autocorrelation). In 2D correlation studies, one array is sequentially shifted relative to the other in both directions; therefore, the correlation arrays are two-dimensional in terms of the k and q shifts:

$$cc_{k,q}^{<x,z>} = \sum_v \sum_u x_{u,v} \cdot z_{u+k,v+q} \tag{4.10}$$

A similar expression is written for autocorrelation $\underline{\underline{ac}}^{<x>}$. Two-dimensional correlation is a computationally intensive operation. In many applications, the image $\underline{\underline{z}}$ is a small image and it is shifted only within some predefined subregion of $\underline{\underline{x}}$.

4.2.4 Properties of the Cross-correlation and Autocorrelation

The cross-correlation sum is not commutative. In fact, in terms of 1D signals,

$$cc_k^{<x,z>} = cc_{-k}^{<z,x>} \tag{4.11}$$

which means that it is the same to shift signal $\underline{z}$ to the left as to shift signal $\underline{x}$ to the right. A salient relationship between the cross-correlation and autocorrelation operators is

$$(cc_k^{<x,z>})^2 \leq ac_0^{<x>} \cdot ac_0^{<z>} \qquad \text{for all k} \tag{4.12}$$

4.3 THE IMPULSE RESPONSE – SYSTEM IDENTIFICATION

The *impulse response* $\underline{h}$ is the output signal generated by a linear time-invariant (LTI) system when the input signal is an impulse. By definition, the units of h_i are [output/input]. The impulse response contains all needed information about the system.

4.3.1 The Impulse Response of a Linear Oscillator

Let us develop these ideas within the context of a single degree of freedom (DoF) oscillator with mass m supported on a spring k and a dashpot c (Figure 4.10a). The single DoF system is an LTI system. This model can be used to simulate or analyze a wide range of dynamic systems, from ionic polarization at the molecular level (Figure 4.10b), to the response of experimental devices such as isolation tables and resonant instruments (Figure 4.10c), and the vibration of a trailer and the seismic response of buildings (Figure 4.10d).

The equation of motion represents the dynamic balance of participating forces acting on the mass when subjected to forced vibrations x(t):

$$m \cdot \ddot{y} + c \cdot \dot{y} + k \cdot y = x \tag{4.13}$$

where

x is the time history of the input force,
y is the time history of the displacement response, and
dots on y denote first and second derivatives (velocity and acceleration).

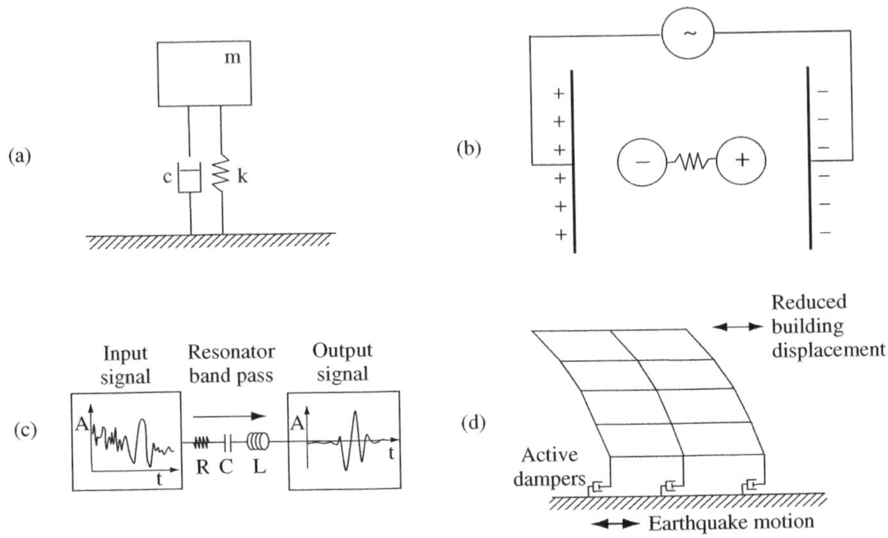

Figure 4.10 The equivalent single DoF oscillator is commonly used in the analysis of engineering and science systems: (a) damped linear oscillator; (b) ionic polarization in AC field; (c) electrical RCL amplifier; (d) seismic response of buildings with active dampers

Imagine impacting the mass of the oscillator with an "impulse". The oscillator will be set in motion and the amplitude of the impulse response h_i at discrete time $t_i = i \cdot \Delta t$ will be (for underdamped systems, $D < 1.0$):

$$h_i = \Delta t \cdot \frac{e^{-D \cdot \omega_n \cdot i \cdot \Delta t}}{m \cdot \omega_n \cdot \sqrt{1-D^2}} \cdot \sin\left(\omega_n \cdot \sqrt{1-D^2} \cdot i \cdot \Delta t\right) \qquad (4.14)$$

where system damping D and natural frequency ω_n are determined by the oscillator parameters m, k and c:

$$D = \frac{c}{2 \cdot \sqrt{m \cdot k}} \qquad \textit{damping} \qquad (4.15)$$

$$\omega_n = \sqrt{\frac{k}{m}} \qquad \textit{natural angular frequency} \qquad (4.16)$$

The natural period T_n is related to the angular frequency ω_n as $T = 2\pi/\omega_n$. Impulse responses $\underline{h}$ for single DoF systems with the same natural frequency but varying damping levels are shown in Figure 4.11.

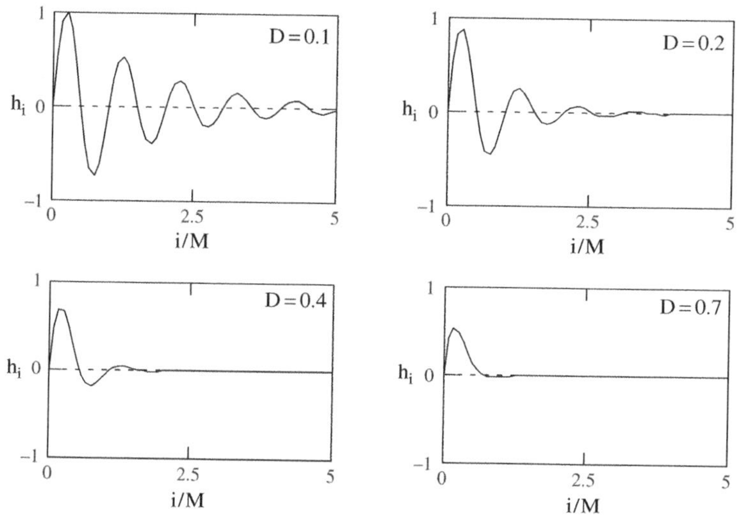

Figure 4.11 Single DoF oscillator. Impulse response $\underline{h}$ for different values of the damping coefficient D. The sampling rate is selected so that $T_n/\Delta t = 12$

The units of the impulse response for the single DoF system are [output/input], that is [displacement/force]. This agrees with the units of the impulse response predicted in Equation 4.14: [time^2/mass].

Note that the amplitude of the impulse response in discrete time $\underline{h}$ is not unique, but it is proportional to the selected sampling interval Δt (Equation 4.16 – for comparison, the impulse response in continuous time is presented in the footnote below).[3] The justification becomes apparent when the role of the impulse response in discrete time convolution is recognized in the next section. For now, let us just say that the impulse response will be repeated at every discrete time t_i so that a smaller sampling interval Δt means more frequent repetitions of smaller h_i values and the effects cancel out in the summation. Thus, convolution results will be independent of the selected Δt. (Note: Δt must satisfy the Nyquist criterion, Chapter 3.)

[3] The response of a single DoF system to an impulse in continuous time is

$$h(t) = \frac{e^{-D \cdot \omega_n \cdot t}}{m \cdot \omega_n \cdot \sqrt{1-D^2}} \cdot \sin\left(\omega_n \cdot \sqrt{1-D^2} \cdot t\right)$$

The impulse response h(t) corresponds to an ideal impulse signal that lasts $\Delta t \rightarrow 0$ and has infinite amplitude, yet its area is 1. The function h(t) only depends on ω_n and D. The units of h(t) are [output/(input·time)].

4.3.2 Determination of the Impulse Response

The impulse and impulse response are mathematical constructs, and they depart form physical reality and immediate experimental determination. Difficulties are overcome by measuring input and output signals, and processing them in the frequency domain (Chapter 6) or through formal deconvolution in the time domain (Chapters 8 and 9). Still, simple tests can be performed to gain insight into the characteristics of the system and the nature of its impulse response.

The operational definition of an impulse-type excitation implies a signal with much shorter duration than any inherent timescale in the system under study. From the point of view of discrete signals, the duration of a physical impulse should be about the sampling interval Δt. Once these conditions are satisfied, the amplitude of the response must be related to the energy delivered to the system through the applied physical impulse.

An alternative and often simpler approach is to use a step function $\underline{u}$ as input. The impulse $\underline{\delta}$ is the derivative of a step, $\delta_i = u_{i+1} - u_i$ (Chapter 3). Therefore, system linearity implies that the impulse response $\underline{h}$ is the derivative of the step response $\underline{sr}$, and $h_i = sr_{i+1} - sr_i$. There may be some subtleties in this approach. For example, consider the implementation of this method with a single DoF oscillator; there are two possibilities:

- A positive step: bring a small mass m^* onto contact with the oscillator mass m and release it at once. (Note: the mass m^* should not be dropped!) The amplitude of the step is $g \cdot m^*$.
- A negative step: enforce a quasi-static deformation (a constant force is applied), then release the system at once.

In both cases, the mass m will oscillate about the final position, gradually converging to it. The time-varying displacement normalized by the amplitude of the step is the step response $\underline{sr}$. Note that the effective mass of the oscillator is $(m + m^*)$ in the first case and (m) in the second case. Then, different damping values and resonant frequencies will be determined in these two tests according to Equations 4.15 and 4.16. Therefore, proper system identification requires analytical correction in the positive step approach.

4.3.3 System Identification

The impulse response completely characterizes the LTI system. If there is an adequate analytical model for the system under study, the measured impulse response $\underline{h}$ can be least squares fitted to determine the system parameters. This is the inverse problem.

Following the example of a single DoF oscillator, let us assume that the system under consideration resembles a single DoF. In this case, Equation 4.14 is fitted to the measured impulse response to determine m, ω_n, and D. Some system characteristics can be identified with point estimators. For example, the decrement of peak amplitudes $h^{<peak>}$ in the measured impulse response $\underline{h}$ can be used to compute the damping D:

$$D = \frac{1}{2\pi} \cdot \ln\left(\frac{h^{<peak>}}{h^{<next\ peak>}}\right) \tag{4.17}$$

where $h^{<peak>}$ and $h^{<next\ peak>}$ are two consecutive peaks (see Figure 4.11). If damping is low ($D < 0.1$), the time between consecutive peaks is the period T_n of the single DoF system. Thus, the natural frequency ω_n is

$$\omega_n = 2\pi \cdot \frac{1}{t^{<next\ peak>} - t^{<peak>}} \tag{4.18}$$

4.4 CONVOLUTION: COMPUTING THE OUTPUT SIGNAL

The system output signal $\underline{y}$ is a convolution between the input $\underline{x}$ and the system impulse response $\underline{h}$. The mathematical expression for convolution logically follows from these observations:

- The impulse response $\underline{h}$ fully characterizes the LTI system.
- A signal $\underline{x}$ can be decomposed into scale and time-shifted impulses δ_{i-k}, where the scaling factor at the discrete time $i = k$ is the signal value x_k (Chapter 3):

$$x_i = \sum_k x_k \cdot \delta_{i-k} \tag{4.19}$$

- The generalized superposition principle applies to causal LTI systems; therefore, "*the sum of scaled and time-shifted impulses → the sum of equally scaled and time-shifted impulse responses*".

Then, the LTI system output $\underline{y}$ to an input $\underline{x}$ can be obtained by replacing the shifted impulse δ_{i-k} by the shifted impulse response h_{i-k} in Equation 4.19:

$$y_i = \sum_k x_k \cdot h_{i-k} \tag{4.20}$$

This is the *convolution sum.*[4] A graphical demonstration is shown in Figure 4.12. The convolution operator is denoted with an asterisk

$$\underline{y} = \underline{x} * \underline{h} \tag{4.21}$$

If the input signal was decomposed into step signals, the convolution sum would be obtained following a similar procedure, starting from the equation of signal decomposition into steps, and replacing the step for the step response.

Dimensional homogeneity in these equations is preserved because the input signal is decomposed in terms of values x_k with dimensions of [input], whereas the shifted impulses are dimensionless. On the other hand, the discrete impulse response $\underline{h}$ carries the dimensions of the transformation [output/input]. Thus the output has dimensions of [output] and the dimensional homogeneity of Equations 4.20 or 4.21 is satisfied.

4.4.1 Properties of the Convolution Operator – Combination of Subsystems

The convolution operator has several important properties (see exercises at the end of the chapter):

- commutative : $$\underline{x} * \underline{h} = \underline{h} * \underline{x} \tag{4.22}$$
- associative : $$\left(\underline{x} * \underline{h}^{<1>}\right) * \underline{h}^{<2>} = \underline{x} * \left(\underline{h}^{<1>} * \underline{h}^{<2>}\right) \tag{4.23}$$
- distributive : $$\left(\underline{x} * \underline{h}^{<1>}\right) + \left(\underline{x} * \underline{h}^{<2>}\right) = \underline{x} * \left(\underline{h}^{<1>} + \underline{h}^{<2>}\right) \tag{4.24}$$

The numbers shown as superscripts in angular brackets <> indicate two different impulse responses. The associative property can be used to compute the response of a system that consists of two subsystems in series with known impulse response: $\underline{h}^{<global>} = \underline{h}^{<1>} * \underline{h}^{<2>}$. On the other hand, if a system consists of two subsystems in parallel, the impulse response of the system can be computed from the impulse response of the individual subsystems as prescribed by the distributive law: $\underline{h}^{<global>} = \underline{h}^{<1>} + \underline{h}^{<2>}$. These results can be generalized to systems with any combination of series and parallel subsystems.

---

[4] The convolution sum in continuous time is defined as

$$y(t) = \int_{-\infty}^{\infty} x(\tau) \cdot h(t-\tau) \cdot d\tau$$

This integral is also known as "Duhamel's integral" and it first appeared in the early 1800s. Note that the integration includes the timescale in $d\tau$ whereas the summation in discrete time does not, which is in agreement with differences in units between h(t) and $\underline{h}$ discussed earlier.

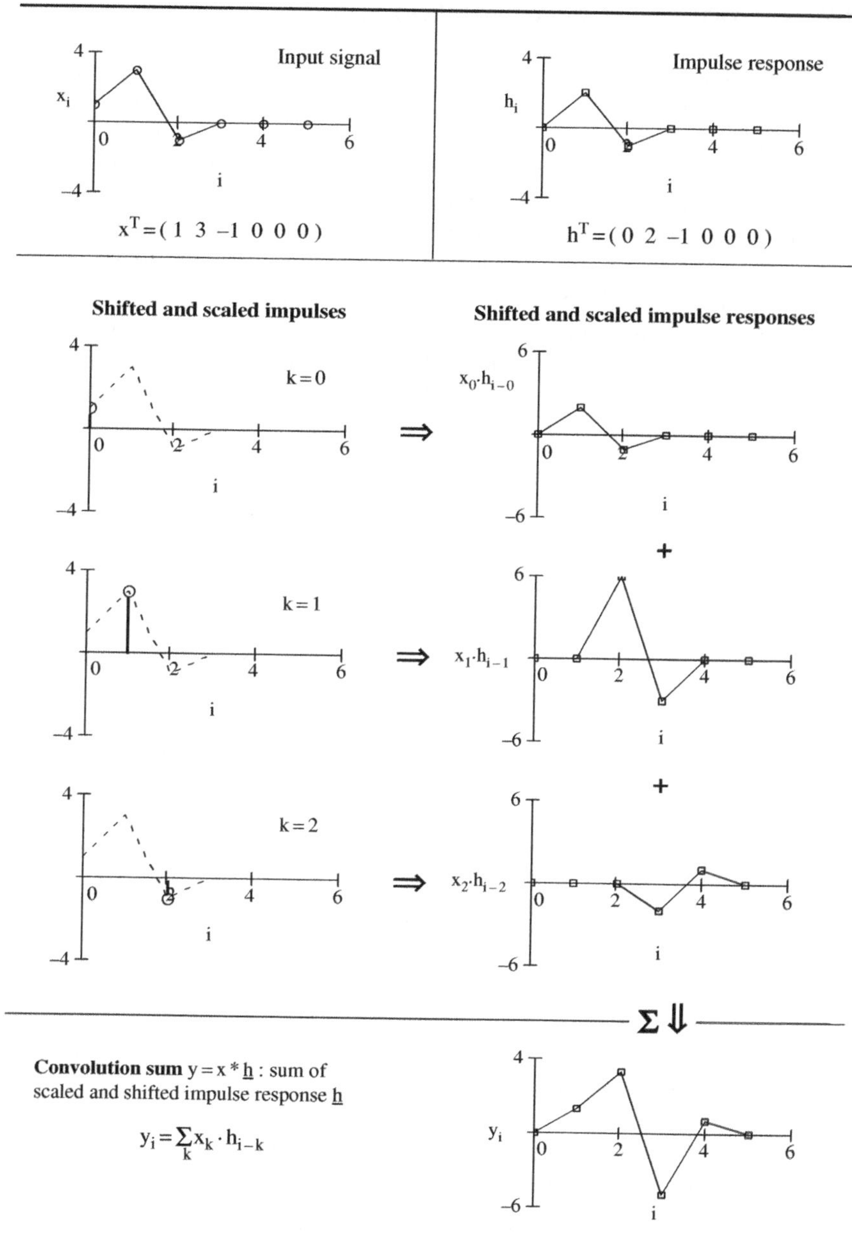

Figure 4.12 Graphical demonstration of the convolution sum

4.4.2 Computing the Convolution Sum

The implementation of the convolution operation in discrete time is demonstrated in the form of a spreadsheet computation in Figure 4.13. Each row shows a scaled and time-shifted impulse response. For example, the row corresponding to $k = 2$ shows the array for the impulse response $(h_0, h_1, h_2, h_3, \ldots)$ shifted two places and scaled by x_2. Entries in every column i are summed to compute the value of the output y_i corresponding to discrete time $t_i = i \cdot \Delta t$.

Convolution can be easily programmed in any algorithmic programming language (FORTRAN, C, Basic), with spreadsheets such as the one shown in Figure 4.13, or with mathematical software. Implementation Procedure 4.3 summarizes the algorithm for computation of the convolution sum and Figure 4.14 presents an example of the system response as computed with the convolution sum.

	$i = 0$	$i = 1$	$i = 2$	$i = 3$	$i = 4$	i
k	$\mathbf{x_0 \cdot h_0}$	$\mathbf{x_k \cdot h_{1-k}}$	$\mathbf{x_k \cdot h_{2-k}}$	$\mathbf{x_k \cdot h_{3-k}}$	$\mathbf{x_k \cdot h_{4-k}}$	$\mathbf{x_k \cdot h_{i-k}}$
0	$x_0 \cdot h_0$	$x_0 \cdot h_1$	$x_0 \cdot h_2$	$x_0 \cdot h_3$	$x_0 \cdot h_4$	...
1		$x_1 \cdot h_0$	$x_1 \cdot h_1$	$x_1 \cdot h_2$	$x_1 \cdot h_3$	...
2			$x_2 \cdot h_0$	$x_2 \cdot h_1$	$x_2 \cdot h_2$	...
3				$x_3 \cdot h_0$	$x_3 \cdot h_1$	...
4					$x_4 \cdot h_0$	...
7						...
Σ	$\Downarrow$	$\Downarrow$	$\Downarrow$	$\Downarrow$	$\Downarrow$	
y_i	$x_0 \cdot h_0$	$\sum_k x_k \cdot h_{1-k}$	$\sum_k x_k \cdot h_{2-k}$	$\sum_k x_k \cdot h_{3-k}$	$\sum_k x_k \cdot h_{4-k}$	$\sum_k x_k \cdot h_{i-k}$

Figure 4.13 Convolution. Computation spreadsheet

Implementation Procedure 4.3 Convolution sum

1. Determine the array that characterizes the system impulse response $\underline{h} = (h_0, h_1, h_2, \ldots h_i, \ldots)$ in discrete time $t_i = i \cdot \Delta t$.
2. Digitize the input signal with the same sampling interval Δt to produce the array $\underline{x} = (x_0, x_1, x_2, \ldots, x_i, \ldots)$. The number of points in arrays $\underline{h}$ and $\underline{x}$ does not need to be the same.

3. For a given value of i, perform the multiplications indicated in the following equation, and sum all values to obtain y_i. The summation is in k.

$$y_i = \sum_k x_k \cdot h_{i-k}$$

4. Repeat for next i, until the last element of the input signal x_{N-1} is reached. The resulting array $\underline{y}$ is the output obtained from the convolution of $\underline{x}$ and $\underline{h}$.

Example

Consider a conveyor belt. The impulse response $\underline{h}$ of a support is determined with a sledgehammer (Figure 4.14a). The predicted time history $\underline{x}$ of the repetitive forcing input is shown in Figure 4.14b. The estimated response of the support is computed by the convolution operation $\underline{y} = \underline{x} * \underline{h}$, and it is shown in Figure 4.14c.

Note: A more efficient convolution algorithm is presented in Chapter 6.

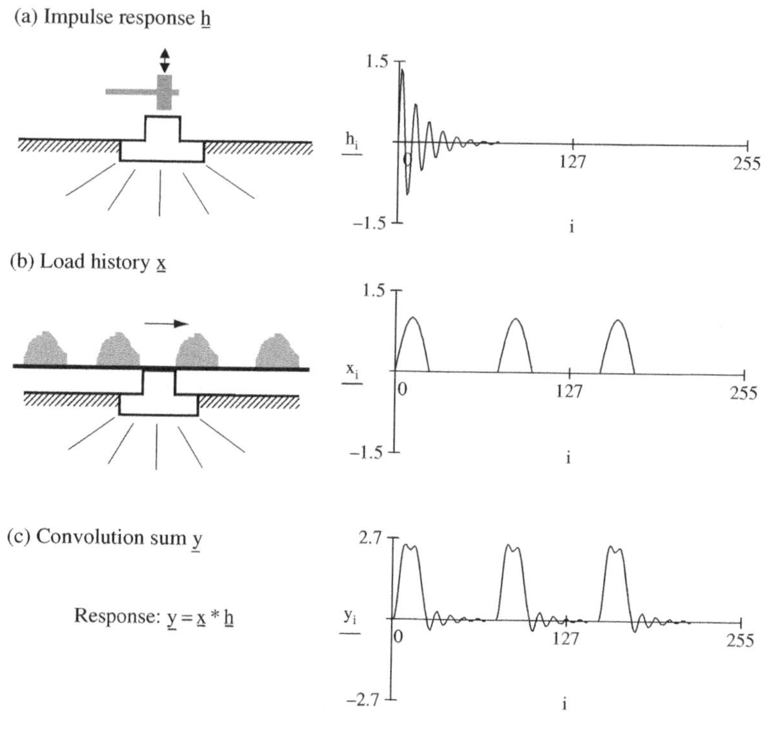

Figure 4.14 Convolution example. The dynamic response at the support of a belt conveyer

4.4.3 Revisiting Moving Kernels and Cross-correlation

Signal processing with moving kernels (Equation 4.4), cross-correlation (Equation 4.8) and convolution (Equation 4.20) share similar mathematical expressions:

moving kernel	*cross-correlation*	*convolution*
$y_i = \sum_{p=1}^{p=m} K_p \cdot x_{\left(i-\frac{m-1}{2}\right)+p}$	$cc_k^{<x,z>} = \sum_i x_i \cdot z_{i+k}$	$y_i = \sum_k x_k \cdot h_{i-k}$

Therefore, these three operations are classified as convolutions. The similarity between cross-correlation and convolution requires careful consideration. Compare the columns in the respective computation sheets (Figures 4.7 and 4.13). The two sheets are the same if: (1) both signals are of the same length N, (2) the signal $\underline{x}$ is tail-reversed in the cross-correlation operation, and (3) circularity applies so that the signal $\underline{h}$ repeats before and after, or it "wraps around". In this case,

$$\underline{cc}^{<x,z>} = \underline{z} * \text{rev}(\underline{x}) \tag{4.25}$$

The tail-reversed version of array $\underline{x} = [2, 4, 6, 5, 3, 1]$ is $\text{rev}(\underline{x}) = [2, 1, 3, 5, 6, 4]$. Note that the first element is x_0 in both arrays. While convolution is commutative, cross-correlation is not (Equation 4.11, $cc_k^{<x,z>} = cc_{-k}^{<z,x>}$) and this is properly accounted for by tail reversal in Equation 4.25.

4.5 TIME DOMAIN OPERATIONS IN MATRIX FORM

Convolution operations, including moving kernels and cross-correlation, are *summations of binary products*. This is analogous to matrix multiplication (Chapter 2).

The binary products involved in the convolution operation can be reproduced in matrix form by creating a matrix $\underline{\underline{h}}$ where each column is a shifted impulse response (Figure 4.15): the k-th column in matrix $\underline{\underline{h}}$ is the array $\underline{h}$ shifted down k places. The signal $\underline{x}$ is an N×1 vector, and the convolution operation in matrix form becomes

$$\underline{y} = \underline{\underline{h}} \cdot \underline{x} \tag{4.26}$$

Convolution is commutative; therefore, convolution in matrix form can be expressed in terms of the matrix $\underline{\underline{x}}$ made of vertically shifted copies of the array $\underline{x}$,

$$\underline{y} = \underline{\underline{x}} \cdot \underline{h} \tag{4.27}$$

The definition of convolution in the time domain does not require both arrays $\underline{x}$ and $\underline{h}$ to have the same number of elements; therefore, the matrix $\underline{\underline{h}}$ may not

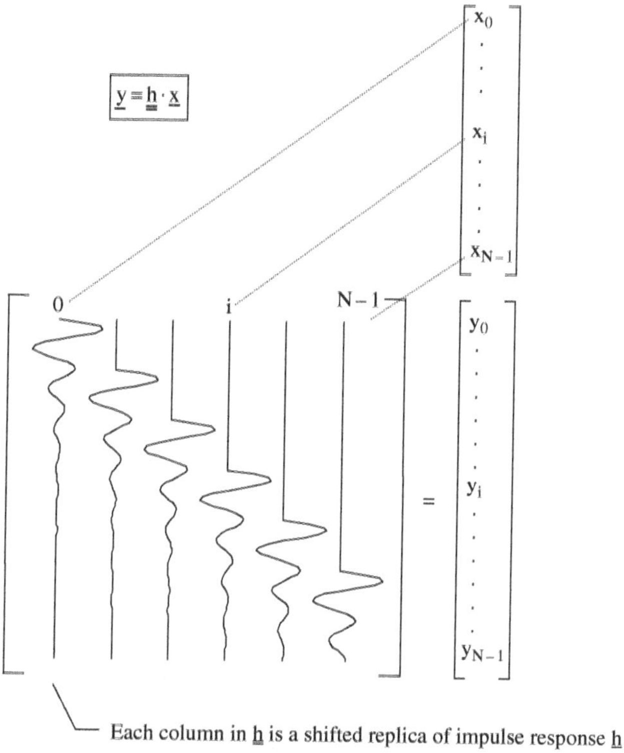

Figure 4.15 Convolution sum in matrix form. Matrix multiplication involves the summation of binary products. These are the operations required to implement convolution

be square. If the matrix $\underline{\underline{h}}$ in Equation 4.26 were invertible, the input $\underline{x}$ to a system could be inferred from the output $\underline{y}$ as $\underline{x} = \underline{\underline{h}}^{-1} \cdot \underline{y}$. This is *deconvolution*. Likewise, if the matrix $\underline{\underline{x}}$ in Equation 4.27 were invertible, the system impulse response $\underline{h}$ could be determined knowing the input and the output, $\underline{h} = \underline{\underline{x}}^{-1} \cdot \underline{y}$. This is *system identification*. These two inverse problems will be addressed in Chapter 8.

Although convolution operations in the time domain can be readily expressed in matrix form, higher computational efficiency is achieved when these operations are performed in the frequency domain. Time domain operations may still be of interest in some applications, such as deconvolution of data streams in real time. Furthermore, time domain operations avoid inherent assumptions made in the transformation to the frequency domain that lead to circular convolution (Chapters 5 and 6).

4.6 SUMMARY

- The decomposition of signals into scaled and shifted impulses leads to the analysis of signals and systems in the time domain.
- The first and most advantageous strategy to control noise is a proper experimental design.
- Detrending techniques remove low-frequency noise.
- Signal stacking is a robust alternative to control noise effects during signal recording. Signal stacking leads to increased signal-to-noise ratio, resolution, and dynamic range.
- Moving kernels permit implemention of a wide range of signal processing procedures. In particular, moving kernels can be used to remove high-frequency noise in recorded one-of-a-kind signals.
- The similarity between two signals is assessed with cross-correlation. Cross-correlation is useful in discovering replicas of a signal in the presence of noise and in identifying selected frequency components. Stationary noise with zero mean cancels out in the cross-correlation sum.
- The impulse response $\underline{h}$ fully characterizes the LTI system. It is a mathematical construct and its direct experimental determination is inappropriate or inconvenient in most cases.
- The output signal $\underline{y}$ is the convolution of the input signal $\underline{x}$ with the system impulse responses $\underline{h}$. Operationally, convolution is the sum of shifted impulse responses $\underline{h}$, scaled by the corresponding amplitude of the input signal.
- Signal processing with moving kernels and cross-correlation operations are convolutions.
- Convolution operations can be expressed in matrix form.

FURTHER READING

Komo, J. J. (1987). Random Signal Analysis in Engineering Systems. Academic Press, Inc., Orlando. 302 pages.

Meyers, D. G. (1990). Digital Fourier Transform: Efficient Convolution and Fourier Transforms Techniques. Prentice-Hall, New York. 355 pages.

Oppenhein, A. V., Willsky, A. S, and Young, I. T. (1983). Signals and Systems. Prentice-Hall, Inc., Englewood Cliffs. 796 pages.

Webster, G. M. (ed.) (1978). Deconvolution – Collection of Papers. Geophysics Reprint Series No. 1. Society of Exploration Geophysics, Tulsa, Okia. 492 pages.

SOLVED PROBLEMS

P4.1 *Application: longitudinal wave propagation.* A cylindrical aluminum rod is suspended by two strings. The rod is impacted at one end and signals are collected with an accelerometer mounted on the opposite end (see sketch: rod length $L = 2.56\,m$). Captured signals record multiple reflections in the rod. Giving an ensemble of 20 signals: (a) detrend the data, (b) stack, and (c) calculate the travel time between reflections using the autocorrelation function. Calculate the wave velocity in the rod.

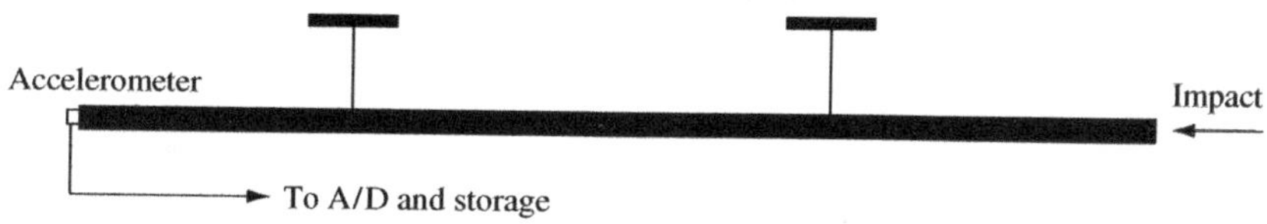

Solution: Twenty signals are collected with a sampling rate of 500 kHz so that the sampling interval is $\Delta t = 2 \cdot 10^{-6}\,s$ (data by J. Alvarellos and J. S. Lee – Georgia Tech). A segment of one record is shown next:

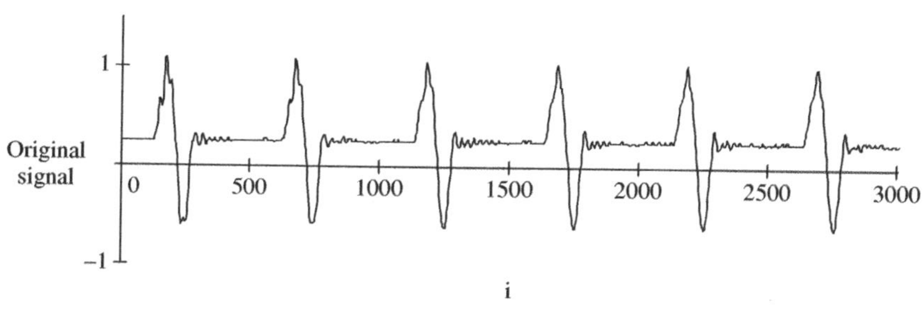

Detrend each signal. Calculate the DC component for each signal and subtract it $z_i^{<detrended>} = z_i - DC$. Repeat for all 20 records.

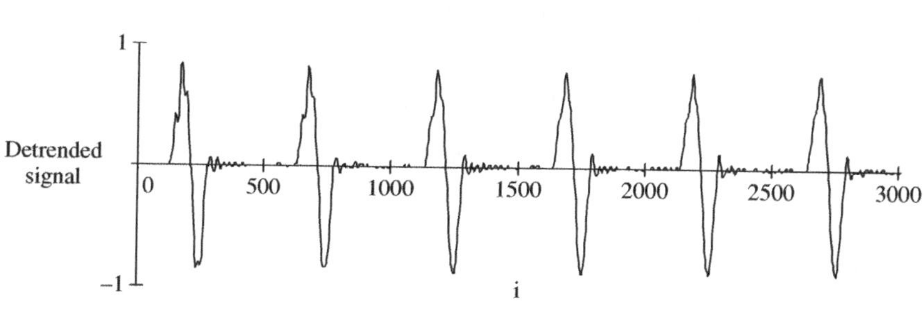

Signal stacking. Implement stacking to improve SNR. Note: given the high signal-to-noise ratio in the original signal, and the inherent noise cancellation in autocorrelation, stacking is not needed in this case.

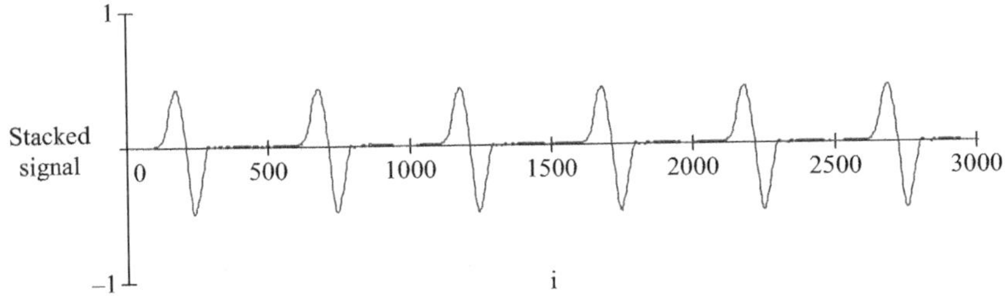

Autocorrelation. Calculate the autocorrelation using the stacked signal:

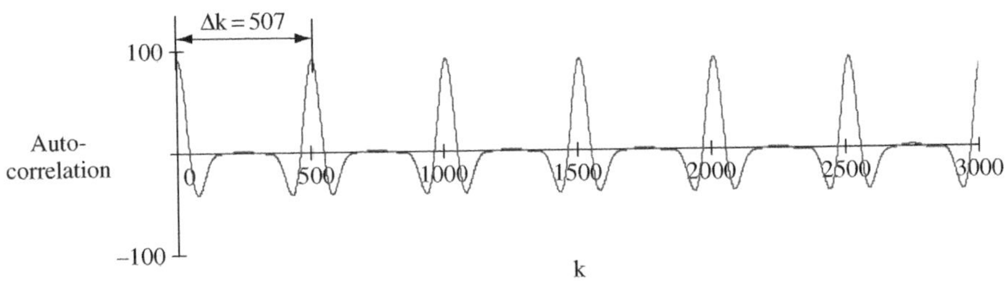

The time difference between consecutive peaks is the travel time tt between reflections:

$$tt = \Delta k \cdot \Delta t = 507 \cdot 2 \cdot 10^{-6} s = 1.014 \cdot 10^{-3} s$$

The wave velocity in the aluminum rod is

$$V = \frac{2L}{tt} = \frac{2 \cdot 2.56 m}{1.014 \cdot 10^{-3} s} = 5049 \frac{m}{s}$$

This is the longitudinal wave velocity in a rod (Note: it is lower than the P-wave velocity in an infinite body $V_P = 6400\,m/s$).

P4.2 *Convolution.* Consider the following short signals and demonstrate the associative property of the convolution operator.

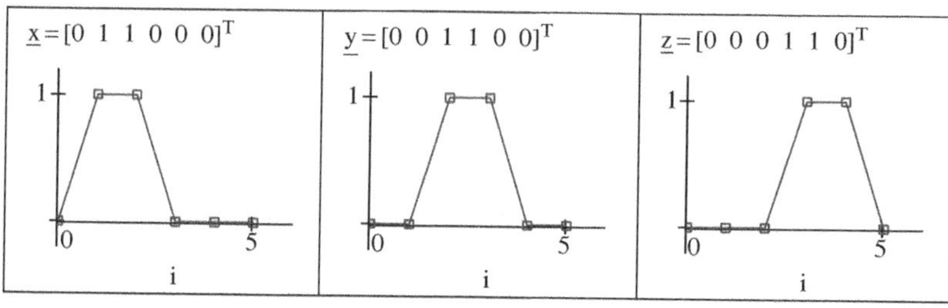

Solution:

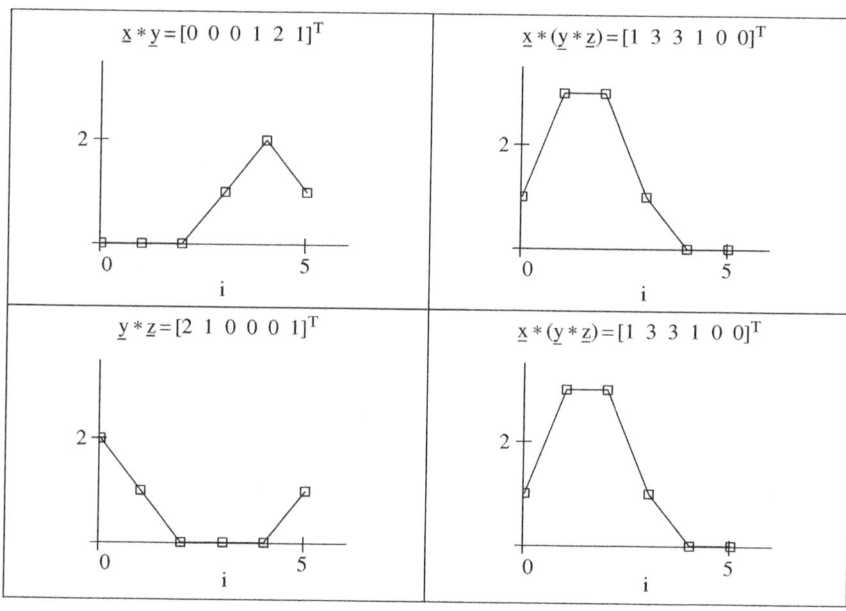

The signals on the right-hand column verify $(\underline{x} * \underline{y}) * \underline{z} = \underline{x} * (\underline{y} * \underline{z})$.

ADDITIONAL PROBLEMS

P4.3 *Noise control in the time domain.* Noise control by stacking in time or in frequency domains is based on statistical principles related to the distribution of the mean. (a) Prove by numerical simulation the relationship between the mean of the means and the population mean, and the standard deviation of the sample means in relation to the population standard deviation. (b) Derive the equation to compute the number of signals required so that the mean $x_i^{<avr>}$ at discrete time i is within predefined bounds, with probability p.

P4.4 *Noise control.* Is smoothing with moving average a linear operation? Is median filtering a linear operation? (Hint: start with the definition of a linear system, Section 3.5.)

P4.5 *Stacking and resolution.* Use numerical simulation to explore the effect of stacking on resolution and dynamic range for different noise levels. Discuss.

P4.6 *Impulse response.* Demonstrate that if the input $\underline{x}$ is a step function, the derivative of the step response is the system impulse response. (Hint: link the concept of numerical derivative with the modified superposition principle for LTI systems.)

P4.7 *Convolution operator.* Verify that convolution satisfies the distributive property (numerically or in close form). Does the demonstration require the assumption of linearity?

P4.8 *Convolution and cross-correlation operators.* Given: $x = [0, 1, 0, -2, 0, 0, 0, 0]$ and $h = [0, 0, 0, 10, 10, 10, 0, 0]$, compute $y = x * h$ and $cc^{<x,h>}$ (no computer needed!)

P4.9 *Convolution and cross-correlation.* Prepare a detailed flowchart for programming the convolution and cross-correlation operators. Program the two algorithms. Use simulated signals to compare results computed with the cross-correlation algorithm and using the convolution algorithm with tail reversal. Compute the autocorrelation of background noise.

P4.10 *Convolution in matrix form.* Prove that convolution in matrix form can be implemented by writing either the impulse response or the input as the transformation matrix. In other words, show that Equations 4.26 and 4.27 lead to the same result.

P4.11 *Convolution in matrix form.* Write the matrix convolution operator $\underline{\underline{h}}$ for an underdamped single DoF system excited by a transient at its base. Is the matrix invertible?

P4.12 *Application: optimal design of speed bumps.* Consider a car as a single DoF system excited at its base, with damping D and resonant frequency ω. Use this model to design speed bumps to promote cruising speeds lower than a preselected speed V_{max}.

P4.13 *Application: pavement monitoring system.* Utilize the concepts developed in Problem 4.13 to design a pavement monitoring system. The goal is to determine the pavement surface profile from acceleration records taken with a small wheel towed behind a car. Design an experimental procedure to calibrate the system and discuss possible nonlinearities related to wheel diameter and the geometry of surface features. Note that this is a deconvolution-type inverse problem!

5

Frequency Domain Analysis of Signals (Discrete Fourier Transform)

Discrete time signals can be analyzed or decomposed into a series of sines and cosines. This representation is called the discrete Fourier transform (DFT) of the signal; it is reversible and no information is lost. The DFT underlies most signal processing strategies, facilitates the interpretation of signals, enhances the characterization of systems, and improves the efficiency of algorithms. However, there are several inherent assumptions and limitations in this transformation.

Why are sines and cosines selected to analyze signals and systems? There are two reasons. First, sines and cosines are *orthogonal functions* and form a base for the analysis of signals, as discussed in this chapter. Second, sines and cosines are *eigenfunctions* for LTI systems; this will be the starting point for Chapter 6.

5.1 ORTHOGONAL FUNCTIONS – FOURIER SERIES

Two functions are orthogonal in the interval [a, b] if

$$\int_a^b f_u(t) \cdot \bar{f}_v(t)\, dt = \begin{cases} 0 & \text{if } u \neq v \\ c & \text{if } u = v \end{cases} \tag{5.1}$$

Discrete Signals and Inverse Problems J. C. Santamarina and D. Fratta

where f_u and f_v are functions with real and imaginary components, $\bar{f}$ indicates complex conjugate of the function f, and c is any number different than zero.

Given a sinusoid of circular frequency $\omega = 2\pi/T$, its u-th harmonic is another sinusoid with circular frequency $\omega = u \cdot (2\pi/T)$, where u is an integer. Harmonics fulfill the orthogonality property; therefore, the following relations hold:

$$\int_0^T \sin\left(\frac{2\pi}{T}t\right) \cdot \sin\left(u\frac{2\pi}{T}t\right) \cdot dt = \begin{cases} 0 & \text{if } u \neq 1 \\ \frac{T}{2} & \text{if } u = 1 \end{cases} \tag{5.2}$$

$$\int_0^T \cos\left(\frac{2\pi}{T}t\right) \cdot \cos\left(u\frac{2\pi}{T}t\right) \cdot dt = \begin{cases} 0 & \text{if } u \neq 1 \\ \frac{T}{2} & \text{if } u = 1 \end{cases} \tag{5.3}$$

$$\int_0^T \sin\left(\frac{2\pi}{T}t\right) \cdot \cos\left(u\frac{2\pi}{T}t\right) \cdot dt = 0 \qquad \text{for all } u \tag{5.4}$$

Invoking Euler's identities (Chapter 2), these equations show that complex exponentials are orthogonal as well (see solved problem at the end of this Chapter):

$$\int_0^T e^{j\cdot\left(\frac{2\pi}{T}t\right)} \cdot e^{-j\cdot\left(u\frac{2\pi}{T}t\right)} \cdot dt = \begin{cases} 0 & \text{if } u \neq 1 \\ T & \text{if } u = 1 \end{cases} \tag{5.5}$$

The integral equation used to determine the orthogonality of two functions is equivalent to the equation used to determine the value of cross-correlation for zero time shift ($\tau = 0$ in continuous time). Hence, orthogonality concepts support the utilization of cross-correlation to identify frequency similarity between two signals (Chapter 4).

5.1.1 Fourier Series

The orthogonality of harmonics suggests that these functions form a *base* in the open interval [0, T[. Then, a continuous periodic function f(t) with period T can be expressed as a linear combination of sinusoids with frequencies that are multiples of the fundamental circular frequency $2\pi/T$. The summation is known as Fourier series. The value at discrete time t_i is

$$f_i = \sum_{u=-\infty}^{\infty} \left[a_u \cdot \cos\left(u\frac{2\pi}{T}t_i\right) + b_u \cdot \sin\left(u\frac{2\pi}{T}t_i\right) \right] \tag{5.6}$$

where the coefficients a_u and b_u are real.

5.1.2 An Intuitive Preview of the Fourier Transform

Imagine N points in the x–t Cartesian coordinates $\underline{x} = [x_0, x_1, x_2, x_3, \ldots]$. If a polynomial is least squares fitted through these points $p(t) = a + bx + cx^2 + dx^3 + \ldots$, we could call the set of coefficients $\underline{p} = [a, b, c, d, \ldots]$ the "polynomial transform of $\underline{x}$".

Likewise, one can least squares fit the Fourier series (Equation 5.6) to the signal $\underline{x}$. The set of coefficients $\underline{X} = [a_0, b_0, a_1, b_1, a_2, b_2, a_3, b_3 \ldots]$ is called the Fourier transform of $\underline{x}$, which is herein denoted with a capital letter. The subset made of all a-coefficients is called the "real" part $Re(\underline{X})$, whereas the subset of b-coefficients is called the "imaginary" part $Im(\underline{X})$. Each subset plotted versus the frequency counter u provides important information about the signal $\underline{x}$:

- The u-th value in $Re(\underline{X})$ is the amplitude of the cosine with frequency $u(2\pi/T)$ that is needed to form the signal $\underline{x}$.
- The u-th value in $Im(\underline{X})$ is the amplitude of the sine with frequency $u(2\pi/T)$ that is needed to form the signal $\underline{x}$.

where the fundamental period T as the length of the time window, so that the fundamental circular frequency is $2\pi/T$. Figure 5.1 shows a collection of simple signals and the corresponding real and imaginary parts of their Fourier transform obtained by fitting the Fourier series to the signals. The signals are simple and Fourier transforms are identified by visual inspection and comparison with Equation 5.6. A few important observations follow:

- A constant signal, $x_i =$ constant for all i, has no oscillations; therefore, all terms for $u > 0$ are null: $a_{u>0} = 0$ and $b_{u>0} = 0$. For $u = 0$, $\cos(0) = 1$, and the first real coefficient a_0 takes the value of the signal. On the other hand, $\sin(0) = 0$, and any value for the first imaginary coefficient b_0 could be used; however, $b_0 = 0$ is typically assumed. For example, fitting the Fourier series to $\underline{x} = [7, 7, 7, 7, 7, \ldots]$ results in $Re(X) = [7, 0, 0, 0, 0, \ldots]$ and $Im(X) = [0, 0, 0, 0, 0, \ldots]$, as shown in Figure 5.1a.
- The Fourier transform of a single frequency cosine signal is an impulse in Re(X), whereas the transform of a single frequency sine is an impulse in Im(X). For example, if a sine signal with amplitude 7 fits three times in the time window, then the Fourier transform obtained by fitting Equation 5.6 is an impulse corresponding to the third harmonic in the imaginary component, $b_3 = 7$, and $Re(X) = [0, 0, 0, 0, 0, 0, \ldots]$ and $Im(X) = [0, 0, 0, 7, 0, 0, \ldots]$ as shown in Figure 5.16.
- Because the Fourier series is a summation, superposition is implied, and the cases in Figure 5.1d and e are readily computed.

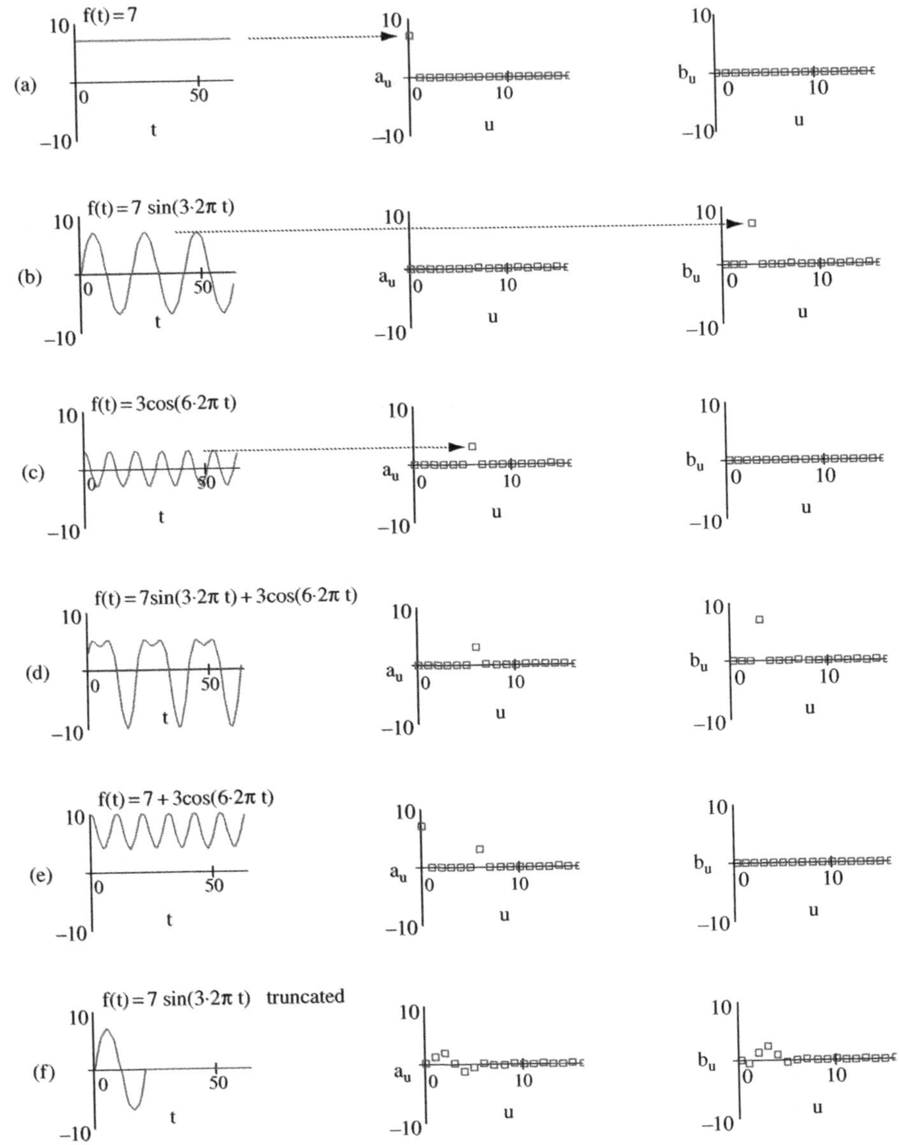

Figure 5.1 Simple signals and the corresponding real (cosine) and imaginary (sine) components of the fitted Fourier series. Note that the truncated sinusoid requires additional frequency components to synthesize the signal

- What is the Fourier transform of a signal duration T with a one-cycle sinusoid duration T/3, shown in Figure 5.1f? A good initial guess is to assume that b_3 will not be zero. Furthermore, there must be other nonzero real and imaginary components; otherwise, the sinusoid would be present throughout the duration of the signal.

This intuitive preview suggests a robust interpretation of the Fourier transform: it is curve fitting the signal a series of cosines (*real part*) and sines (*imaginary part*). However, there are several subtleties. For example, note that the signal $\underline{x}$ exists from time zero to T, that is $0 \leq t_i < T$. However, the sinusoids that are used to fit the signal $\underline{x}$ exist from "the beginning of time till the end of time, all the time", that is $-\infty < t < +\infty$. The implications of discretization are explored in the next sections.

5.2 DISCRETE FOURIER ANALYSIS AND SYNTHESIS

There are four types of Fourier time-frequency transforms according to the continuous or discrete representation of the information in each domain: continuous-continuous, continuous-discrete, discrete-continuous, and discrete-discrete. Current engineering and science applications invariably involve discrete time and frequency representations. Consequently, only the case of discrete-discrete transformation is considered.

There is an immediate and most relevant consequence of selecting discrete time and frequency representations: *The discrete time and frequency Fourier transform presumes periodic signals.* In other words, any aperiodic signal $\underline{x}$ with N points $[x_0, \ldots, x_{N-1}]$ is automatically assumed periodic with fundamental period $T = N \cdot \Delta t$. A schematic representation is shown in Figure 5.2.

5.2.1 Synthesis: The Fourier Series Rewritten

The Fourier series in Equation 5.6 is rewritten to accommodate the discrete nature of the signals in time and frequency domains, and the inherent periodicity associated with the discrete representation. The sequence of changes is documented next.

Change #1: Exponentials

Sines and cosines are replaced for complex exponentials by means of Euler's identities with complex coefficients X_u (Chapter 2):

$$f(t) = \sum_{u=-\infty}^{+\infty} X_u \cdot e^{j \cdot \left(u \frac{2\pi}{T} t\right)} \tag{5.7}$$

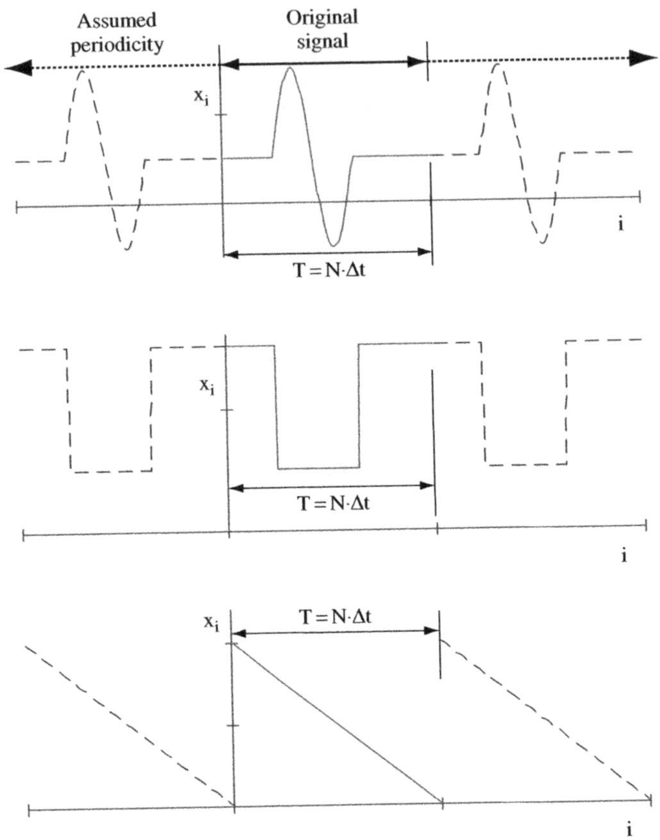

Figure 5.2 The discrete time and frequency Fourier transform assumes periodicity. Therefore, aperiodic signals are converted to periodic signals. The continuous line represents the captured signal. The dotted lines are the presumed periodic repetition from time $-\infty$ to $+\infty$

Change #2: Discrete Time

The inherent periodicity of a discrete time signal $\underline{x}$ is $T = N \cdot \Delta t$ and discrete time time is $t_i = i\Delta t$. Then, Equation 5.7 becomes

$$x_i = \sum_{u=-\infty}^{+\infty} X_u \cdot e^{j \cdot \left(u \frac{2\pi}{N \cdot \Delta t} i \cdot \Delta t\right)} = \sum_{u=-\infty}^{+\infty} X_u \cdot e^{j \cdot \left(u \frac{2\pi}{N} i\right)} \tag{5.8}$$

Change #3: Nyquist Criterion

The highest frequency that can be resolved from a discrete time signal is the Nyquist frequency $1/(2 \cdot \Delta t)$, as shown in Chapter 3. Therefore, the highest frequency of any harmonic to be included in the series is $u_{max} \cdot (1/T) = 1/(2 \cdot \Delta t)$. Replacing $T = N \cdot \Delta t$, the discrete time Fourier series need not extend beyond $u_{max} = N/2$. Keeping N summation terms, from $-N/2$ to $(N/2) - 1$,

$$x_i = \sum_{u=-\frac{N}{2}}^{\frac{N}{2}-1} X_u \cdot e^{j \cdot \left(u \frac{2\pi}{N} i\right)} \tag{5.9}$$

Change #4: Shift in Summation Limits

The complex exponential does not change if either u or $u + N$ appear in the exponent because $e^{j2\pi} = e^{j2\pi N} = 1$. Then the summation limits are shifted while keeping N-terms in the summation. In particular, Equation 5.9 can be written as

$$x_i = \sum_{u=0}^{N-1} X_u \cdot e^{j \cdot \left(u \frac{2\pi}{N} i\right)} \tag{5.10}$$

where negative frequencies are avoided. The fact that the summation limit goes above N/2 does not imply that higher frequencies are extracted from the discrete signal. This is just a mathematical effect that will be discussed further in the text. An important conclusion from this analysis is that the *Fourier series for discrete time periodic signals is finite*.

5.2.2 Analysis: Computing the Fourier Coefficients

Fourier coefficients X_u can be obtained by least squares fitting the signal $\underline{x}$ with the Fourier series in Equation 5.10: given the array $\underline{x}$, identify each coefficient X_u so that the total square error E between measured values x_i and predicted values $x_i^{<pred>}$ is minimized, $\min[E = \Sigma(x_i - x_i^{<pred>})^2]$. When the fitting is complete, the residual is $E = 0$. (There may be some numerical noise. See solved problems in Chapter 3.)

A better alternative is to call upon the orthogonality property of harmonics to identify how much the signal $\underline{x}$ (N points sampled with an interval Δt) resembles a given sinusoid of frequency $\omega_u = u \cdot 2\pi/(N \cdot \Delta t)$. Following this line of thought, the Fourier coefficients are computed as the zero-shift value of the cross-correlation:[1]

$$X_u = \sum_{i=0}^{N-1} x_i \cdot e^{-j \cdot \left(u \frac{2\pi}{N} i\right)} \tag{5.11}$$

Note that Equation 5.11 is a summation in the time index i, whereas Equation 5.10 is a summation in the frequency index u. The Fourier coefficient $Re(X_0) = \sum x_i$ captures the *static component* of the signal (zero-offset or DC-offset) and the zero frequency imaginary coefficient is assumed $Im(X_0) = 0$. The array $\underline{X}$ formed with the complex Fourier coefficients X_u is the "discrete Fourier transform" or frequency domain representation of the discrete time signal $\underline{x}$. The magnitude of the Fourier coefficient X_u relates to the magnitude of the sinusoid of frequency $\omega_u = u \cdot 2\pi/T$ that is contained in the signal with phase ϕ_u

$$|X_u| = \sqrt{[Re(X_u)]^2 + [Im(X_u)]^2} \qquad \textit{amplitude} \tag{5.12}$$

$$\varphi_u = \tan^{-1}\left(\frac{Im(X_u)}{Re(X_u)}\right) \qquad \textit{phase} \tag{5.13}$$

5.2.3 Selected Fourier Pair

The analysis equation and its corresponding synthesis equation form a "Fourier pair". From Equations 5.10 and 5.11,

$$X_u = \sum_{i=0}^{N-1} x_i \cdot e^{-j \cdot \left(u \frac{2\pi}{N} i\right)} \qquad \textit{analysis equation: time} \rightarrow \textit{frequency} \tag{5.14}$$

$$x_i = \frac{1}{N} \sum_{u=0}^{N-1} X_u \cdot e^{j \cdot \left(u \frac{2\pi}{N} i\right)} \qquad \textit{synthesis equation: frequency} \rightarrow \textit{time} \tag{5.15}$$

The normalization factor 1/N is added in the synthesis equation to maintain energy content in time → frequency → time transformations.

There are different Fourier pairs available in computer software and invoked in the literature. This Fourier pair is notably convenient in part owing to the

---

[1] The Fourier transform and the Laplace transform in continuous time are:

$$\text{Fourier: } X(\omega) = \int_{-\infty}^{\infty} x(t) \cdot e^{-j \cdot \omega \cdot t} dt \qquad \text{Laplace: } X(s) = \int_{-\infty}^{\infty} x(t) \cdot e^{-s \cdot t} dt$$

where s is the complex variable $s = \sigma + j \cdot \omega$. When $\sigma = 0$, the Laplace transform becomes the Fourier transform. The z-transform is the discrete time equivalent of the Laplace transform.

parallelism between the analysis and the synthesis equations. (Other advantages will be identified in Chapter 6.) If the DFT is implemented with a given analysis equation, the inverse DFT (IDFT) must be computed with the corresponding synthesis equation in the pair. Table 5.1 summarizes the Fourier pair and related expressions.

The DFT of a one-dimensional (1D) signal in time involves the frequency $\omega = 2\pi/T$ and its harmonics. If the parameter being monitored varies along a spatial coordinate ℓ, the wave number $\kappa = 2\pi/\lambda$ is used instead. Analogous to signals in time, the maximum wavelength λ that is captured in the discrete record depends on the sampling interval $\Delta\ell$ and the number of points N so that $\lambda = N \cdot \Delta\ell$, and the exponent $u \cdot \omega \cdot t$ in the complex exponential becomes

$$u \cdot \frac{2\pi}{\lambda} \cdot \ell = u \cdot \frac{2\pi}{N \cdot \Delta\ell} \cdot i \cdot \Delta\ell = u \cdot \frac{2\pi}{N} \cdot i \qquad \textit{in space} \qquad (5.16)$$

Therefore, the formulation presented earlier is equally applicable to spatial variables.

Table 5.1 Summary: discrete Fourier transform pair and related expessions

Analysis (*from time → to frequency*)	**Synthesis** (*from frequency → to time*)
$X_u = \sum_{i=0}^{N-1} x_i \cdot e^{-j\left(u\frac{2\pi}{N}i\right)}$	$x_i = \frac{1}{N}\sum_{u=0}^{N-1} X_u \cdot e^{j\cdot\left(u\frac{2\pi}{N}i\right)}$
Static component:	$X_0 = \sum_i x_i$
Magnitude:	$\lvert X_u\rvert = \sqrt{[\mathrm{Re}\,(X_u)]^2 + [\mathrm{Im}\,(X_u)]^2}$
Phase:	$\varphi_u = \tan^{-1}\left[\frac{\mathrm{Im}\,(X_u)}{\mathrm{Re}\,(X_u)}\right]$
Parseval's identity:	$\sum_{i=0}^{N-1} x_i^{\,2} = \frac{1}{N} \cdot \sum_{u=0}^{N-1} \lvert X_u\rvert^2$

The following expressions are worth highlighting:

$t_i = i \cdot \Delta t$	$T = N \cdot \Delta t$
$f_{min} = \frac{1}{T}$	$f_{max} = \frac{1}{2 \cdot \Delta t}$
$f_u = u\frac{1}{T} = u\frac{1}{N \cdot \Delta t}$	$\omega_u = 2\pi f_u = u\frac{2\pi}{T} = u\frac{2\pi}{N \cdot \Delta t}$

Note:
The physical dimensions are the same in both domains.
Summations in "u" can be reduced to (N/2)+1 terms by recalling the symmetry and periodicity properties. When the summation is extended from $u = 0$ to $u = N - 1$ the operation is called "double sided". When the summation is extended from $u = 0$ to N/2, the operation is called "single sided".

5.2.4 Computation - Example

In 1965, J. Tukey and J. Cooley published an algorithm for the efficient implementation of the DFT. This algorithm and other similar ones developed since are known as the "fast Fourier transform" (FFT). Maximum computational efficiency is attained when the signal length is a power of 2, $N = 2^r$, where r is an integer.

Signal analysis and synthesis are demonstrated in Figure 5.3. The aperiodic tooth signal in Figure 5.3a is transformed to the frequency domain. (Recall that the discrete time and frequency representation presumes this signal repeats itself.) Both real and imaginary components are shown in Figures 5.3b and c. Observe that the static component is equal to Σx_i. The synthesis of the signal is incrementally computed by adding increasingly more terms in the Fourier series. Figures 5.2d–k show the evolution of the synthesized signal. The last synthesized signal in Figure 5.2k is identical to the original signal $\underline{x}$.

5.3 CHARACTERISTICS OF THE DISCRETE FOURIER TRANSFORM

The most important properties of the DFT are reviewed in this section. Exercises at the end of this chapter suggest the numerical verification of these properties.

5.3.1 Linearity

The Fourier transform is a sum of binary products, thus, it is distributive. Therefore, given two discrete time signals $\underline{x}$ and $\underline{y}$, and their Fourier transforms $\underline{X}$ and $\underline{Y}$

$$\left(a \cdot \underline{x} + b \cdot \underline{y}\right) \xrightarrow{\text{DFT}} (a \cdot \underline{X} + b \cdot \underline{Y}) \tag{5.17}$$

5.3.2 Symmetry

The cosine is an even function $\cos(u\theta) = \cos(-u\theta)$, whereas sine is odd $\sin(u\theta) = -\sin(-u\theta)$. Therefore, it follows from Euler's identities (Chapter 2) that the Fourier coefficient for the frequency index u is equal to the complex conjugate of the Fourier coefficient for $-u$

$$X_u = \overline{X_{-u}} \tag{5.18}$$

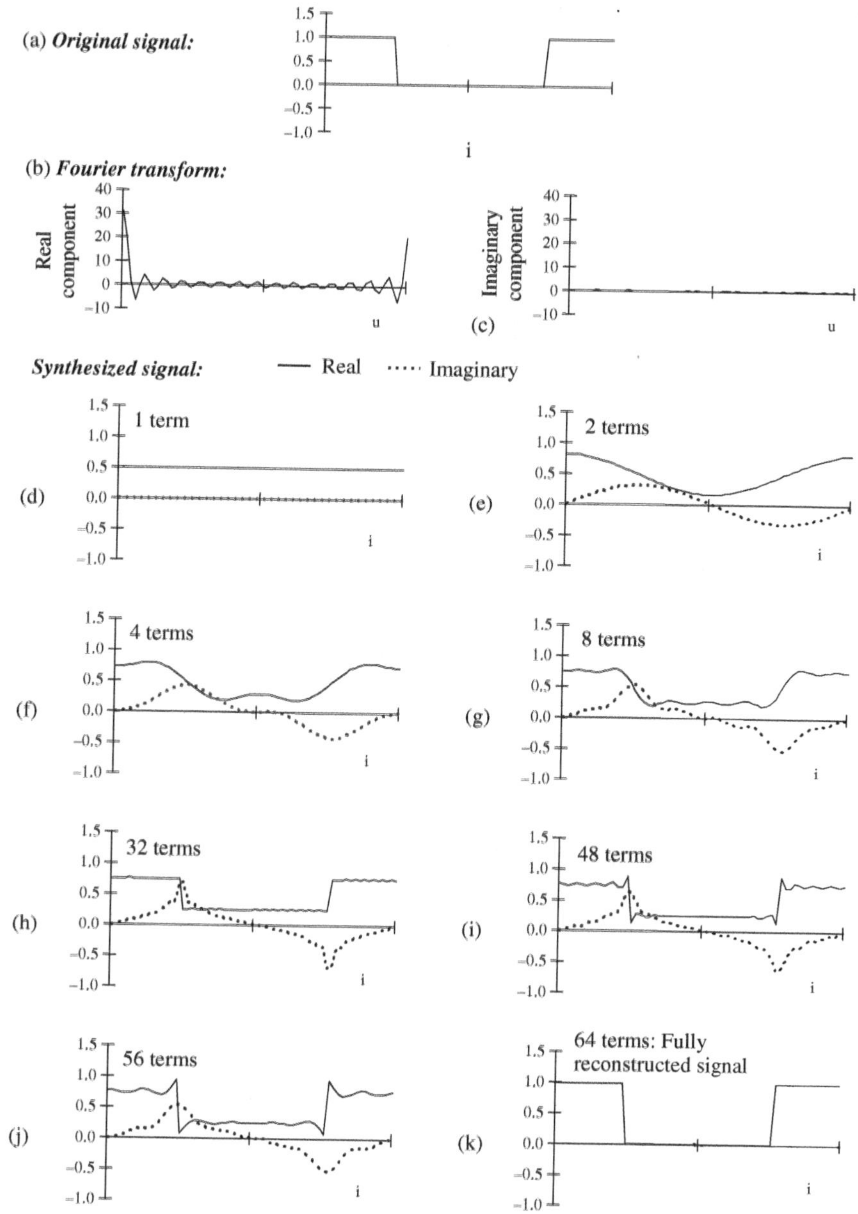

Figure 5.3 Analysis and synthesis: (a) original signal, N = 64; (b) and (c) analysis: real and imaginary components of the DFT; (d)–(k) synthesis: incremental reconstruction of the signal by adding an increasingly higher number of Fourier components

5.3.3 Periodicity

As invoked earlier in relation to Equation 5.10, the complex exponential for frequency $\omega_u = (u+N) \cdot (2\pi/N \cdot \Delta t)$ has the same values at discrete times t_i as an exponential with lower frequency $\omega_u = u(2\pi/N \cdot \Delta t)$. Therefore,

$$X_u = X_{u+N \cdot k} \tag{5.19}$$

where k is an integer. Therefore, *the discrete time and frequency domain assumption inherently implies a periodic signal in time and in frequency*, and the corresponding arrays in each domain repeat with periodicity:

$$T = N \cdot \Delta t \text{ (in time domain)} \qquad N\frac{2\pi}{T} = \frac{2\pi}{\Delta t} \text{ (in frequency domain)} \tag{5.20}$$

Figure 5.4 presents a discrete signal $\underline{x}$ and its discrete transform $\underline{X}$, and highlights the periodicities in time domain and frequency domains.

5.3.4 Convergence – Number of Unknown Fourier Coefficients

It would appear that there are N complex coefficients X_u; hence, $2 \cdot N$ unknowns. However, the periodicity and symmetry properties of the Fourier transform guarantee that $X_u = \overline{X_{N-u}}$, where the bar indicates complex conjugate. Furthermore, X_0 and $X_{N/2}$ are real. Then, the number of unknowns is reduced to N. Indeed, this must be the case: each value x_i permits writing one equation like Equation 5.15, and given that complex exponentials form a base, the number of unknown Fourier coefficients must be equal to the number of equations N. The following numerical example verifies these observations. Consider the time series $\underline{x} = [1, 0, 1, 1, 0, 1, 1, 2]$ with $N = 8$ elements. The DFT of $\underline{x}$ is obtained using Equation 5.14:

u	0	1	2	3	4	5	6	7
X_u	7	$1+j \cdot \sqrt{2}$	$-1+j \cdot \sqrt{2}$	$1+j \cdot \sqrt{2}$	-1	$1-j \cdot \sqrt{2}$	$-1-j \cdot \sqrt{2}$	$1-j \cdot \sqrt{2}$

Note that the array $\underline{X}$ fulfills the relation $X_u = \overline{X_{N-u}}$, and that $X_0 = 7$ and $X_{N/2} = -1$ are real; therefore, there are only N unknowns.

The fact that N values in the time domain are fitted with N Fourier coefficients in the frequency domain implies that there will be no convergence difficulties in the DFT of discrete time signals. (Convergence problems develop in continuous

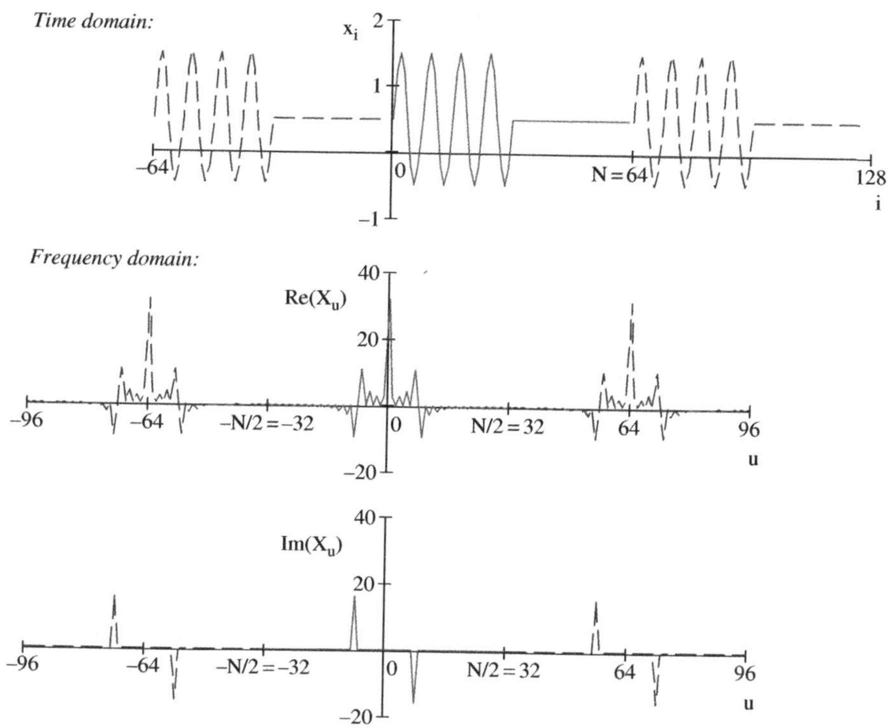

Figure 5.4 The DFT presumes the signal is periodic both in the time and the frequency domains. Observe the symmetry properties of real and imaginary components. The time series $\underline{x}$ has a DC offset, thus $\mathrm{Re}(X_0) \neq 0$.

time signals around discontinuities. This is Gibb's phenomenon, and it manifests as ripples and overshoots near discontinuities.) In addition, the N information units available in the time domain are preserved in the frequency domain, as confirmed by the fact that $\underline{x} = \mathrm{IDFT}[\mathrm{DFT}(\underline{x})]$, indicating that there is no loss of information going from time to frequency and vice versa.

5.3.5 One-sided and Two-sided Definitions

The DFT was defined for the frequency index u that ranges from $u = -N/2$ to $(N/2) - 1$ or from $u = 0$ to $u = N - 1$. These are called *two-sided definitions*. Yet, there is no need to duplicate computations when $X_u = \overline{X_{N-u}}$: one does not physically measure negative frequencies, and cannot resolve above the Nyquist frequency. Therefore, *one-sided definitions* are established between $u = 0$ and

$u = N/2$. Two-sided definitions are advantageous in analytical derivations. However, one-sided definitions are computationally efficient. (See exercise at the end of this chapter.) To avoid confusion, derivations, computations, and examples in this text are obtained with two-sided definitions.

5.3.6 Energy

The energy in a signal $\underline{x}$ is the sum of the square of the amplitude of the signal at each point. Each Fourier coefficient X_u indicates the amplitude of the sinusoid of frequency $\omega_u = u \cdot 2\pi/T$ that is contained in the signal. Therefore, the energy in the signal is also computed from the Fourier coefficients, as prescribed in Parseval's identity,

$$\sum_{i=0}^{N-1} x_i^2 = \frac{1}{N} \cdot \sum_{u=0}^{N-1} |X_u|^2 \tag{5.21}$$

The plot of $|X_u|^2$ versus frequency is the *autospectral density* of the signal, also known as power spectral density. (Spectral values in one-sided computations are twice those corresponding to the two-sided definition except for the zero-frequency term.)

5.3.7 Time Shift

Consider a wave train propagating along a rod. The signal is detected with two transducers. If the medium is not dispersive or lossy, and the coupling between the transducers and the rod are identical, then the only difference between the signal $\underline{x}$ detected at the first transducer and the signal $\underline{y}$ detected at the second transducer is the wave travel time between the two points $r \cdot \Delta t$. For a single frequency ω sinusoid,

$$\begin{aligned} &\text{if} && x_i = e^{j\omega i\Delta t} \\ &\text{and} && y_i = x_{i-r} = e^{j\omega(i-r)\Delta t} = x_i e^{-j\omega r\Delta t} \\ &\text{then} && Y_u = e^{-j\left(u\frac{2\pi}{N}r\right)} \cdot X_u \end{aligned} \tag{5.22}$$

For the given travel time, the higher the frequency signal, the higher the phase shift. When phase is measured, computed arctan values can only range between $[\pi/2, -\pi/2]$, and proper "phase unwrapping" is required (Chapter 6).

5.3.8 Differentiation

The derivative of a continuous time sinusoid $x(t) = A \cdot \sin(\omega \cdot t)$ is $y(t) = \omega \cdot A \cdot \cos(\omega \cdot t)$. In words, the derivative of a sinusoid implies a linear scaling of the amplitude by the frequency and a $\pi/2$ phase shift. The first derivative in discrete time $\underline{y}$ can be approximated by finite differences. The corresponding DFT is obtained by invoking the time shift property (Equation 5.22):

$$y_i = \frac{x_i - x_{i-1}}{\Delta t} \quad \text{then} \quad Y_u = \frac{1 - e^{-j\left(u\frac{2\pi}{N}\right)}}{\Delta t} X_u \tag{5.23}$$

The magnitude of the coefficient that multiplies X_u increases with u. Thus, this result predicts the magnification of high-frequency components when a differentiation transformation is imposed. This is in agreement with observations in the time domain whereby the derivative of a signal is very sensitive to the presence of high-frequency noise.

5.3.9 Duality

The parallelism between the analysis and synthesis equations in a Fourier pair (Equations 5.14 and 5.15, Table 5.1) leads to the property of duality. Before proceeding, notice that the exponents have the opposite sign in the Fourier pair; this means opposite phase: one is turning clockwise and the other counterclockwise, or in terms of time series, one is the tail-reverse version of the other. (For clarity, replace the exponentials for their trigonometric identities: a tail-reverse cosine is the same cosine; however, a tail-reversed sine becomes [−]sine, thus opposite phase.)

Now, consider the signal $\underline{x}$ shown in Figure 5.5a. The DFT of signal $\underline{x}$ computed with Equation 5.14 is shown in Figures 5.5b and c. Then, the analysis Equation 5.14 is used again to compute a second DFT but this time of $\underline{X}$, that is DFT[DFT($\underline{x}$)]. Figure 5.5c shows that the result is the original signal but in reversed order and scaled by N. In mathematical terms,

$$(x_0, x_{N-1} \ldots x_1) = \frac{1}{N} \cdot \text{DFT}\left[\text{DFT}\left(x_0, x_1 \ldots x_{N-1}\right)\right] \tag{5.24}$$

Duality is a useful concept in the interpretation of time and frequency domain operations and properties.

5.3.10 Time and Frequency Resolution

The time resolution is defined as the time interval between two consecutive discrete times; this is the sampling interval $\Delta t = t_{i+1} - t_i$. Likewise, frequency

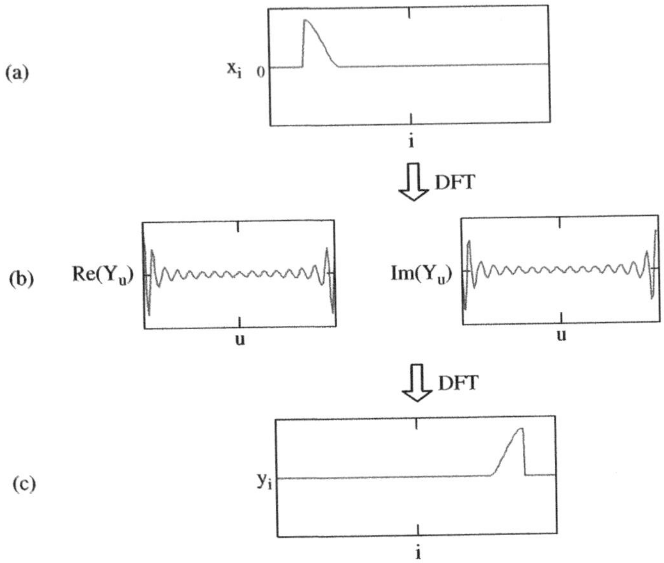

Figure 5.5 The duality property of the DFT: (a) the original signal $\underline{x}$; (b) its discrete Fourier transformed to the frequency domain $\underline{X}$; (c) the *forward* (not inverse) discrete Fourier transformation of $\underline{X}$ sends the series back to the time domain, but the signal appears tail-reversed

resolution is the frequency interval between two consecutive discrete frequencies $\Delta f = f_{u+1} - f_u$, where each frequency f_u is the u-th harmonic of the first frequency $f_u = u \cdot f_1 = u/(N \cdot \Delta t)$. Then $\Delta f = f_{u+1} - f_u = (u + 1 - u)/(N \cdot \Delta t)$:

$$\Delta f = \frac{1}{N \cdot \Delta t} \quad \text{that is} \quad N = \frac{1}{\Delta f \cdot \Delta t} \tag{5.25}$$

This is known as the "uncertainty principle" in signal processing: *when limited to N pieces of information, the resolution in frequency can only be improved at the expense of the resolution in time* (see solved example at the end of this Chapter).

5.3.11 *Time and Frequency Scaling*

The length N of the array $\underline{x}$ can be reduced by decimation (removal of intermediate points) or increased by interpolation. Similar effects are obtained by varying the sampling interval Δt during A/D conversion: down-sampling or up-sampling. In either case, the total time duration of the signal remains the same. Consider a stationary continuous signal x(t) sampled with two different sampling rates Δt and $\alpha \cdot \Delta t$:

Discrete time	Signal	Discrete frequency	DFT
$t_i = i \cdot \Delta t$	y_i	$\omega_u = u \dfrac{2\pi}{N \cdot \Delta t}$	Y_u
$t_k = k \cdot (\alpha \cdot \Delta t)$	z_k	$\omega_v = v \dfrac{2\pi}{M \cdot \alpha \cdot \Delta t}$	Z_v

The values z_i and z_k are equal at the same discrete time $t_i = t_k$; therefore, $i = k \cdot \alpha$. Likewise, the values of Y_u and $\alpha \cdot Z_v$ are equal at the same discrete frequency $\omega_u = \omega_v$; therefore, $u = v/\alpha$. Thus,

$$\text{if} \quad y_{k \cdot \alpha} \qquad \text{then} \quad \frac{1}{\alpha} \cdot Y_{\left(\frac{v}{\alpha}\right)} \tag{5.26}$$

This result shows the inherent inverse relation between time and frequency. The factor $1/\alpha$ in the frequency domain reflects the selected Fourier pair. Down-sampling is restricted by the Nyquist frequency.

5.4 COMPUTATION IN MATRIX FORM

The summation of binary products in analysis and synthesis equations is equivalent to matrix multiplication, and the transformation $\underline{X} = \mathrm{DFT}(\underline{x})$ implied in Equation 5.14 can be computed as:

$$\underset{[N,1]}{\underline{X}} = \underset{[N,N]}{\underline{\underline{F}}} \cdot \underset{[N,1]}{\underline{x}} \qquad \textit{Time} \rightarrow \textit{Frequency} \tag{5.27}$$

where each row in the Fourier transform matrix $\underline{\underline{F}}$ is the array of values that represents a complex exponential. In other words, the i-th element in the u-th row of $\underline{\underline{F}}$ is

$$F_{u,i} = e^{-j \cdot \left(u \frac{2\pi}{N} i\right)} \tag{5.28}$$

Note that u and i play the same roles in the exponent; therefore, the element $F_{u,i}$ is equal to the element $F_{i,u}$ and the matrix is symmetric $\underline{\underline{F}}^T = \underline{\underline{F}}$.

Similarly, the implicit operations in matrix multiplication apply to the synthesis equation or inverse Fourier transform. The elements in the inverse Fourier matrix $\underline{\underline{InvF}}$ have positive exponent, and the following equality holds (see Equation 5.15):

$$InvF_{u,i} = \frac{1}{N} e^{j \cdot \left(u \frac{2\pi}{N} i \right)} = \frac{1}{N} \overline{e^{-j \cdot \left(u \frac{2\pi}{N} i \right)}} = \frac{1}{N} \overline{F_{u,i}} \tag{5.29}$$

where the bar indicates complex conjugate. (Note: this is in agreement with the duality property, where the conjugate implies reversal.) Therefore, the inverse Fourier transform is

$$\underline{x} = \frac{1}{N} \cdot \overline{\underline{\underline{F}}} \cdot \underline{X} \qquad \textit{Frequency} \rightarrow \textit{Time} \tag{5.30}$$

Matrix $\overline{\underline{\underline{F}}}$ is the Hermitian adjoint of $\underline{\underline{F}}$ (Chapter 2). It follows from Equations 5.27 and 5.30 that $\underline{x} = 1/N \cdot \overline{\underline{\underline{F}}} \cdot \left(\underline{\underline{F}} \cdot \underline{x} \right)$. Then

$$\underline{\underline{I}} = \frac{1}{N} \cdot \overline{\underline{\underline{F}}} \cdot \underline{\underline{F}} \tag{5.31}$$

Implementation Procedure 5.1 outlines the implementation of Fourier transform operations in matrix form.

Implementation Procedure 5.1 Fourier analysis in matrix form

1. Digitize the signal x(t) with a sampling interval Δt to generate the array $\underline{x}$ [N × 1].
2. Create the Fourier transformation matrix $\underline{\underline{F}}$:

$$F_{u,i} = e^{-j \cdot \left(u \frac{2\pi}{N} i \right)}$$

for i and u that range between [0. . . N − 1]. The matrix is symmetric.

3. The DFT of the signal $\underline{x}$ is $\underline{X} = \underline{\underline{F}} \cdot \underline{x}$.
4. The magnitude and the phase of each frequency component are

$$\text{Magnitude}: |X_u| = \sqrt{[\mathrm{Re}(X_u)]^2 + [\mathrm{Im}(X_u)]^2}$$

$$\text{Phase}: \varphi_u = \tan^{-1}\left[\frac{\mathrm{Im}(X_u)}{\mathrm{Re}(X_u)} \right]$$

for corresponding frequency: $f_u = u \frac{1}{N \cdot \Delta t}$ or $\omega_u = u \frac{2\pi}{N \cdot \Delta t}$

5. Conversely, given a signal in the frequency domain $\underline{X}$, its IDFT is the time domain signal $\underline{x}$,

$$\underline{x} = \frac{1}{N}\overline{\underline{\underline{F}}} \cdot \underline{X} \quad \text{where} \quad \overline{F_{u,i}} = \text{complex conjugate of } F_{u,i}$$

Note: The fast Fourier transform (FFT) is preferred for large signals. The FFT algorithm is included in all commercially available mathematical software and in public domain codes at numerous internet sites.

5.5 TRUNCATION, LEAKAGE, AND WINDOWS

Short duration transients can be adequately recorded from beginning to end. Some A/D converters even permit pretriggering to gather the background signal prior to the transient. However, long-duration or ongoing signals are inevitably truncated and we only see a finite "window of the signal".

The effects of truncation are studied with a numerical example in Figure 5.6. The sinusoid is truncated when six cycles are completed (Figure 5.6a). The autospectral density is shown in Figure 5.6b. Given that this is a single-frequency sinusoid, the autospectral density is an impulse at the frequency of the signal. Figure 5.6c shows a similar signal truncated after 5.5 cycles. The autospectral density is shown in Figure 5.6d. In contrast to the previous case, energy has "leaked" into other frequencies.

Leakage is the consequence of two inherent characteristics of the DFT. The first one is the *unmatched harmonic* effect whereby the sinusoid frequency f* in Figure 5.6c is not a harmonic of $f_{min} = 1/(N \cdot \Delta t)$; therefore, the DFT cannot produce an impulse at f*. Instead, the DFT "curve-fits" the signal with harmonically related sinusoids at frequencies $f_u = u/(N \cdot \Delta t)$. The second cause for leakage results from the *presumed periodicity* in the DFT: the signal in Figure 5.6c is effectively considered the periodic signal in Figure 5.6e. The resulting sharp discontinuities at the end of the signal require higher-frequency components; in addition, the lack of complete cycles leads to a nonzero static component.

The window imposed on the analog signal during A/D conversion into a finite record is a sharp-edged off–on–off window and magnifies discontinuity effects. Leakage is reduced by "windowing the signal" with gradual window arrays $\underline{w}$. The windowed signal $\underline{x}^{<win>}$ is obtained multiplying the signal $\underline{x}$ with the window $\underline{w}$ point by point:

$$x_i^{<win>} = x_i \cdot w_i \tag{5.32}$$

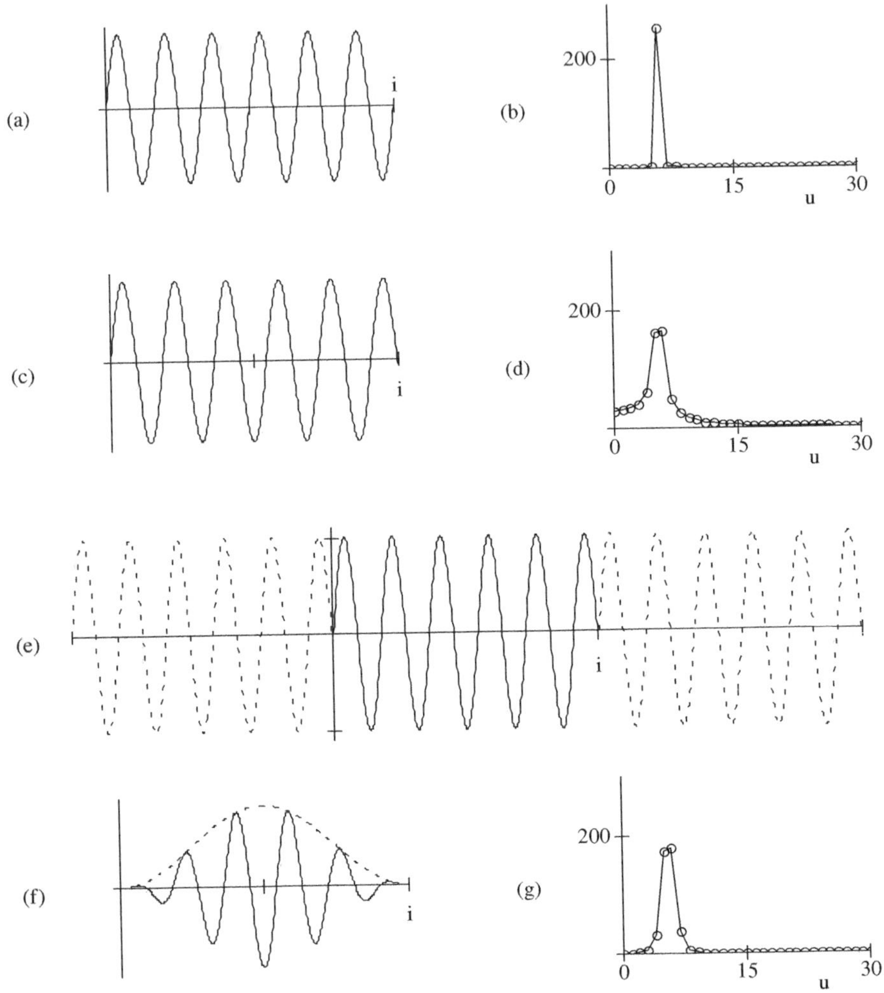

Figure 5.6 Truncation and windowing: (a, b) the DFT of a single frequency sinusoid is an impulse if it completes an integer number of cycles in the duration of the signal T; (c, d) this signal has an incomplete number of cycles; its DFT is not an impulse and has a static component; (e) periodic assumption in the DFT; (f) signal in frame 'c' but windowed with smooth transition towards zero ends; (g) autospectrum of the windowed signal

The Hanning and Hamming windows are two common windowing functions:

$$\textit{Hanning} \qquad w_i = \begin{cases} \frac{1}{2} + \frac{1}{2} \cdot \cos\left[\frac{2\pi}{E}(i-M)\right] & |i-M| \le \frac{E}{2} \\ 0 & \text{otherwise} \end{cases} \tag{5.33}$$

$$\textit{Hamming} \qquad w_i = \begin{cases} 0.54 + 0.46 \cdot \cos\left[\frac{2\pi}{E}(i-M)\right] & |i-M| \le \frac{E}{2} \\ 0 & \text{otherwise} \end{cases} \tag{5.34}$$

These windows are centered around $i = M$ and have a time width $E \cdot \Delta t$. In this format, the rectangular window becomes

$$\textit{Rectangular} \; w_i = \begin{cases} 1 & |i-M| \le \frac{E}{2} \\ 0 & \text{otherwise} \end{cases} \tag{5.35}$$

Figure 5.6f shows the signal in Figure 5.6c when the Hanning window is used. Finally, Figure 5.6g shows the autospectral density of the windowed signal.

The energy available in the windowed signal is reduced by windowing. The ratio of the energy in the original signal $\underline{x}$ and the windowed signal $\underline{x}^{<win>}$ can be computed in the time domain:

$$\beta = \sqrt{\frac{\sum_{i=0}^{N-1} x_i^2}{\sum_{i=0}^{N-1} \left(x_i^{<win>}\right)^2}} \tag{5.36}$$

5.6 PADDING

A longer duration $N \cdot \Delta t$ signal renders a better frequency resolution $\Delta f = 1/(N \cdot \Delta t)$. Therefore, a frequently used technique to enhance the frequency resolution of a stored signal length N consists of "extending" the signal by appending values to a length $M > N$. This approach requires careful consideration.

There are various "signal extension" strategies. Zero padding, the most common extension strategy, consists of appending zeros to the signal. Constant padding extends the signal by repeating the last value. Linear padding extends the signal while maintaining the first derivative at the end of the signal constant. Finally, periodic padding uses the same signature for padding.

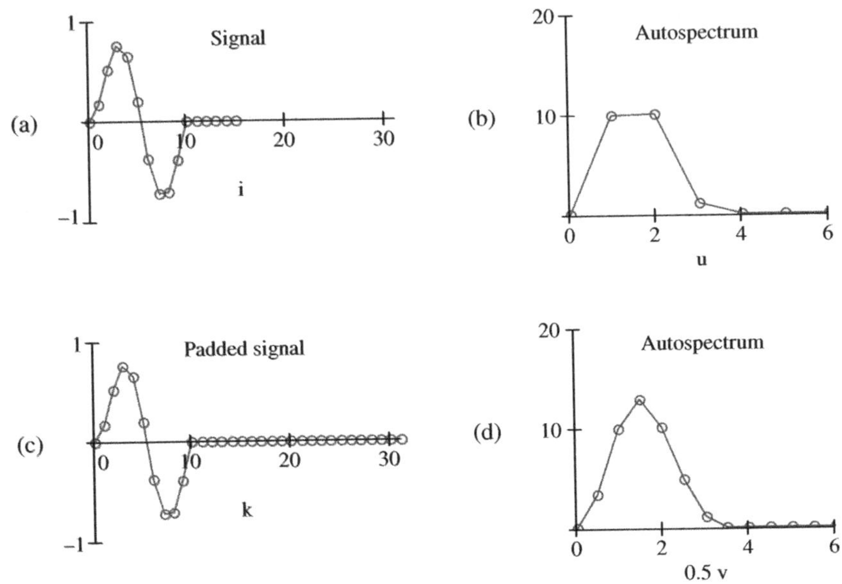

Figure 5.7 Time and frequency resolution: (a, b) original N = 16 signal and its auto spectrum; (c, d) zero-padded signal with N = 32 and its auto spectrum. Padding increases frequency resolution. The peak in the autospectral density of the original signal is absent because there is no corresponding harmonic. (Note: the time interval Δt is kept constant, the number of points N is doubled, and the frequency interval is halved.)

Figure 5.7 presents an example of zero padding. The signal length is $N = 16$ and the DFT decomposes it into harmonics $f_u = u/(16\Delta t)$, while the padded signal is length $M = 32$ and the associated harmonics are $f_v = v/(32\Delta t)$. The sinusoid duration is $11 \cdot \Delta t$; thus, its main frequency is $f^* = 1/(11 \cdot \Delta t)$. Therefore, the harmonic for $v = 3$ in the DFT of the padded signal is quite close to f*, but there is no harmonic in the DFT of the original signal near f*.

The following observations follow from this example and related analyses:

- Signal extension is not intended to add information. Therefore, there is no new information in the frequency domain if the same information is available in the time domain.
- The real effect of padding is to create harmonic components that better "fit" the signal.
- Zero and periodic padding may create discontinuities; plot the signal in the time domain to verify continuity.

- The negative effects of padding are reduced when signals are properly detrended and windowed first.
- The signal length can be increased by adding zeros at the front of the signal; however, this implies a time shift in all frequencies, and a frequency-dependent phase shift, as predicted in Equation 5.22.
- Signal extension to attain a signal length $N = 2^r$ allows the use of more computationally efficient Fast Fourier transform algorithms. However, harmonics may be lost: for example, a sinusoid with period $450 \cdot \Delta t$ in a signal length $N = 900$ has a harmonic at $u = 2$, but it has no harmonic when the signal is zero padded to $M = 2^{10} = 1024$.
- When the main frequency in the signal under study is a known value f*, then record length N and sample interval Δt are selected so that f* is one the harmonics $f_u = u/(N\Delta t)$ in the discrete spectrum.
- The DFT presumes the signal is periodic with fundamental period $T = N \cdot \Delta t$. Signal extension increases the fundamental period and prevents circular convolution effects in frequency domain computations (Chapter 6).
- The previous observations apply to deterministic signals. In the case of random signals, signal extension must preserve stationary conditions.

Enhanced resolution with harmonics that better "fit" the signal lead to more accurate system identification (review Figure 5.7). Consider a low-damping single degree of freedom oscillator: the narrow resonant peak may be missed when the frequency resolution is low and no harmonic f_u matches the resonant frequency. In this case, the inferred natural frequency and damping of the oscillator would be incorrect.

5.7 PLOTS

A signal in the time domain (time or space) is primarily plotted as x_i versus time $t_i = i \cdot \Delta t$. However, there are several alternatives in the frequency domain to facilitate the interpretation of the information encoded in the signal. Consider the signal in Figure 5.8a, which shows the free vibration of an oscillator after being excited by a very short impulse-like input signal. Various plots of the DFT are shown in Figures 8b–h:

- Figure 5.8b shows the autospectral density versus the frequency index u. The first mode of vibration is clearly seen. When the autospectral density is plotted in log scale, other low-amplitude vibration modes are identified (Figure 5.8c).

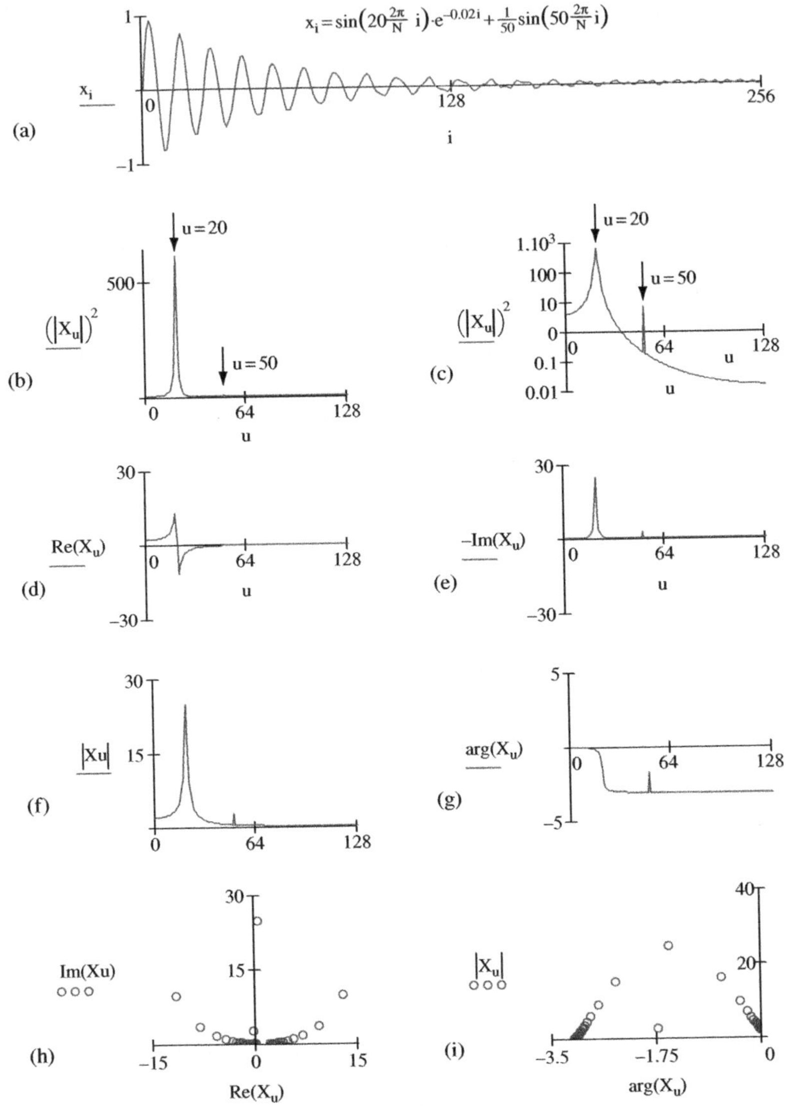

Figure 5.8 Different plots of the DFT of a signal: (a) original signal $\underline{x}$ in time domain; (b, c) autospectral density – normal and log magnitudes; (d, e) real and imaginary components versus frequency index u; (f, g) amplitude and phase versus frequency index u; (h) imaginary versus real component (Cole–Cole plot); (i) amplitude versus phase. Frequency domain data are presented single-sided, for $u = [0, N/2]$

- Figures 5.8d and e show the real $Re(X_u)$ and imaginary $Im(X_u)$ components of the DFT versus the frequency index u.
- Figures 5.8f and g show the amplitude $|X_u|$ and the phase φ_u versus the frequency index u.
- Figure 5.8h shows the imaginary component $Im(X_u)$ versus the real component $Re(X_u)$. This is called the Cole–Cole plot, and it is used to identify materials that show relaxation behavior (e.g. response of a viscoelastic material); a relaxation defines a semicircle in these coordinates.
- Figure 5.8i shows a plot of amplitude versus phase.

Any frequency is readily recovered from the frequency counter u as $f_u = u/(N\Delta t)$. In particular, the frequency associated with the peak response is the oscillator resonant frequency. The oscillator damping is reflected in both time and frequency domains: low damping is denoted by multiple oscillations in the time domain (Figure 5.8) and a narrow peak in the frequency domain (Chapter 4).

5.8 THE TWO-DIMENSIONAL DISCRETE FOURIER TRANSFORM

A 2D signal x(p, q) captures the variation of a parameter in two dimensions p and q. During A/D conversion, the signal is digitized along a grid made of M discrete values in p and N discrete values in q. The discrete 2D signal is a matrix $\underline{\underline{x}}$ where entry $x_{i,k}$ corresponds to location $p = i \cdot \Delta p$ and $q = k \cdot \Delta q$. The 2D signal may involve data gathered in any two independent dimensions, such as a digital picture or a sequence of time series obtained at different positions in space.

The DFT $\underline{\underline{X}}$ of $\underline{\underline{x}}$ is also a matrix; each entry $X_{u,v}$ corresponds to frequencies $f_u = u/(M \cdot \Delta p)$ and $f_v = v/(N \cdot \Delta q)$. The 2D Fourier transform pair is

$$X_{u,v} = \sum_{i=0}^{M-1} \left[\sum_{k=0}^{N-1} x_{i,k} \cdot e^{-j\left(v \cdot \frac{2\pi}{N} \cdot k\right)} \right] \cdot e^{-j\left(u \cdot \frac{2\pi}{M} \cdot i\right)} \qquad \text{2D } \textit{Analysis} \qquad (5.37)$$

$$x_{i,k} = \frac{1}{M} \cdot \sum_{u=0}^{M-1} \left[\frac{1}{N} \cdot \sum_{v=0}^{N-1} X_{u,v} \cdot e^{j\cdot\left(v \cdot \frac{2\pi}{N} \cdot k\right)} \right] \cdot e^{j\cdot\left(u \cdot \frac{2\pi}{M} \cdot i\right)} \qquad \text{2D } \textit{Synthesis} \qquad (5.38)$$

The 2D DFT can be computed with 1D algorithms in two steps. First, an intermediate matrix $\underline{\underline{INT}}$ is constructed where each row is the DFT of the corresponding row of $\underline{\underline{x}}$. The columns of the final 2D Fourier transform $\underline{\underline{X}}$ are obtained by computing the DFT of the corresponding columns in $\underline{\underline{INT}}$.

Analysis and synthesis operations can be expressed in matrix form, in analogy to the case of 1D signals. In particular, if the discrete signal is square $M = N$, the 2D Fourier transform of $\underline{\underline{x}}$ is

$$\underline{\underline{X}} = \left[\underline{\underline{F}} \cdot \left(\underline{\underline{F}} \cdot \underline{\underline{x}} \right)^T \right]^T = \underline{\underline{F}} \cdot \underline{\underline{x}} \cdot \underline{\underline{F}} \quad \textit{from } p\text{–}q \textit{ to } f_p\text{–}f_q \tag{5.39}$$

where the second equality follows from $\underline{\underline{F}}^T = \underline{\underline{F}}$. The k-th element in the v-th row of $\underline{\underline{F}}(N \times N)$ is

$$F_{v,k} = e^{-j \cdot \left(v \cdot \frac{2\pi}{N} \cdot k \right)} \tag{5.40}$$

Because $N \cdot \underline{\underline{I}} = \overline{\underline{\underline{F}}} \cdot \underline{\underline{F}}$ (Equation 5.31), the synthesis equation in matrix form is

$$\underline{\underline{x}} = \frac{1}{N^2} \cdot \overline{\underline{\underline{F}}} \cdot \underline{\underline{X}} \cdot \overline{\underline{\underline{F}}} \qquad \textit{from } f_p - f_q \textit{ to } p - q \tag{5.41}$$

Other concepts such as resolution, truncation and leakage, discussed in relation to 1D signals, apply to 2D signals as well.

Examples of 2D DFT are presented in Figure 5.9 (see solved example at the end of this Chapter). The following observations can be made (analogous to the 1D DFT Figure 5.1). The DFT of a uniform 2D signal has only the real DC component at $u = 0$, $v = 0$ (Figure 5.9a). The DFT of the linear combination of 2D signals is the linear combination of DFT of the individual signals (Figure 5.9b). A single frequency sinusoid becomes an impulse in the frequency domain in the same direction as the signal in the time domain (Figure 5.9c); if there is leakage, it manifests parallel to the u and v axes.

5.9 PROCEDURE FOR SIGNAL RECORDING

The most robust approach to signal processing is to *improve the data at the lowest possible level* (review Section 4.1.5). Start with a proper experimental design: explore various testing approaches, select the transducers that are best fitted to sense the needed parameter under study, match impedances, reduce noise by proper insulation (electromagnetic, mechanical, thermal, chemical, and biological) and use quality peripheral electronics. If the signal is still poor, then the option of signal stacking should be considered before analog filters are included in the circuitry.

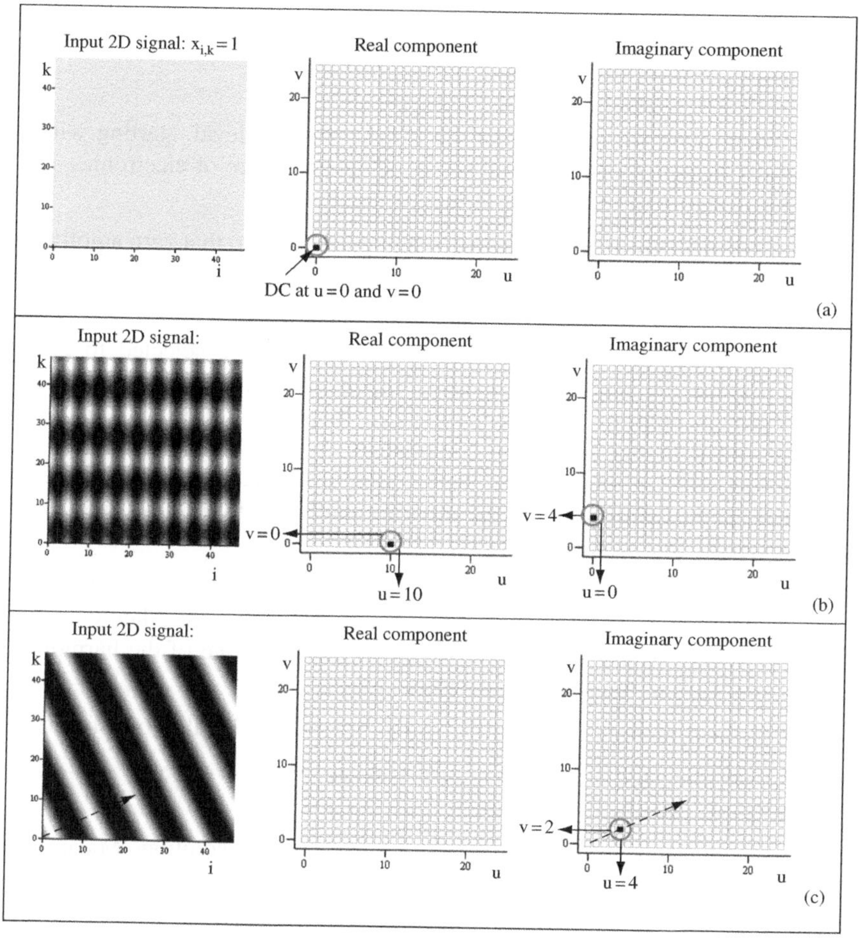

Figure 5.9 The 2D-DFT: (a) constant-value signal: the only nonzero value in 2D-DFT is the DC component; (b) the signal $x_{i,k} = \cos(10\frac{2\pi}{N}i) + \sin(4\frac{2\pi}{N}i)$ has one peak in the real part and one peak in the imaginary components of the 2D-DFT – note the direction in each case relative to the image; (c) the 2D-DFT of the single frequency sinusoid $x_{i,k} = \sin\left[4\frac{2\pi}{N}(i + 0.5k)\right]$ is aligned in the same direction as the oscillations in the signal

Once these recommendations have been taken into consideration, start planning the signal digitization and storage. Concepts discussed in this and the previous chapters permit outlining of common guidelines for signal recording that are applicable to most situations. When signal processing involves DFTs, data gathering must consider signal length, truncation and leakage, windowing, and frequency resolution. Guidelines are summarized in the Implementation Procedure 5.2.

Implementation Procedure 5.2 Recommended procedure for signal recording

1. The signal must be improved at the lowest possible level, starting with a carefully designed experimental setup, adequate choice of electronics, and proper isolation of the system to reduce noise.
2. It is advantageous to extend the recording duration T so that zero amplitude is recorded at the front and tail ends of the signal. This is possible in short-duration events.
3. The sampling interval or time resolution Δt must be selected to properly digitize the highest-frequency component of interest f_{max}, fulfilling the Nyquist criterion. It is recommended that $\Delta t \sim 1/(10 \cdot f_{max})$ be used. If unwanted higher frequency components are expected, they should be removed with an analog filter before digitalization. Many A/D systems include antialiasing filters at the input to automatically remove frequency components that would be aliased otherwise.
4. The total number of points to be recorded is estimated as $N = T/\Delta t$. If you know the main frequency in the signal under study f*, then combine record length N and sample interval Δt so that f* is one of the harmonics $f_u = u/(N\Delta t)$ in the discrete spectrum.
5. Detrend and remove spikes in the signal before the signal is transformed.
6. Window truncated signals to reduce leakage. Windowing and zero-offset corrections may be repeated.
7. Extend the recorded signal to increase frequency resolution. Make sure that there is a harmonic in the padded signal that corresponds to the main component f* in the signal under study.

5.10 SUMMARY

- Harmonically related sinusoids and complex exponentials are orthogonal functions in the open interval [0, T[. Therefore, they form a base that can be used to express any other function as a linear combination. This is the foundation for the DFT.
- For a robust interpretation of the DFT of a signal length N, remember that: (1) the DFT is equivalent to fitting the signal with a series of cosines and sines and storing the amplitudes in the “real” and “imaginary” arrays, (2) the signal

is assumed periodic with period equal to the duration of the signal $T = N \cdot \Delta t$, and (3) only harmonically related sinusoid frequencies $f_u = u/(N \cdot \Delta t)$ are used.

- The DFT has a finite number of terms. In fact, there are N information units in a signal length N, both in the time domain and in the frequency domain. There are no convergence difficulties in the DFT of discrete time signals.
- The parallelism between analysis and synthesis relations in a Fourier pair leads to the duality of the DFT.
- Resolution in time is inversely proportional to resolution in frequency. Signal extension or padding decreases the frequency interval between consecutive harmonics.
- The truncation of ongoing signals produces leakage. Leakage effects are reduced by windowing signals with smooth boundary windows.
- The DFT can be applied to signals that vary along more than one independent variable, such as 2D images or data in space-time coordinates.
- The signal must be improved at the lowest possible level, starting with careful experimental setup, adequate choice of electronics, and proper isolation of the system under study to reduce noise. While planning analog-to-digital conversion, the experimenter must take into consideration the system under study and the mathematical implications of digitization and DFT operations.

FURTHER READING AND REFERENCES

Bracewell, R. N. (1986). The Fourier Transform and Its Applications. McGraw-Hill Book Company, New York. 474 pages.

Cooley, J. W. and Tukey, J. W. (1965). An Algorithm for the Machine Computation of Complex Fourier Series. Math. Comput. Vol. 19, pp. 297–301.

Jackson, L. B. (1991). Signals, Systems, and Transforms. Addison-Wesley Publishing Co., Reading, Mass. 482 pages.

Orfanidis, S. J. (1996). Introduction to Signal Processing. Prentice-Hall, Inc., Englewood Cliffs, NJ. 798 pages.

Wright, C. P. (1995). Applied Measurement Engineering. Prentice-Hall, Englewood Cliffs, NJ. 402 pages.

SOLVED PROBLEMS

P5.1 *Fourier series.* Demonstrate that:

$$\int_0^T e^{j\left(\frac{2\pi}{T}t\right)} \cdot e^{-j\left(u\frac{2\pi}{T}t\right)} \cdot dt = \begin{cases} 0 & \text{if } u \neq 1 \\ T & \text{if } u = 1 \end{cases}$$

Solution: Using Euler's identities

$$f(T) = \int_0^T \left[\cos\left(\frac{2\pi}{T}t\right) + j \cdot \sin\left(\frac{2\pi}{T}t\right)\right] \cdot \left[\cos\left(u\frac{2\pi}{T}t\right) - j \cdot \sin\left(u\frac{2\pi}{T}t\right)\right] \cdot dt$$

$$f(T) = \int_0^T \left[\begin{array}{l} \cos\left(\frac{2\pi}{T}t\right) \cdot \cos\left(u\frac{2\pi}{T}t\right) - j \cdot \cos\left(\frac{2\pi}{T}t\right) \cdot \sin\left(u\frac{2\pi}{T}t\right) \\ \\ +j \cdot \sin\left(\frac{2\pi}{T}t\right) \cdot \cos\left(u\frac{2\pi}{T}t\right) + \sin\left(\frac{2\pi}{T}t\right) \cdot \sin\left(u\frac{2\pi}{T}t\right) \end{array}\right] \cdot dt$$

$$f(T) = \int_0^T \left[\cos\left(\frac{2\pi}{T}t\right) \cdot \cos\left(u\frac{2\pi}{T}t\right)\right] \cdot dt - j \cdot \int_0^T \left[\cos\left(\frac{2\pi}{T}t\right) \cdot \sin\left(u\frac{2\pi}{T}t\right)\right] \cdot dt$$

$$+j \cdot \int_0^T \left[\sin\left(\frac{2\pi}{T}t\right) \cdot \cos\left(u\frac{2\pi}{T}t\right)\right] \cdot dt + \int_0^T \left[\sin\left(\frac{2\pi}{T}t\right) \cdot \sin\left(u\frac{2\pi}{T}t\right)\right] \cdot dt$$

Invoking Equation 5.4, the previous equation simplifies to

$$f(T) = \int_0^T \left[\cos\left(\frac{2\pi}{T}t\right) \cdot \cos\left(u\frac{2\pi}{T}t\right)\right] \cdot dt + \int_0^T \left[\sin\left(\frac{2\pi}{T}t\right) \cdot \sin\left(u\frac{2\pi}{T}t\right)\right] \cdot dt$$

And, from Equations 5.3 and 5.4:

$$f(T) = \underbrace{\int_0^T \left[\cos\left(\frac{2\pi}{T}t\right) \cdot \cos\left(u\frac{2\pi}{T}t\right)\right] \cdot dt}_{\begin{array}{lll} 0 & \text{if} & u \neq 0 \\ \frac{T}{2} & \text{if} & u \neq 0 \end{array}} + \underbrace{\int_0^T \left[\sin\left(\frac{2\pi}{T}t\right) \cdot \sin\left(u\frac{2\pi}{T}t\right)\right] \cdot dt}_{\begin{array}{lll} 0 & \text{if} & u \neq 0 \\ \frac{T}{2} & \text{if} & u \neq 0 \end{array}}$$

P5.2 *Digitization.* Given a sampling interval $\Delta t = 10^{-3}$s and a record length $T = 0.5$s, compute: (a) frequency resolution, (b) frequency corresponding to the frequency counter $u = 13$, (c) the shortest time shift compatible with a phase shift $\Delta\phi = \pi$ for the frequency component that corresponds to $u = 10$.

Solution:

(a) The frequency resolution is $\Delta f = \frac{1}{T} = \frac{1}{0.5s} = 2Hz$

(b) The frequency corresponding to $u = 13$ is $f_{13} = u \cdot \Delta f = f = 13 \cdot 2\,Hz = 26\,Hz$

(c) Phase and time shifts are related as $\dfrac{\Delta\phi_u}{2\pi} = \dfrac{\delta t}{T_u}$

The time shift is $\delta t = \dfrac{\Delta\phi_u}{2\pi}\dfrac{T}{u} = 0.025\,s$

P5.3 *2D-Fourier transform.* Create a 2D image $\underline{\underline{x}}$ to represent ripples on a pond. Calculate the discrete Fourier transform $\underline{\underline{X}}$. Analyze the results.

Solution: Definition of function $\underline{\underline{x}}$ ($N \times N$ elements where $N = 64$)

Distance from the center of the pond: $r_{i,k} = \sqrt{\left(i - \frac{N}{2}\right)^2 + \left(k - \frac{N}{2}\right)^2}$

Displacement function: $x_{i,k} = \dfrac{\sin\left(10\dfrac{2\pi}{\max(r)} r_{i,k}\right)}{r_{i,k} + 10}$

Discrete Fourier transform matrix: $F_{u,i} = e^{-ju\frac{2\pi}{N}i}$

2D discrete Fourier transform: $\underline{\underline{X}} = \underline{\underline{F}} \cdot \underline{\underline{x}} \cdot \underline{\underline{F}}$

Magnitude: $|X_{u,v}| = X_{u,v} \cdot \overline{X_{u,v}}$

where the spatial indices i and k range from 0 to $N - 1$ and the frequency indices u and v range from 0 to $N - 1$. Time and frequency domain plots are presented next. Only one quadrant of the 2D-DFT is shown:

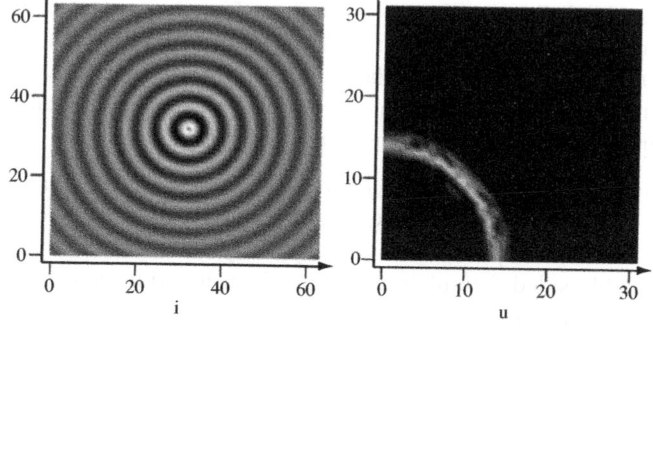

Interpretation: There are 15 ripples in both i and k directions along the center of the plot. That is the location of peak energy along the u and v axis.

Explore this solution further. What happens if you shift the center of the ripples away from the center of the image? What is the 2D-DFT of a signal with elliptical ripples?

ADDITIONAL PROBLEMS

P5.4 *Fourier series.* Compute the fourth-order discrete Fourier series ($u = 0, 1, 2, 3, 4$) that best approximates an odd square wave. Repeat for an even square wave. Compare the coefficients for sine and cosine components in both cases. What can be concluded about the decomposition of even and odd signals?

P5.5 *Discrete Fourier transform pairs.* There are various Fourier pairs besides the one presented in Table 5.1; for example:

$$\text{Analysis:}\quad a_u = \frac{1}{N}\sum_{i=0}^{N-1} x_i \cdot \cos\left(u\frac{2\pi}{N}i\right) \text{ and } b_u = \frac{1}{N}\sum_{i=0}^{N-1} x_i \cdot \sin\left(u\frac{2\pi}{N}i\right)$$

$$\text{Synthesis:}\quad x_i = \sum_{u=0}^{N-1}\left[a_u \cdot \cos\left(u\frac{2\pi}{N}i\right) + j \cdot b_u \cdot \sin\left(u\frac{2\pi}{N}i\right)\right]$$

Determine the relationship between this Fourier pair and the one presented in Table 5.1. Explicitly state the relationship between a_u and b_u, and X_u.

P5.6 *Properties of the discrete Fourier transform.* Demonstrate the following properties of the DFT of discrete periodic signals: linearity, periodicity, differentiation, Parseval's relation, time shift, and $N \cdot \underline{\underline{I}} = \underline{\underline{F}} \cdot \overline{\underline{\underline{F}}}$ (matrix operations). Is the magnification of high-frequency components linear with frequency in Equation 5.23?

P5.7 *Single-sided discrete Fourier transform.* Use the properties of the DFT to show that the computation of the DFT can be reduced to coefficients $u = 0$ to $u = N/2$. Rewrite the synthesis equation to show this reduced summation limits. Corroborate your results using numerical simulation. Compare the autospectral density in both cases.

P5.8 *Discrete Fourier transform of a complex exponential.* What is the DFT of a complex exponential? Consider both positive and negative exponents. Solve this problem both analytically and numerically. (Important: use double sided formulation, that is, from $u = 0$ to $N - 1$; this exercise is revisited in Chapter 7.)

P5.9 *Padding*. Generate a $N = 300$ points sinusoid $x_i = \sin\left(8 \cdot \frac{2\pi}{N} \cdot i\right)$. Consider different padding criteria to extend the signal to $N = 512$ points and compute the DFT in each case. Analyze spectra in detail and draw conclusions.

P5.10 *Application: signal recording and preprocessing*. Capture a set of signals within the context of your research interests. Follow the recommendations outlined in the Implementation Procedure 5.3. For each signal:

- Detrend the signal.
- Window the signal with a Hamming window (test different widths E).
- Compute the DFT and plot results in different forms to highlight the underlying physical process.
- Infer the characteristics of the system (e.g. damping and resonance if testing a single DoF system).
- Double the number of points by padding, compute the DFT and compare the spectra with the original signals.
- Repeat the exercise varying parameters such as sampling interval Δt, number of stored points N, and signal amplitude.

P5.11 *Application: sound and octave analysis*. The *octave* of a signal frequency f is the first harmonic 2f. In "octave analysis", frequency is plotted in logarithmic scale. Therefore, the central frequency of each band increases logarithmically, and bins have constant log-frequency width; that is, the frequency width of each bin increases proportionally to the central frequency. Systems that operate with octave analysis include filters with upper-to-lower frequency ratio 2n, where n is either 1, 1/2, 1/6, or 1/12. This type of analysis is preferred in studies of sound and hearing. Create a frequency sweep sinusoid $\underline{x}$ with frequency increasing linearly with time. Plot the signal. Compute $\underline{X} = \text{DFT}(\underline{x})$. and plot the magnitude versus linear and logarithmic frequency. Draw conclusions.

P5.12 *Application: Walsh series*. A signal can be expressed as a sum of square signals with amplitude that ranges between +1 and −1. In particular, the Walsh series is orthogonal, normalized, and complete. Research the Walsh series and:

1. Write the Walsh series in matrix form (length $N = 16$).
2. Study the properties of the matrix. Is it invertible?

3. Apply the Walsh's decomposition to a sinusoidal signal, a stepped signal (e.g. transducer with digital output), and to a small digital image.
4. Analyze your results and compare with Fourier approaches (see also the Hadamard transform).

6

Frequency Domain Analysis of Systems

The discrete Fourier transform brings a signal from the time domain to the frequency domain by fitting the discrete time signal with a finite series of harmonically related sinusoids (Chapter 5). This chapter shows that the system response $\underline{y}$ to an input signal $\underline{x}$ can be readily computed using frequency domain operations. The first question to be addressed is whether sinusoids offer any advantage in the study of systems.

Because of the equivalence with convolution, cross-correlation and filtering are reviewed in this context. Procedures presented in this chapter presume that systems are linear time-invariant (LTI). Therefore, the generalized superposition principle applies.

6.1 SINUSOIDS AND SYSTEMS – EIGENFUNCTIONS

Consider a single degree of freedom (DoF) system. When this system is excited with an impulse, it responds with a characteristic signature known as the impulse response $\underline{h}$. The impulse response has all the information needed to characterize the LTI system (Chapter 4).

What is the response of an LTI system when it is excited with a single frequency sinusoidal forcing function? Consider the single DoF oscillator analyzed in Chapter 4. Deformation compatibility at the boundary is required to maintain a linear response; therefore, the mass displacement will also be a sinusoidal function of the same frequency as the input force, and with some amplitude and phase (Figure 6.1; also Section 4.4). Euler's identities allow us to extend this observation to complex exponentials. This conclusion extends to all LTI systems

Discrete Signals and Inverse Problems J. C. Santamarina and D. Fratta

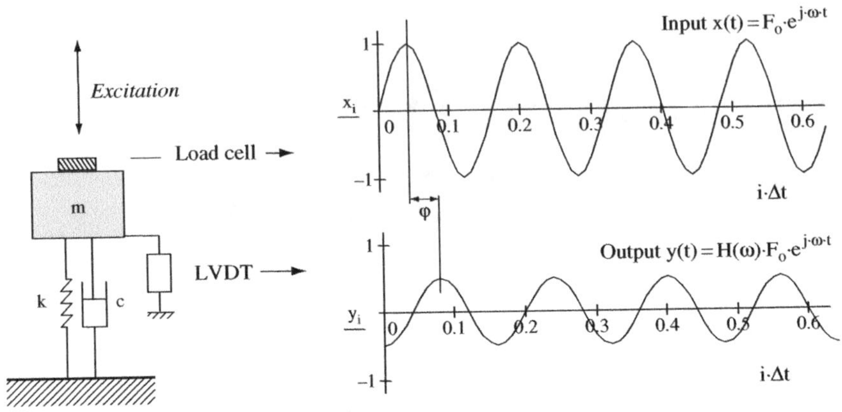

Figure 6.1 A single DoF oscillator excited with a single-frequency sinusoidal forcing function – Amplitude and phase response

and states that

$$\begin{aligned} &\text{if} \quad x(t) = \text{sinusoid or a complex exponential} \\ &\text{then} \quad y(t) = H_u \cdot x(t) \quad \text{in an LTI system} \end{aligned} \tag{6.1}$$

where the complex number H_u conveys amplitude and phase information corresponding to the excitation frequency ω_u. This situation resembles eigenvectors in matrix algebra: given the transformation $\underline{y} = \underline{\underline{a}} \cdot \underline{x}$, a vector $\underline{x}$ is an eigenvector of $\underline{\underline{a}}$ if the outcome $\underline{y}$ can be expressed as the product of a scalar λ times the vector $\underline{x}$ (Section 2.2.3),

$$\begin{aligned} &\text{if} \quad \underline{x} \text{ is an eigenvector} \\ &\text{then} \quad \underline{y} = \lambda \cdot \underline{x} \end{aligned} \tag{6.2}$$

where λ is the corresponding eigenvalue. In other words, the output $\underline{y}$ is like the input $\underline{x}$ but scaled by λ, which may be a complex number. By analogy, sinusoids and complex exponentials are "eigenfunctions" for LTI systems, and H_u are the corresponding eigenvalues.

6.2 FREQUENCY RESPONSE

A series of complex numbers H_u can be determined by exciting the system at different frequencies ω_u. The array of H_u values is the system *frequency response* $\underline{H}$. Implementation Procedure 6.1 outlines a possible experiment to determine the frequency response $\underline{H}$.

Implementation Procedure 6.1 Determination of the frequency response $\underline{H}$ – Sinusoidal sweep method

1. Connect a frequency-controlled sinusoidal output source to the system under consideration.
2. Monitor the system response. Verify that transducers have an operating frequency compatible with the frequency range of interest.
3. Tune the source to a circular frequency ω_u. Measure the time history of the source $\underline{x}$ and the response $\underline{y}$. Determine the amplitude of $\underline{x}$ and $\underline{y}$, and compute

$$|H_u| = \frac{\text{amplitude of } \underline{y} \text{ at frequency } \omega_u}{\text{amplitude of } \underline{x} \text{ at frequency } \omega_u}$$

4. Measure the relative phase φ between $\underline{x}$ and $\underline{y}$ (Figure 6.1): φ_u
5. Repeat for different circular frequencies $\omega_u = 2\pi \cdot f_u$.
6. Assemble the array $\underline{H}$ of complex numbers H_u:

$$H_u = |H_u| \cdot \cos(\varphi_u) + j \cdot |H_u| \cdot \sin(\varphi_u)$$

7. Transducers and peripheral electronics transform the signal. Determine their frequency response through calibration with known specimens and correct the measured response $\underline{H}$ to determine the true system frequency response (see Implementation Procedures 6.5 and 6.6).

Notes

- The proper assembly of the array $\underline{H}$ requires that entries H_u are obtained at equal frequency spacing Δf for a total of N/2 readings. Then, this assembly must be repeated with the tail reserve of its complex conjugate to obtain the double-sided form of $\underline{H}$. The complete array is used to evaluate the system response to an N-point input array $\underline{x}$ that is captured with $\Delta t = 1/(N \cdot \Delta f)$.
- The log-linear plot of the magnitude of $\underline{H}$ versus Δ_u helps identify the presence of higher modes (see Figure 5.8b).

This method is recommended when the SNR is very low. When the SNR is adequate, the use of broadband signals leads to more efficient determination of $\underline{H}$*; the required signal processing methods are presented later in Implementation Procedure 6.6.*

6.2.1 Example: A Single Degree-of-freedom Oscillator

The frequency response of simple systems can be mathematically computed in closed form. Consider once again the single DoF oscillator analyzed in Chapter 4 (also Figure 6.1). The equation of motion is

$$m \cdot \ddot{y} + c \cdot \dot{y} + k \cdot y = x(t) \tag{6.3}$$

If the forcing function is a complex exponential, $x(t) = F_o \cdot e^{j \cdot \omega \cdot t}$, Equation 6.3 can be written as

$$m \cdot \ddot{y} + c \cdot \dot{y} + k \cdot y = F_o \cdot e^{j \cdot \omega \cdot t} \tag{6.4}$$

where F_o is the amplitude of the forcing function. Because a complex exponential is an eigenfunction of the system, Equation 6.1 predicts the mass displacement y(t) to be

$$y(t) = H(\omega) \cdot \left[F_o \cdot e^{j \cdot \omega \cdot t}\right] \qquad \textit{for excitation frequency } \omega \tag{6.5}$$

This equation is substituted into Equation 6.4. After computing the derivatives, the following expression for $H(\omega)$ is obtained:

$$H(\omega) = \frac{1}{k} \cdot \left[\frac{1}{1 + j \cdot 2D \cdot \frac{\omega}{\omega_n} - \left(\frac{\omega}{\omega_n}\right)^2} \right] \tag{6.6}$$

This is the oscillator frequency response $H(\omega)$. The coefficient D is the damping ratio $D = c/(2 \cdot m \cdot \omega_n)$, and ω_n is the oscillator natural frequency $\omega_n = \sqrt{k/m}$. Figure 6.2 shows the amplitude $|H(\omega)|$ and the phase $\varphi = \tan^{-1}\{Im[H(\omega)]/Re[H(\omega)]\}$ as a function of the excitation frequency. Results are presented in dimensionless form in terms of $[H(\omega) \cdot k]$ and ω/ω_n.

6.2.2 Frequency Response and Impulse Response

The frequency response of the single DoF system, Equation 6.6, is a function of all the characteristics of the system. This can be generalized to all LTI systems: *an LTI system is completely characterized by its frequency response* $\underline{H}$.

(a)

(b)

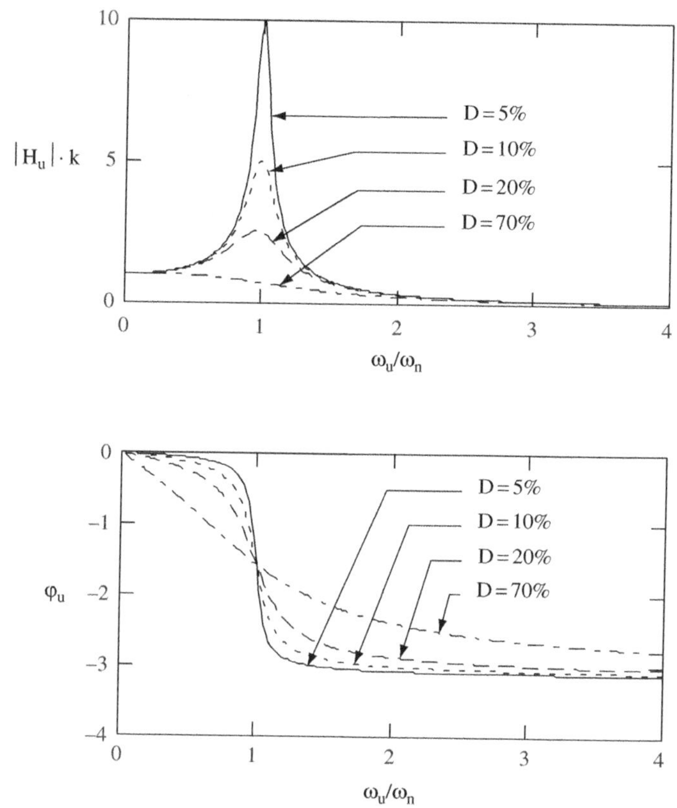

Figure 6.2 Frequency response of a single DoF system for different damping ratios: (a) dimensionless amplitude $|H_u| \cdot k$; (b) phase. Frequency is presented in dimensionless form ω_u/ω_n

A similar observation was made in the time domain about the impulse response $\underline{h}$ (Section 4.3). Therefore, the impulse response and the frequency response have the same information content and must be mathematically related. Indeed, this is the case: *for an LTI system, the frequency response* $\underline{H}$ *is the DFT of the impulse response* $\underline{h}$:

$$\underline{H} = \text{DFT}(\underline{h})$$

$$\text{or} \qquad H_u = \sum_{i=0}^{N-1} h_i \cdot e^{-j\left(u \frac{2\pi}{N} i\right)} \tag{6.7}$$

Once again, the DFT preserves the information in the array (see also Section 5.1).

6.3 CONVOLUTION

It has been shown that a signal can be expressed as a linear combination of harmonically related complex exponentials. In particular, the input signal $\underline{x}$ can be expressed as

$$x_i = \frac{1}{N} \cdot \sum_{u=0}^{N-1} X_u \cdot e^{j \cdot \left(u \frac{2\pi}{N} i\right)} \tag{6.8}$$

If $\underline{x}$ acts on an LTI system, each of the complex exponentials in the summation will cause a scaled and phase-shifted exponential in the response, as prescribed in Equation 6.1. Applying the superposition principle "sum of the causes → sum of the effects", the output $\underline{y}$ of the LTI system to the input signal $\underline{x}$ can be computed from Equation 6.8 by replacing the input sinusoids by their response (Equation 6.1):

$$y_i = \frac{1}{N} \cdot \sum_{u=0}^{N-1} H_u \cdot \left[X_u \cdot e^{j \cdot \left(u \frac{2\pi}{N} i\right)}\right] \tag{6.9}$$

This time domain result y_i is computed with frequency domain coefficients X_u and H_u. Regrouping coefficients,

$$y_i = \frac{1}{N} \cdot \sum_{u=0}^{N-1} [H_u \cdot X_u] \cdot e^{j \cdot \left(u \frac{2\pi}{N} i\right)} \tag{6.10}$$

This is the discrete Fourier synthesis equation for $\underline{y}$. Hence, the coefficients in square brackets must be the u-th term of the DFT of $\underline{y}$,

$$Y_u = H_u \cdot X_u \tag{6.11}$$

where:

X_u is the u-th element of the DFT of the input signal $\underline{x}$;
H_u is the u-th element of the frequency response $\underline{H}$ which is $\underline{H} = \text{DFT}(\underline{h})$;
Y_u is the u-th element of the DFT of the output signal $\underline{y}$; and
$\omega_u = 2\pi f_u = u \cdot 2\pi/(N \cdot \Delta t)$ is the frequency of the u-th harmonic.

Therefore, the convolution sum in the time domain $\underline{y} = \underline{h} * \underline{x}$ becomes a point-by-point multiplication in the frequency domain $Y_u = H_u \cdot X_u$. An alternative demonstration is presented under solved problems at the end of this chapter.

A signal has the same dimensions in both time and frequency domains. Therefore, the units of the frequency response $\underline{H}$ must be [units of output]/[units of input], according to Equation 6.11. For example, the units of $H(\omega)$ in Equation 6.6 for the single DoF oscillator are [units of deformation]/[units of force].

6.3.1 Computation

Because of the efficient computation of the DFT with fast Fourier transform algorithms, convolution and other operations in the time domain are often executed in the frequency domain. The computation of convolution in the frequency domain is outlined in Implementation Procedure 6.2.

Implementation Procedure 6.2 Convolution in the frequency domain

1. Determine the values H_u that define the system frequency response $\underline{H}$ for the harmonically related frequencies $\omega_u = u \cdot 2\pi/(N \cdot \Delta t)$. The frequency sweep method can be used, as described in Implementation Procedure 6.1. Note: alternative procedures are presented later in this Chapter.
2. Compute the DFT of the input signal $\underline{X} = \text{DFT}(\underline{x})$.
3. Obtain the output signal in the frequency domain $\underline{Y}$ by multiplying point by point the two arrays in the frequency domain:

$$Y_u = H_u \cdot X_u$$

   Values X_u and H_u are complex numbers; consequently, the values Y_u are complex numbers as well.
4. Compute the response $\underline{y}$ in the time domain as $\underline{y} = \text{IDFT}(\underline{Y})$.

Example

A numerical example is presented in Figure 6.3. The sawtoothed input $\underline{x}$ is convolved with the system impulse response $\underline{h}$ using frequency domain operations.

Note: Some Fourier pairs require the multiplication of the convolution by a normalizing factor, such as N or $\sqrt{N}$. The Fourier pair selected in this book is compatible with the approach outlined in this Implementation Procedure. To test the available DFT algorithm, implement the following computation with $N = 8$,

define: $\underline{h} = (0, 1, 2, 0, -1, 0, 0, 0)$ and $\underline{x} = (1, 0, 0, 0, 0, 0, 0, 0)$

compute: $\underline{H} = \text{DFT}(\underline{h})$

$$\underline{X} = \mathrm{DFT}(\underline{x})$$

$$Y_u = H_u \cdot X_u \qquad \text{for all u}$$

$$\underline{y} = \mathrm{IDFT}(\underline{Y})$$

result $\underline{y} = (0, 1, 2, 0, -1, 0, 0, 0)$

The computed $\underline{y}$ *should be equal to* $\underline{h}$. *Otherwise, the computation of the convolution operator with frequency domain operations must be corrected for the proper normalization factor.*

A numerical example is shown in Figure 6.3. The sawtooth input signal $\underline{x}$ and the system impulse response $\underline{h}$ are known (Figures 6.3b, c). Convolution is computed in the frequency domain. The output signal $\underline{y}$ is displayed in Figure 6.3d.

6.3.2 Circularity

The DFT presumes the signal is periodic, with period equal to the signal duration $T = N \cdot \Delta t$ (Section 5.3.3). This can produce misleading results when the convolution operation is implemented in the frequency domain.

Consider a low-damping single DoF system. The "assumed periodic" input $\underline{x}$ and the impulse response $\underline{h}$ are shown in Figures 6.4a and b; the known signals are indicated by continuous lines and the presumed signals are represented as dashed lines. The convolution computed in the frequency domain renders the output $\underline{y}$ shown in Figure 6.4c; the computed result is the continuous line. The tail on the left-hand side of the response $\underline{y}$ is caused by the "prior excitation" in the presumed periodic input signal $\underline{x}$. This effect is known as "circular convolution".

The effects of circular convolution are minimized or cancelled when the signal length is extended by padding, $M > N$, so that the signal presumed period $M \cdot \Delta t$ increases (see Chapter 5).

6.3.3 Convolution in Matrix Form

The point-by-point operation $Y_u = H_u \cdot X_u$ can be captured in matrix form by assembling a diagonal matrix $\underline{\underline{diagH}}$ whose entries in the main diagonal are $diagH_{u,u} = H_u$ and other entries are zero ($diagH_{u,v} = 0$ for $u \neq v$). Then,

$$\underline{Y} = \underline{\underline{diagH}} \cdot \underline{X} \tag{6.12}$$

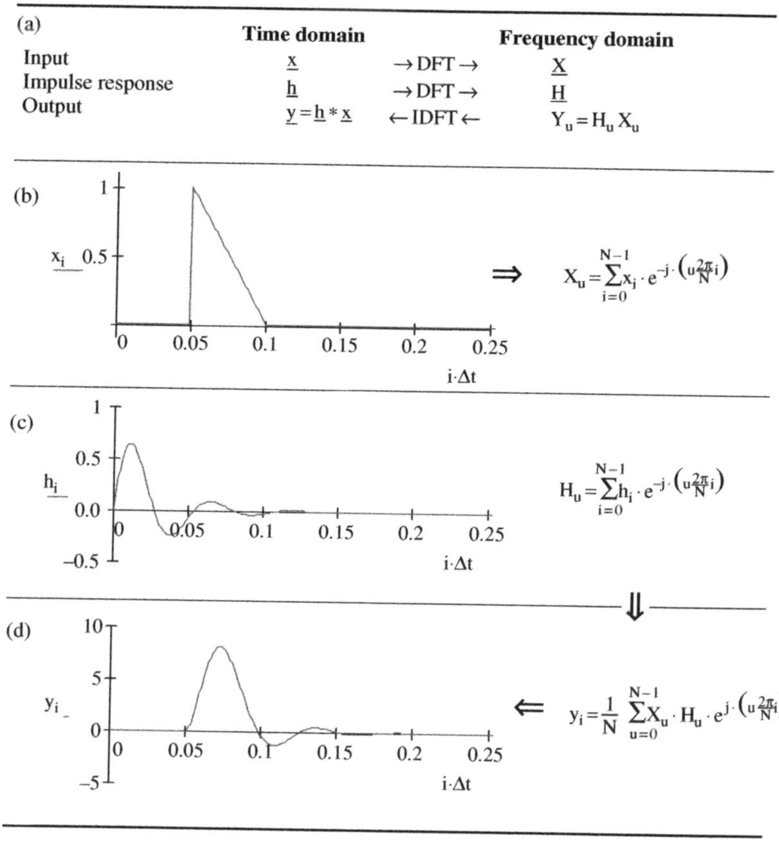

Figure 6.3 Convolution in the frequency domain. The sawtooth input signal $\underline{x}$ and the impulse response of the system $\underline{h}$ are known. (a) Sequence of calculations; (b) input signal $\underline{x}$; (c) impulse response $\underline{h}$; (d) output signal $\underline{y}$

Expressing the DFT in matrix form: $\underline{Y} = \underline{\underline{F}} \cdot \underline{y}$ and $\underline{X} = \underline{\underline{F}} \cdot \underline{x}$ (Section 5.4). Equation 6.11 becomes

$$\underline{\underline{F}} \cdot \underline{y} = \underline{\underline{\text{diagH}}} \cdot \underline{\underline{F}} \cdot \underline{x} \tag{6.13}$$

On the other hand, the inverse of $\underline{\underline{F}}$ is $\underline{\underline{F}}^{-1} = (1/N) \cdot \overline{\underline{\underline{F}}}$, where the entries in matrix $\overline{\underline{\underline{F}}}$ are the complex conjugates of the entries in $\underline{\underline{F}}$ (Section 5.4). Premultiplying both sides by $\underline{\underline{F}}^{-1}$,

$$\underline{y} = \frac{1}{N} \cdot \overline{\underline{\underline{F}}} \cdot \underline{\underline{\text{diagH}}} \cdot \underline{\underline{F}} \cdot \underline{x} \tag{6.14}$$

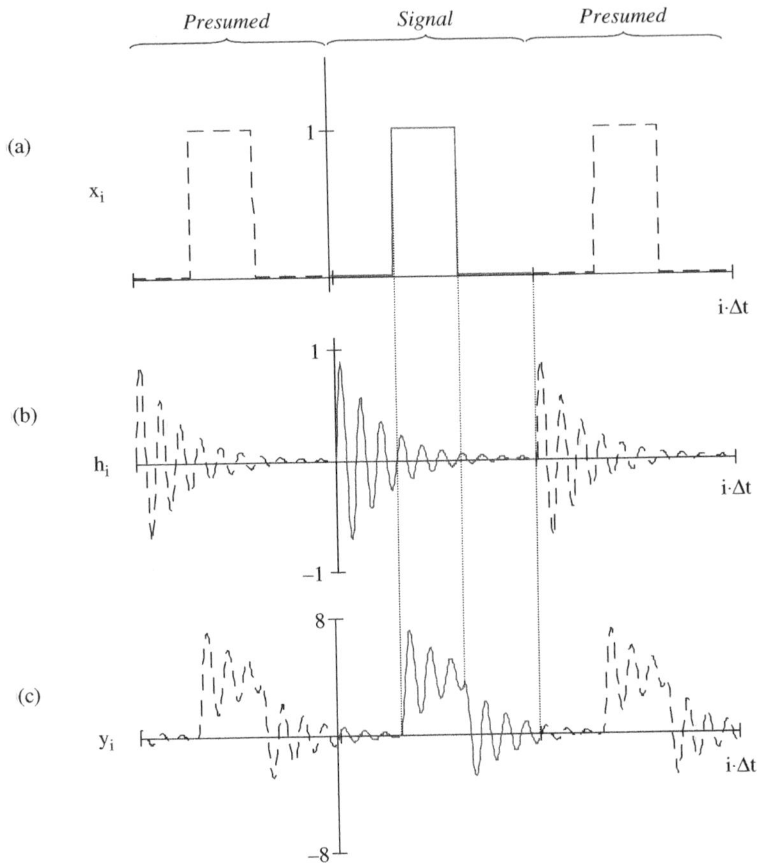

Figure 6.4 Circularity. The example corresponds to a single DoF system with low damping. The continuous trace shows the signals used in the computations. The dashed lines represent the periodicity assumption of the DFT. (a) Input signal $\underline{x}$; (b) impulse response $\underline{h}$; (c) output signal $\underline{y}$ obtained by implementing the convolution of $\underline{x}$ and $\underline{h}$ with frequency domain operations. The system "responds" before the input is applied!

This equation must be equivalent to the matrix form of the convolution in the time domain, $\underline{y} = \underline{\underline{h}} \cdot \underline{x}$, where the columns in matrix $\underline{\underline{h}}$ are shifted impulse responses $\underline{h}$ (Section 4.5). Hence

$$\underline{\underline{h}} = \frac{1}{N} \cdot \overline{\underline{\underline{F}}} \cdot \underline{\underline{\text{diagH}}} \cdot \underline{\underline{F}} \tag{6.15}$$

Example-Verification

Equation 6.15 is numerically verified in Figure 6.5. The different frames show (a) an impulse response vector $\underline{h}$ with $N = 8$ elements; (b) the DFT of $\underline{h}$ computed as $\underline{H} = \underline{\underline{F}} \cdot \underline{h}$; (c) the DFT matrix $\underline{\underline{F}}$; (d) the matrix $\underline{\underline{h}}$ computed in the time domain where each column contains the impulse response but shifted one time increment; and (e) the matrix $\underline{\underline{h}}$ computed with Equation 6.15. This matrix shows the effect of circular convolution: the lower tails of the shifted impulse responses appear "wrapped" at the top of each column.

6.4 CROSS-SPECTRAL AND AUTOSPECTRAL DENSITIES

The cross-correlation operation was introduced in the time domain to identify similarities between two signals (Section 4.3). The similarity between the computational procedures for convolution and cross-correlation led to the conclusion that the cross-correlation $\underline{cc}$ of $\underline{x}$ and $\underline{z}$ is equivalent to the convolution "$*$" of $\underline{z}$ with the tail-reversed $\underline{x}$ (see Section 4.4.3):

$$\underline{cc}^{<x,z>} = \underline{z} * \text{rev}(\underline{x}) \tag{6.16}$$

Tail reversal is equivalent to measuring the phase in the opposite direction: a tail-reversed cosine is the same cosine; however, a tail-reverse sine is [−]sine. Therefore, if $\underline{X} = \text{DFT}(\underline{x})$, the conjugate of $\underline{X}$ is the DFT of rev($\underline{x}$). Applying the DFT to both sides of Equation 6.16,

$$\begin{aligned} \text{DFT}(\underline{cc}^{<x,z>}) &= \text{DFT}[\underline{z} * \text{rev}(\underline{x})] \\ CC_u^{<x,z>} &= Z_u \cdot \{\text{DFT}[\text{rev}(\underline{x})]\}_u \\ CC_u^{<x,z>} &= Z_u \cdot \overline{X_u} \end{aligned} \tag{6.17}$$

Likewise the DFT of the autocorrelation is $\underline{AC}^{<x>} = \text{DFT}(\underline{ac}^{<x>})$

$$\begin{aligned} AC_u^{<x>} &= X_u \cdot \overline{X_u} \\ &= [\text{Re}(X_u)]^2 + [\text{Im}(X_u)]^2 \end{aligned} \tag{6.18}$$

The cross-correlation and autocorrelation arrays in the frequency domain are called the *cross-spectral* and the *autospectral densities*, respectively.

(a) Consider the impulse response $\underline{h}$:

$$\underline{h} = \begin{bmatrix} 0 \\ 1 \\ 2 \\ 0 \\ -1 \\ 0 \\ 0 \\ 0 \end{bmatrix}$$

(b) Its Fourier transform is $\underline{H}$=DFT($\underline{h}$)

$$H = \begin{bmatrix} 2 \\ 1.707 - 2.707j \\ -3 - 1j \\ 0.293 + 1.293 \\ 0 \\ 0.293 - 1.293j \\ -3 + 1j1 \\ 1.707 + 2.707j \end{bmatrix}$$

(c) The DFT matrix is computed as $F_{u,i} = e^{-j\left(u\frac{2\pi}{N}i\right)}$

$$F = \begin{bmatrix} 1 & 1 & 1 & 1 & 1 & 1 & 1 & 1 \\ 1 & 0.7-0.7j & -j & -0.7-0.7j & -1 & -0.7+0.7j & j & 0.7+0.7j \\ 1 & -j & -1 & j & 1 & -j & -1 & j \\ 1 & -0.7-0.7j & j & 0.7-0.7j & -1 & 0.7+0.7j & -j & -0.7+0.7j \\ 1 & -1 & 1 & -1 & 1 & -1 & 1 & -1 \\ 1 & -0.7+0.7j & -j & 0.7+0.7j & -1 & 0.7-0.7j & j & -0.7-0.7j \\ 1 & j & -1 & -j & 1 & j & -1 & -j \\ 1 & 0.7+0.7j & j & -0.7+0.7j & -1 & -0.7-0.7j & -j & 0.7-0.7j \end{bmatrix}$$

(d) The impulses response matrix $\underline{\underline{h}}$ in the time domain is

$$\begin{bmatrix} 0 & 0 & 0 & 0 & 0 & 0 & 0 & 0 \\ 1 & 0 & 0 & 0 & 0 & 0 & 0 & 0 \\ 2 & 1 & 0 & 0 & 0 & 0 & 0 & 0 \\ 0 & 1 & 1 & 0 & 0 & 0 & 0 & 0 \\ 0 & 0 & 0 & 0 & 0 & 0 & 0 & 0 \\ -1 & 0 & 2 & 1 & 0 & 0 & 0 & 0 \\ 0 & -1 & 0 & 2 & 0 & 0 & 0 & 0 \\ 0 & 0 & -1 & 0 & 2 & 1 & 0 & 0 \\ 0 & 0 & 0 & -1 & 0 & 2 & 1 & 0 \end{bmatrix}$$

(e) The $\underline{h}$ matrix computed as $\underline{\underline{h}} = \left[\frac{1}{N}\underline{\underline{F}} \cdot \underline{\underline{\text{diag}H}} \cdot \underline{\underline{F}}\right]$

$$\begin{bmatrix} 0 & 0 & 0 & 0 & -1 & 0 & 2 & 1 \\ 1 & 0 & 0 & 0 & 0 & -1 & 0 & 2 \\ 2 & 1 & 0 & 0 & 0 & 0 & -1 & 0 \\ 0 & 2 & 1 & 0 & 0 & 0 & 0 & -1 \\ -1 & 0 & 2 & 1 & 0 & 0 & 0 & 0 \\ 0 & -1 & 2 & 1 & 0 & 0 & 0 & 0 \\ 0 & -1 & 0 & 2 & 1 & 0 & 0 & 0 \\ 0 & 0 & -1 & 0 & 2 & 1 & 0 & 0 \\ 0 & 0 & 0 & -1 & 0 & 2 & 1 & 0 \end{bmatrix}$$

circularity

Figure 6.5 Convolution as matrix multiplication. Circular convolution affects the matrix $\underline{\underline{h}}$ computer with frequency domain operations. Compare frames d and e – upper triangle

Cross-correlation is not commutative (Section 4.2.4). However, circularity in the frequency domain renders $\underline{cc}^{<z,x>} = \text{rev}(\underline{cc}^{<x,z>})$. The computation of cross- and autocorrelations in the frequency domain is summarized in Implementation Procedure 6.3. A simple numerical example is presented in Figure 6.6. This algorithm is more efficient than the computation of correlations in the time domain because it involves fewer multiplications. Nevertheless, one must be aware of the periodicity assumption that underlines the DFT.

Caution. The values of the spectra in one-sided computations are twice those corresponding to the two-sided definition. Two-sided definitions are used in this text.

Implementation Procedure 6.3 Cross-correlation and autocorrelation

1. Given $\underline{x}$, compute the DFT: $\underline{X} = \text{DFT}(\underline{x})$.
2. Given $\underline{z}$, compute the DFT: $\underline{Z} = \text{DFT}(\underline{z})$.
3. Determine the complex conjugate for each value $\overline{X_u} = \text{Re}(X_u) - j \cdot \text{Im}(X_u)$.
4. Perform the following point-by-point multiplication:

$$CC_u^{\langle x,z\rangle} = Z_u \cdot \overline{X_u}$$

5. Compute the inverse Fourier transform of $\underline{CC}^{<x,z>}$ to determine the cross-correlation of $\underline{x}$ and $\underline{z}$ in the time domain

$$\underline{cc}^{<x,z>} = \text{IDFT}(\underline{CC}^{<x,z>})$$

6. The same procedure applies to autocorrelation $\underline{AC}^{<x>}$, but $\underline{z}$ and $\underline{Z}$ should be replaced by $\underline{x}$ and $\underline{X}$.

Example

Cross-correlation and autocorrelation are used to find similarities between and within signals (see Section 4.2). A numerical example of cross-correlation is presented in Figure 6.6.

The computation of the cross-correlation in the frequency domain is affected by the underlying assumption of periodicity in the DFT (circularity, Section 6.3.2).

(a)

	Time domain		Frequency domain
Input signal 1	$\underline{x}$	$\Rightarrow$ DFT $\Rightarrow$	$\underline{X}$
Input signal 2	$\underline{z}$	$\Rightarrow$ DFT $\Rightarrow$	$\underline{Z}$
Cross-correlation	$\underline{cc}^{<x,z>}$	$\Leftarrow$ IDFT $\Leftarrow$	$CC_u^{\langle x,z\rangle} = \overline{X_u} \cdot Z_u$

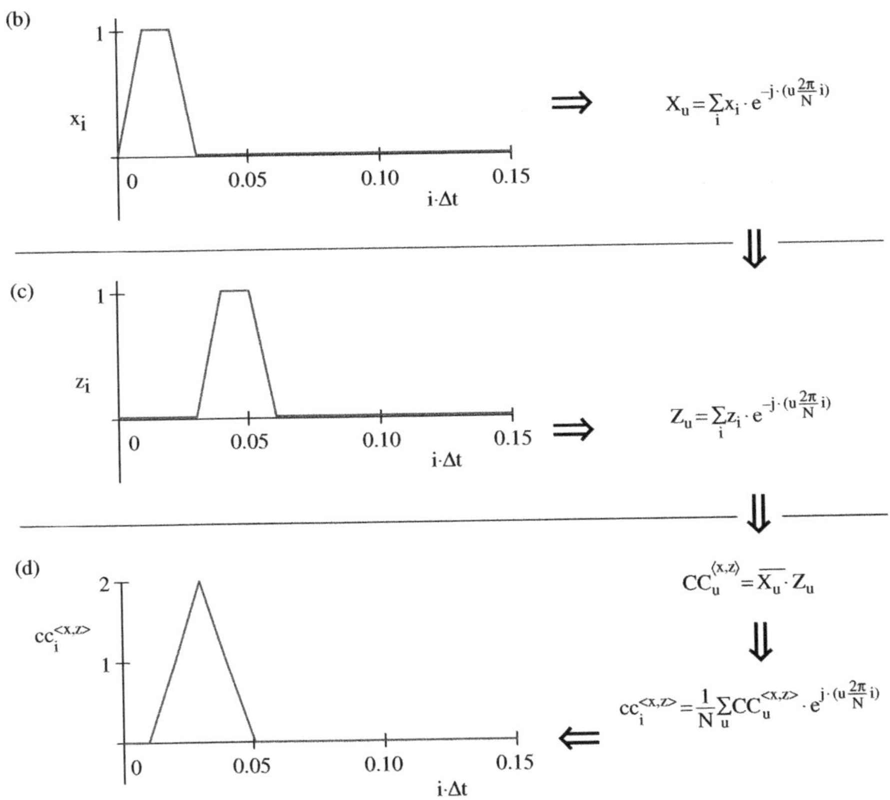

Figure 6.6 Cross-correlation in the frequency domain: (a) calculations; (b) signal $\underline{x}$; (c) signal $\underline{z}$ is a shifted version of $\underline{x}$; (d) cross-correlation. The peak in the cross-correlation array indicates the time shift between the two signals

6.4.1 Important Relations

If $\underline{y}$ is the output signal of an LTI system excited with the input $\underline{x}$, so that $\underline{y} = \underline{h} * \underline{x}$, then $Y_u = H_u \cdot X_u$, and

$$CC_u^{\langle x,y\rangle} = \overline{X_u} \cdot Y_u = \overline{X_u} \cdot H_u \cdot X_u = H_u \cdot AC_u^{\langle x\rangle} \tag{6.19}$$

$$AC_u^{\langle y\rangle} = \overline{Y_u} \cdot Y_u = \overline{H_u} \cdot \overline{X_u} \cdot H_u \cdot X_u = |H_u|^2 \cdot AC_u^{\langle x\rangle} \tag{6.20}$$

Although the DFT of the sum of two signals is the sum of the DFT of the signals, $DFT(\underline{x} + \underline{y}) = DFT(\underline{x}) + DFT(\underline{y})$, the additivity rule does <u>not</u> apply to the cross-spectra or the autospectra. Therefore, $AC^{<x+y>} \neq AC^{<x>} + AC^{<y>}$ (see exercises at the end of this chapter).

6.5 FILTERS IN THE FREQUENCY DOMAIN - NOISE CONTROL

A filter in the frequency domain is a "window" $\underline{W}$ that passes certain frequency components X_u and rejects others. This is a point-by-point multiplication

$$Y_u = X_u \cdot W_u \tag{6.21}$$

The modified array $\underline{Y}$ is transformed back to the time domain. Filters can alter the amplitude spectrum, the phase spectrum, or both, depending on the filter coefficients W_u.

Many signal processing operations can be implemented as filters, including noise control. The primary recommendation still remains: improve the signal-to-noise ratio at the lowest possible level, starting with a proper experimental design; then consider signal stacking if signals are repeatable (Section 4.1.5).

6.5.1 Filters

The filter coefficients W_u at frequencies $f_u = u/(N \cdot \Delta t)$ determine the filter performance. The most common filters are:

- *Low-pass.* A frequency component X_u passes if the corresponding frequency $f_u = u/(N \cdot \Delta t)$ is below the selected cutoff frequency. Frequency components above the cutoff frequency are attenuated or rejected (Figure 6.7b). Low-pass filters are used to suppress high-frequency noise.

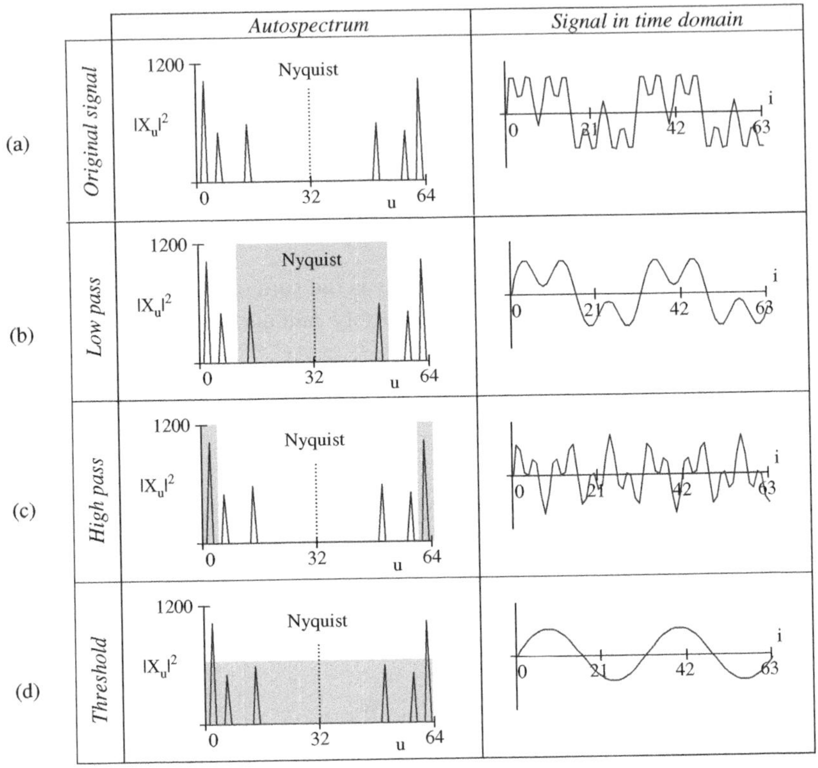

Figure 6.7 Filters – frequency domain. Autospectral densities and corresponding signals. The filtered signals are blocked. Note the double-sided definition of these filters. The original signal is $x_i = \sin(2\frac{2\pi}{N}i) + 0.7\sin(6\frac{2\pi}{N}i) + 0.75 \cdot \sin(14\frac{2\pi}{N}i)$. The filtered frequencies are shaded

- *High-pass*. High-frequency components above the cutoff frequency pass, while low-frequency components are attenuated or rejected (Figure 6.7c). Low-frequency components are common in measurement and applications. Some examples include: uneven illumination in photography, low-frequency bench vibration during ultrasonic testing, and the 60 Hz of power lines with respect to radio signals.
- *Band-pass*. These filters are a combination of low- and high-pass filters. The intent is to keep a frequency band. The opposite effect is achieved with *band-reject* filters, where a selected band of frequencies is rejected. A *notch filter* is a narrow band-reject filter.

- *All-pass*. This filter is used for phase control only. The magnitude response is 1.0 across the spectrum and the phase response is designed to cause a frequency-dependent phase shift in the signal. Typically, all-pass filters are used to correct the phase shift imposed by other filters in a series of filters.

If the transition region from "pass" to "reject" is gradual, the *cutoff frequency* corresponds to a reduction in the signal magnitude of $-3\,\text{dB}$, that is $|Y_u| = 0.7|X_u|$.

Phase shift $\Delta\varphi$ and time shift δt are related as $\Delta\varphi/2\pi = \delta t/T$. If the phase shift varies linearly with frequency $\Delta\varphi_u = \alpha f_u$,

$$\delta t_u = \frac{\Delta\varphi_u}{2\pi f_u} = \frac{\alpha}{2\pi} = \text{constant} \tag{6.22}$$

A *linear-phase filter* causes a constant time shift δt_u in all frequency components and it does not distort the waveform (see solved problems at the end of this Chapter.)

6.5.2 Frequency and Time

The point-by-point multiplication in the frequency domain indicated in Equation 6.21 implies a convolution in the time domain between the signal $\underline{x}$ and the inverse discrete Fourier transform (IDFT) of $\underline{W}$. This vector must be the array named "kernel" $\underline{\kappa}$ in Section 4.1.3. Therefore the kernel $\underline{\kappa} = \text{IDFT}(\underline{W})$ is the *filter impulse response*. Conversely, knowing the kernel, one can determine the window (note that real signals in one domain become complex-valued signals in the other domain in most cases). Then, the understanding of filters in the frequency domain helps gain insight into the design of moving kernels in the time domain.

Consider the windows $\underline{W}$ shown in Figure 6.8. The corresponding kernels $\underline{\kappa}$ are also shown in the figure. Kernels obtained from windows with sharp boundaries show excessive ringing. In general, smoothly varying band-pass filters are preferred.

6.5.3 Computation

If the filter is the DFT of a kernel in time, then the filter $\underline{W}$ must satisfy the periodicity property in frequency (Figure 5.2). In addition, if double-sided operations are used, the filter must be defined above the Nyquist frequency (or in the negative frequencies) as well, as shown in Figure 6.7. This annoyance is avoided when single-sided operations are used. The process of filtering a signal in the frequency domain is summarized in Implementation Procedure 6.4.

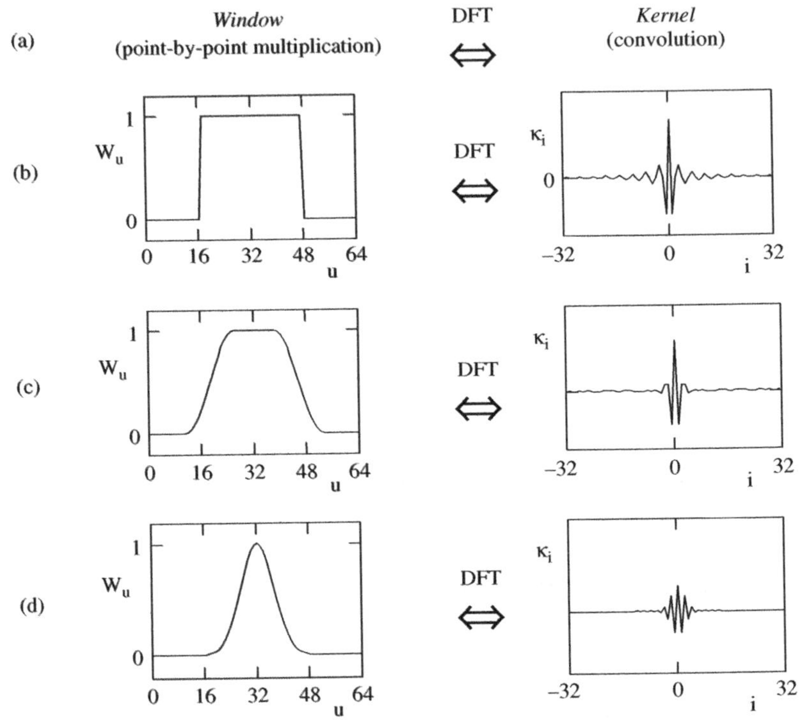

Figure 6.8 Windows $\underline{W}$ and kernels $\underline{\kappa}$: (a) mathematical relation; (b) sharp boundary wide window; (c) gradual boundary wide window; (d) gradual boundary narrow window. Sharp boundaries lead to kernels with high ringing.

Implementation Procedure 6.4 Filtering noise in the frequency domain

1. Given a signal $\underline{x}$, compute its DFT: $\underline{X} = \text{DFT}(\underline{x})$.
2. Plot the magnitude $|X_u|$ vs. f_u to determine the frequency band of interest.
3. Choose the type of filter to be used.
4. Caution: if double-sided DFT is used, then the filter must have a compatible double-sided definition. *Low-pass filter:* removes frequency components below counter u* and above N−1−u*, where u* is the frequency counter for the cutoff frequency. *High-pass filter:* keeps frequency components between the counter at the cutoff frequency u* and N−1−u*.

5. Define the array that represents the filter or window $\underline{W}$. Superimpose a plot of this array onto the spectrum of the signal to confirm the selection of the filter.
6. Apply the window to the signal: multiply point by point $Y_u = X_u \cdot W_u$.
7. Compute the inverse Fourier transform of the filtered signal: $\underline{y} = \text{IDFT}(\underline{Y})$.

Example

The effect of different filters is explored in Figure 6.7.

Note: Electronic filters are frequently used during data acquisition. Antialiasing low-pass filters must be included in the measurement system before the signal is digitized. Analog-to-digital devices often have antialising filters built-in. Because the information content in a signal increases with increasing bandwidth, filtering removes information. Consequently, the design of filters is a critical task in signal recording and postprocessing.

The information content in a signal increases with the bandwidth. Filters reduce the information content and rejected frequencies are irreversibly lost. That is, the convolution of the signal with the filter is a linear transformation, but it is not necessarily invertible.

There are versatile nonlinear signal-enhancement operations in the time domain (Figure 4.5). Likewise, there are nonlinear filters in the frequency domain as well. For example, the band pass of a filter can be implemented by thresholding: if $|X_u| >$ threshold, then $Y_u = X_u$; otherwise, $Y_u = 0$ (Figure 6.7d). The threshold may be established in linear scale or in dB.

6.5.4 Revisiting Windows in the Time Domain

Time windows are used to reduce "leakage" (Section 5.5): the signal $\underline{x}$ in the time domain is multiplied point by point with the window ($x_i \cdot w_i$) before it is transformed to the frequency domain. Multiplying a signal times a window in the time domain is equivalent to the convolution sum of their transforms in the frequency domain (duality property). On the other hand, windows are used for filtering in the frequency domain which is equivalent to convolution with a kernel in the time domain.

At this point, the concepts of "window" and "kernel" have been encountered in both the time and the frequency domains. What is the difference? Typically, kernels are convolved whereas windows are multiplied point by point with the array being processed. Both operations may take place either in the time domain or in the frequency domain, and each operation is the Fourier transform of the other.

	Time domain		*Frequency domain*
Point-by-point (·)	$y_i = x_i \cdot w_i$	DFT	$Y_u = X_u \cdot W_u$
Convolution (∗)	$\underline{y} = \underline{x} * \underline{\kappa}$		$\underline{Y} = \underline{X} * \underline{K}$

6.5.5 *Filters in Two Dimensions (Frequency–Wavenumber Filtering)*

The process of filtering in the frequency domain can be extended to two-dimensional (2D) signals. The original 2D signal is 2D discrete Fourier transformed to the f–k space. A 2D band-pass window is multiplied point by point, keeping only the information of interest. Finally, the windowed spectrum is inverse transformed to the original 2D space of the signal.

For example, consider wave propagation (refer to Figure 6.9). Time series gathered at different aligned locations can be transformed into the frequency–wavenumber space f–k ($f = 1/T, k = 1/\lambda$). In this space, events that emerge with characteristic slopes $f/k = \lambda/T = V$. Unwanted events are conveniently identified and filtered. In geophysical applications, 2D f–k filtering permits removing a coherent component such as surface waves or "ground roll" from signals.

6.6 *DETERMINING $\underline{H}$ WITH NOISELESS SIGNALS (PHASE UNWRAPPING)*

The determination of the impulse response $\underline{h}$ in the time domain is hampered by the mathematical nature of the impulse signal. A more convenient alternative in the time domain is to apply a step, to measure the step response, and to compute its time derivative. Then, the frequency response is $\underline{H} = \text{DFT}(\underline{h})$.

One can also determine the frequency response $\underline{H}$ by exciting the system with single-frequency sinusoids, which are system eigenfunctions (Implementation Procedure 6.1). If needed, the impulse response is computed as $\underline{h} = \text{IDFT}(\underline{H})$.

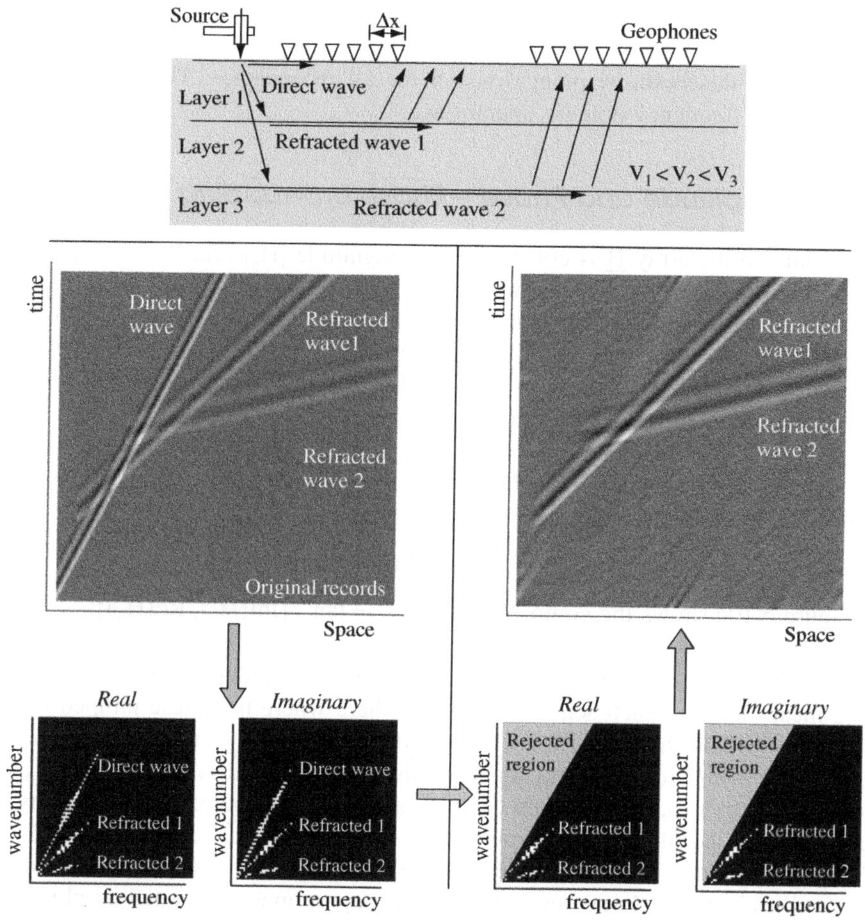

Figure 6.9 2D filters. f–k filtering for seismic applications. The direct wave is removed from the records by rejecting the high wavenumber region. One quadrant shown in frequency-wavenumber space

However, the most effective and versatile approach to determine the frequency response $\underline{H}$ is to use any broadband input signal and to process the data in the frequency domain. Indeed, if convolution in the frequency domain is $Y_u = H_u \cdot X_u$, then the u-th entry in the frequency response array $\underline{H}$ is

$$H_u = \frac{Y_u}{X_u} \quad \text{for frequency} \quad \omega_u = u\frac{2\pi}{N \cdot \Delta t} \tag{6.23}$$

where the arrays $\underline{X}$ and $\underline{Y}$ are the DFTs of the measured input and output signals, $\underline{X} = \text{DFT}(\underline{x})$ and $\underline{Y} = \text{DFT}(\underline{y})$. The frequency response array $\underline{H}$ is obtained by repeating this point-by-point division for all frequencies. This is a salient advantage of frequency domain operations!

6.6.1 Amplitude and Phase – Phase Unwrapping

Each entry in the array $\underline{H}$ is complex. The magnitude $|H_u|$ relates the amplitude of the response sinusoid to the amplitude of a single input sinusoid of the same frequency ω_u; the units of $|H_u|$ are those of [output/input]. The phase between output and input sinusoids is $\varphi_u = \tan^{-1}[\text{Im}(H_u)/\text{Re}(H_u)]$. The analysis of the computed phase often requires an additional step. Consider a system that causes a constant phase shift δt to all frequencies:

- Figure 6.10a shows the true phase shift $\varphi_u = 2\pi(\delta t/T_u) = \delta t \cdot \omega_u$. (Note that this is a linear phase system.)
- Figure 6.10b shows the computed ratio $\text{Im}(H_u)/\text{Re}(H_u) = \tan(\varphi_u)$.
- Figure 6.10c shows the computed phase $\varphi_u = \tan^{-1}[\text{Im}(H_u)/\text{Re}(H_u)]$.

The computed phase appears "wrapped" between $-\pi/2$ and $+\pi/2$. Phase unwrapping means shifting each segment up everywhere where the phase jumped from $-\pi/2$ to $+\pi/2$ as shown in Figure 6.10d. (If the time shift is negative, the phase jumps from $+\pi/2$ to $-\pi/2$ and segments must be shifted down.) The jumps are not always obvious, and in some cases local jumps may be related to the physical nature of the phenomenon, rather than the mathematical effect described in Figure 6.10. Increased frequency resolution may help clarify some apparent discontinuities in phase. In any case, phase unwrapping must be guided by physical insight into the system under consideration.

Implementation Procedure 6.5 summarizes the steps involved in computing and interpreting the frequency response $\underline{H}$ of any LTI system using noiseless signals.

Implementation Procedure 6.5 Determination of the frequency response $\underline{H}$ with a generic broadband signal – No noise

1. Select any broadband source that operates in the frequency range of interest.
2. Select the signal duration $N \cdot \Delta t$, the number of points N and the sampling interval Δt following the guidelines in Implementation Procedure 5.2.

3. Measure the input at the interface between the source and the system. This is the input signal in the time domain $\underline{x}$.

4. Capture the measured response $\underline{y}$.

5. Compute the DFTs of the input and the output: $\underline{X} = DFT(\underline{x})$ and $\underline{Y} = DFT(\underline{y})$.

6. Compute the "measured" frequency response as a point-by-point division:

$$H_u^{<meas>} = \frac{Y_u}{X_u} \qquad \text{for frequency} \qquad \omega_u = u\frac{2\pi}{N\Delta t}$$

7. Transducers, multimeters, analyzers and other peripheral electronics must operate within their frequency range. Each component transforms the signal. Therefore, determine the frequency response of transducers and peripheral electronics $\underline{H}^{<tran>}$ in calibration studies with known specimens.

8. Assuming that the response of transducers and peripheral electronic $\underline{H}^{<tran>}$ is in series with the system response, then the measured response is

$$H_u^{<meas>} = H_u^{<sys>} \cdot H_u^{<tran>}$$

Therefore, the sought system response is

$$H_u^{<sys>} = \frac{H_u^{<meas>}}{H_u^{<tran>}}$$

Note: The system response $\underline{H}^{<sys>}$ is an array of complex numbers. Results are commonly presented as amplitude $|H_u|$ and phase φ_u versus frequency ω_u. The *phase is calculated as* $\varphi_u = \arctan\,[Im(H_u)/Re(H_u)]$ *and it yields values between $-\pi/2$ and $\pi/2$. The phase spectrum is "unwrapped" by accumulating the phase at every jump between $-\pi/2$ and $\pi/2$ (Section 6.6.1).*

6.7 DETERMING $\underline{H}$ WITH NOISY SIGNALS (COHERENCE)

Equation 6.23 is valid for ideal noiseless signals. Yet, noise is always present. In most cases, the input signal $\underline{x}$ can be measured close enough to the system so that

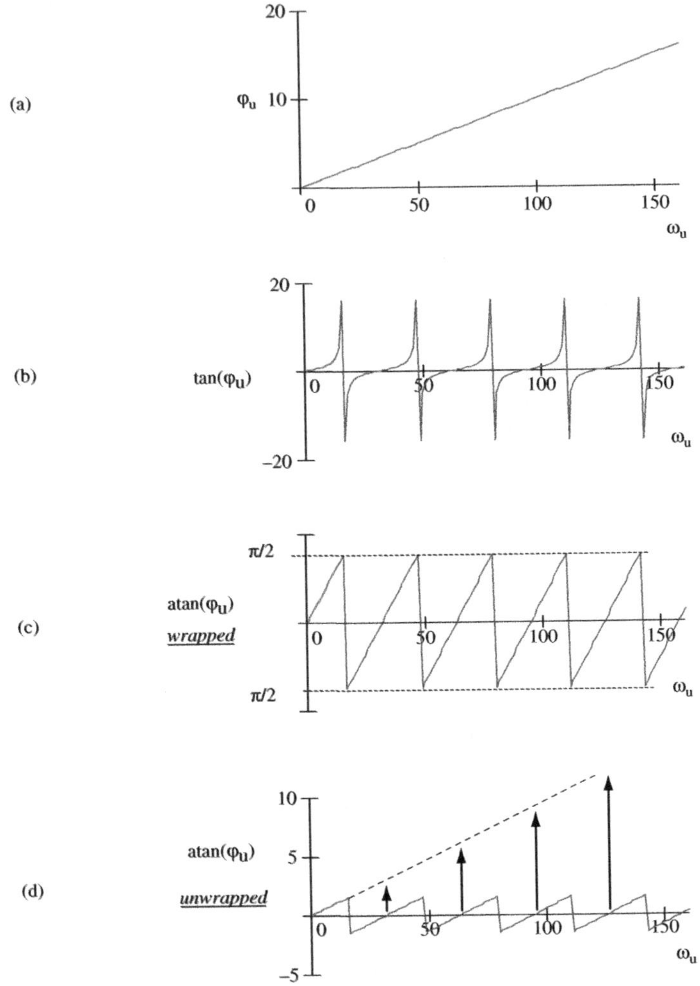

Figure 6.10 Phase unwrapping: (a) consider a process that produces a time shift $\delta t =$ 0.1 between input an output so the phase shift is linear with frequency $\phi_u = \delta t \cdot \omega_u$. (b) However, $\tan(\varphi_u)$ is not a linear but a periodic function of frequency. This is the value computed with input and output data: $\tan(\varphi_u) = \mathrm{Im}(H_u)/\mathrm{Re}(H_u)$. (c) The inferred phase shift $\phi_u = \arctan\,[\mathrm{Im}(H_u)/\mathrm{Re}(H_u)]$ oscillates between $\pi/2$ and $-\pi/2$ in a seesaw function that is characteristic of the wrapped phase. (d) The original phase spectrum is reconstructed by "unwrapping" the phase, adding π at each jump from $\pi/2$ to $-\pi/2$ in the spectrum

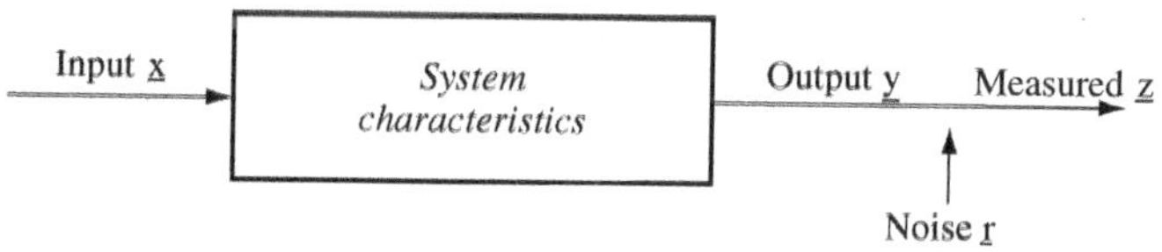

Figure 6.11 System characterization with noisy signals. The measured output signal $\underline{z}$ included noise $\underline{r}$ added at the output

no undetected signal goes into the system, still, noise $\underline{r}$ gets added at the output $\underline{y}$ (see Figure 6.11). Then, the *measured* output $\underline{z}$ in the frequency domain is

$$Z_u = Y_u + R_u = H_u \cdot X_u + R_u \quad \text{for frequency } \omega_u = u\frac{2\pi}{N \cdot \Delta t} \tag{6.24}$$

where $\underline{R}$ is the discrete Fourier transform of the noise $\underline{R} = DFT(\underline{r})$. If the frequency response is computed with Equation 6.22, one obtains

$$H_u^{<Noisy>} = \frac{H_u \cdot X_u + R_u}{X_u} = H_u + \frac{R_u}{X_u} \tag{6.25}$$

Hence, the error in the frequency response depends on the ratio R_u/X_u. The procedure to determine the frequency response must be modified to correct for noise effects. In the new procedure, the input and the output will be measured "M" times so that the effect of noise will be canceled by averaging spectral densities. The proper equation to compute the frequency response is

$$H_u = \frac{(Z_u \cdot \overline{X_u})_{avr}}{(X_u \cdot \overline{X_u})_{avr}} = \frac{(CC_u^{<x,z>})_{avr}}{(AC_u^{<x>})_{avr}} \tag{6.26}$$

where the average spectral densities are

$$\begin{aligned} (CC_u^{<x,z>})_{avr} &= \frac{1}{M} \sum_{\text{all meas}} (CC_u^{<x,z>})_{\text{each meas}} \\ (AC_u^{<x>})_{avr} &= \frac{1}{M} \sum_{\text{all meas}} (AC_u^{<x>})_{\text{each meas}} \end{aligned} \tag{6.27}$$

The values of auto and cross-correlations for each measurement are averaged at each u-th frequency for similar signals to eliminate uncorrelated noise. Note that contrary to signal stacking in the time domain (Section 4.1.2), the various signals $\underline{x}$ need not be the same.

Why does Equation 6.26 work? The average cross-spectral density is

$$\begin{aligned}(CC_u^{<x,z>})_{avr} &= \left(Z_u \cdot \overline{X_u}\right)_{avr} \\ &= \left[(X_u \cdot H_u + R_u) \cdot \overline{X_u}\right]_{avr} \\ &= \left[\left(X_u \cdot \overline{X_u}\right) \cdot H_u\right]_{avr} + \left(R_u \cdot \overline{X_u}\right)_{avr}\end{aligned} \tag{6.28}$$

In the first term on the right-hand side, H_u is a constant and can be factored out of the summation. In the second term, the sum goes to zero because noise and the input signal are uncorrelated. Therefore, the numerator in Equation 6.25 tends to

$$(CC_u^{<x,z>})_{avr} = H_u \cdot (AC_u^{<x>})_{avr} \tag{6.29}$$

and the computational procedure prescribed in Equation 6.26 adequately estimates the frequency response without the effects of random noise. (Notice the parallelism between Equation 6.19 for a single noiseless signal and Equation 6.29 for averaged spectra of an ensemble of noisy signals.)

As an example, consider a simple system with frequency response $H_u = 0.5$ and $\varphi_u = 0$ for all frequencies f_u. Figure 6.12 shows the computed frequency

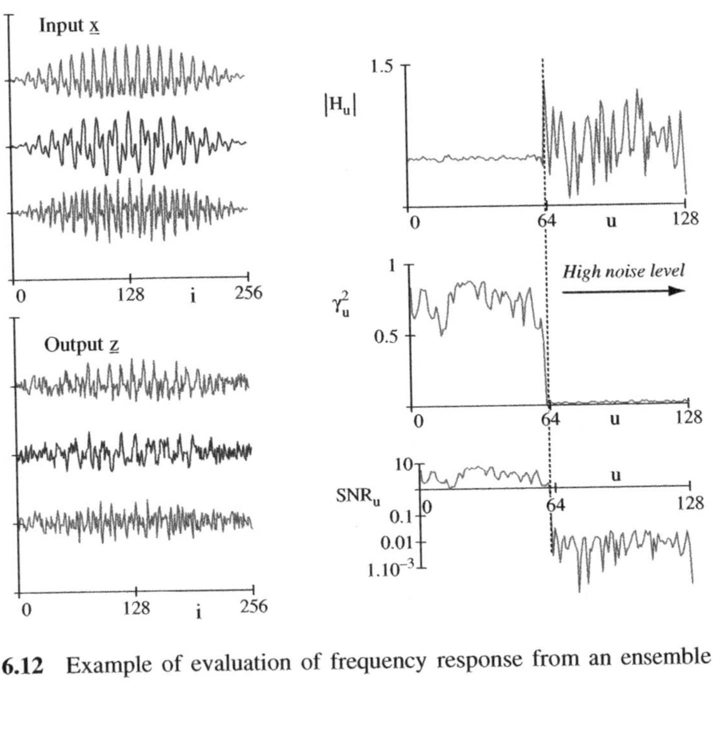

Figure 6.12 Example of evaluation of frequency response from an ensemble of noisy signals

response $\underline{H}$ given a set of input signals and the corresponding output signals "gathered" with additive Gaussian noise.

6.7.1 Measures of Noise – Coherence

The noiseless output $\underline{y}$ is fully caused by the input $\underline{x}$ (Figure 6.11). This is not the case for the measured output $\underline{z}$. The coherence operator yields a real valued array $\underline{\gamma}^2$ where each u-th entry denotes the energy in the measured output that can be justified by the input. In mathematical terms,

$$\gamma_u^2 = \frac{(AC_u^{<y>})_{avr}}{(AC_u^{<z>})_{avr}} \quad \text{for frequency} \quad \omega_u = u\frac{2\pi}{N \cdot \Delta t} \tag{6.30}$$

Coherence quantifies the energy in the measured output that was caused by the input. Coherence $\underline{\gamma}^2$ is properly determined using average spectra for an ensemble of signals:

$$\gamma_u^2 = \frac{\left|(CC_u^{<x,z>})_{avr}\right|^2}{(AC_u^{<x>})_{avr} \cdot (AC_u^{<z>})_{avr}} \quad \text{for frequency } \omega_u \tag{6.31}$$

If only one signal is available, this equation results in $\gamma^2 = 1.0$ for all frequencies. Thus, the value of coherence is meaningful when average spectra for multiple signals are used. The average spectra are computed as indicated in Equation 6.27. It can be shown through mathematical manipulations and arguments similar to those invoked for the derivation of the frequency response that

$$\gamma_u^2 = \frac{|H_u|^2 \cdot (AC_u^{<x>})_{avr}}{|H_u|^2 \cdot (AC_u^{<x>})_{avr} + (AC_u^{<r>})_{avr}} \tag{6.32}$$

where $\underline{AC}^{<r>}$ is the autospectrum of noise $\underline{R}$. This equation agrees with the definition in Equation 6.30 (recall identities in Equations 6.19, 6.20). Furthermore, it shows that if coherence is one (that is, $\gamma_u^2 = 1$ at frequency $\omega_u = u \cdot 2\pi/(N \cdot \Delta T)$, then all the energy in the output is caused by the input.

Coherence is a valuable diagnostic tool. Coherence less than 1.0 indicates one or more of the following situations:

- Noise in the output
- Unaccounted inputs in the system
- Nonlinear system behavior

- Lack of frequency resolution and leakage: a local drop in coherence observed near a resonant peak suggests that the system resonant frequency does not coincide with a harmonic $\omega_u = u \cdot 2\pi/(N \cdot \Delta T)$ in the discrete Fourier transformation (Section 5.6).

The signal-to-noise ratio (SNR) is computed as the ratio between the autospectral density of the signal without noise $\underline{y}$ and the autospectral density of noise $\underline{r}$:

$$SNR_u = \frac{(AC_u^{<y>})_{avr}}{(AC_u^{<r>})_{avr}} \qquad \text{for frequency } \omega_u \tag{6.33}$$

Its proper determination requires ensemble averages, similar to Equation 6.29. From Equation 6.30, the SNR is related to coherence as

$$SNR_u = \frac{\gamma_u^2}{1-\gamma_u^2} \qquad \text{for frequency } \omega_u \tag{6.34}$$

The range for SNR varies from 0 to infinity; the higher the SNR, the stronger the signal is with respect to noise.

Figure 6.12 shows the spectrum of coherence and SNR for the simple system with frequency response $H_{\underline{u}} = 0.5$ and $\varphi_u = 0$ for all frequencies f_u.

6.7.2 Statistical Interpretation

Statistical parameters can be computed for the value of a signal $\underline{x}$ at discrete time t_i, using the ensemble of M signals,

$$\text{mean value} \qquad \mu_i = \frac{1}{M}\sum_k x_i^{<k>} \tag{6.35}$$

$$\text{mean square value} \qquad \psi_i^2 = \frac{1}{M}\sum_k \left(x_i^{<k>}\right)^2 \tag{6.36}$$

$$\text{variance} \qquad \sigma_i^2 = \frac{1}{M}\sum_k \left(x_i^{<k>} - \mu_i^{<k>}\right)^2 = \psi_i^2 - \mu_i^2 \tag{6.37}$$

If the signal is ergodic (Section 3.1), these statistical parameters can be computed using the N-values in one signal. In this case, the mean square value ψ^2 is known as the root mean square *rms* of the signal and it is equal to the value of the autocorrelation for zero-time shift $ac_0^{<x>} = \psi^2 = \sigma^2 + \mu^2$. On the basis of Parseval's identity, ψ^2 is also equal to $1/N^2$ times the area of the autospectral density plot. These observations suggest the mathematical link between ensemble statistics and spectral densities. These concepts are invoked in the next section.

6.7.3 Number of Records – Accuracy in the Frequency Response

The computation of the frequency response is enhanced by increasing the length $N \cdot \Delta t$ of recorded signals and the number of signals M in the ensemble. Longer signals reduce estimate errors in time average operators such as cross-correlation and autocorrelation (recall Figure 4.8). Furthermore, the length $N \cdot \Delta t$ must be much larger than the average time shift δt between input and output to avoid bias.

On the other hand, the higher the number of signals M in the ensemble, the lower the variance in the estimate of the frequency response. Following an analogous discussion to Section 4.1.2, the mean computed from samples size M has a coefficient of variation (cov) proportional to the cov of the population and inversely proportional to $\sqrt{M}$, where the cov is the standard deviation divided by the mean $cov = \sigma/\mu$.

The number of signals M that must be processed to estimate the magnitude of the frequency response $|H_u|$, with an expected cov in the estimate, given a signal-to-noise ratio SNR or coherence γ^2, is (see Bendat and Piersol 1993)

$$M = \frac{1}{2cov^2\,SNR} = \frac{1}{2cov^2}\frac{1-\gamma^2}{\gamma^2} \tag{6.38}$$

Similar statistical arguments were used in the time domain (Section 4.1.2). However, the criterion followed in the time domain was to obtain a good estimate of the signal x_i at a given time t_i. The aim in Equation 6.38 is to obtain a good estimate of the frequency response H_u at frequency ω_u. For clarity, subindices are not included in Equation 6.38.

For example, for an $SNR = 1$ and a desired $cov = 0.01$ (excellent) in the estimate of $\underline{H}$, it would require $M = 5000$ signals, while for $cov = 0.1$ (good) the number $M = 50$. The desired cov can be lowered if single-point estimates of system characteristics are replaced by a more comprehensive consideration of the array $\underline{H}$. For example, estimating the mechanical characteristics of a single DoF oscillator from resonant frequency and the amplitude at resonance is more sensitive to errors than least squares fitting the theoretical response to the measured response.

6.7.4 Experimental Determination of $\underline{H}$ in Noisy Conditions

In summary, the frequency response $\underline{H}$ can be determined by exciting the system with:

- a step or a "quasi-impulse"; noise is controlled by stacking in the time domain the response for the same repeatable input. The result is $\underline{h}$ (or an integral of $\underline{h}$

when a step is used), and the frequency response is computed as $\underline{H} = \text{DFT}(\underline{h})$. This approach was discussed in Chapter 4.

- steady-state single-frequency ω_u sinusoids to determine the values of H_u one at the time, and repeating the measurement at multiple frequencies to form $\underline{H}$. High signal-to-noise ratios may be attained even for signals buried in noise (lock-in amplifiers facilitate this task).
- generic broadband input signals and computing $\underline{H}$ with spectral quantities. Noise is controlled by spectra averaging.

The methodology for the last approach is outlined in Implementation Procedure 6.6 and relevant equations are summarized in Table 6.1.

Implementation Procedure 6.6 Determination of the frequency response $\underline{H}$ using generic broadband signals – Noisy output

1. Start the experiment following the initial guidelines provided in Implementation Procedure 6.5. Conduct preliminary measurements to assess the level of noise. Compute the SNR ratio and estimate the coherence.

2. Determine the number of signals to be stacked for the required cov in the measurement

$$M \approx \frac{1}{2 \cdot \text{cov}^2} \left[\frac{1 - \gamma^2}{\gamma^2} \right]$$

For example, if the coherence at the resonant frequency is $\gamma^2 = 0.8$ and the desired cov of the peak frequency response is cov = 2%, then a total of M = 312 measurements will be needed.

3. Collect input $\underline{x}$ and output $\underline{z}$ signals: acquire the longest possible record to reduce the bias in the estimate of the frequency response.

4. Pre-process the signals by detrending the arrays (Section 4.1.1), applying smooth transition windows (Section 5.5) and extending the recorded time series by zero-padding (Section 5.6) as needed.

5. For each measured input and output signals: compute the following arrays:

$$\underline{X} = \text{DFT}(\underline{x});\ \text{the conjugate of } \underline{X};\ \text{and } \underline{Z} = \text{DFT}(\underline{z})$$

$$\underline{CC}^{<x,z>} \quad \text{where the u-th entry is } (Z_u \cdot \overline{X_u})$$

$$\underline{AC}^{<x>} \quad \text{where the u-th entry is } (X_u \cdot \overline{X_u})$$

6. Use these results for *all measurements* to compute average spectra

$$(\underline{CC}^{<x,z>})_{avr} \quad \text{where the u-th entry is } (CC_u^{<x,z>})_{avr} = \sum_{\text{all meas.}} (Z_u \cdot \overline{X_u})_{meas}$$

$$(\underline{AC}^{<x>})_{avr} \quad \text{where the u-th entry is } (AC_u^{<x>})_{avr} = \sum_{\text{all meas.}} (X_u \cdot \overline{X_u})_{meas}$$

7. Finally, compute the frequency response array $\underline{H}$. The u-th entry is

$$H_u = \frac{(CC_u^{<x,z>})_{avr}}{(AC_u^{<x>})_{avr}} \quad \text{and it corresponds to frequency } \omega_u = u\frac{2\pi}{N \cdot \Delta t}$$

8. The coherence γ^2 and signal-to-noise ratios corresponding to the ensemble of collected signals are

$$\gamma_u^2 = \frac{\left|(CC_u^{<x,z>})_{ave}\right|^2}{(AC_u^{<x>})_{avr} \cdot (AC_u^{<z>})_{avr}} \quad \text{and } SNR_u = \frac{\gamma_u^2}{1-\gamma_u^2} \text{ at } \omega_u = u\frac{2\pi}{N \cdot \Delta t}$$

9. Analyze low coherence values to identify possible causes. Consider noise, nonlinear system behavior, or poor resolution near resonance.

10. Correct the measured $\underline{H}$ for the frequency response of transducers and peripheral electronics (see Implementation Procedure 6.5).

The input signal may be random noise. The average autocorrelation of white random noise is constant $\sim\alpha$ for all frequencies ω_u, and Equation 6.25 becomes $H_u \approx \alpha\ (CC_u^{<x,z>})_{ave}$. Furthermore, systems that preferentially amplify certain frequencies, such as a low damping resonant oscillator, tend to vibrate in that frequency and exhibit an increase in coherence near resonance. These two observations suggest the use of "unmeasured ambient noise" to explore the frequency response of systems that cause sharp amplification in narrow frequency bands.

Table 6.1 Summary of equations

	No noise – ideal signal	Noise added to output
Time domain		
Input signal	$\underline{x}$	$\underline{x}$
Output signal	$\underline{y}$	$\underline{z} = \underline{y} + \underline{r}$
Frequency domain		
Input signal	$\underline{X}$	$\underline{X}$
Output signal	$\underline{Y}$	$\underline{Z} = \underline{Y} + \underline{R}$
Component in DFT	$X_u = \mathrm{Re}\,(X_u) + j \cdot \mathrm{Im}\,(X_u)$	
Complex conjugate	$\overline{X_u} = \mathrm{Re}\,(X_u) - j \cdot \mathrm{Im}\,(X_u)$	
Autospectrum	$AC_u^{\langle x \rangle} = \overline{X_u} \cdot X_u = [\mathrm{Re}\,(X_u)]^2 + [\mathrm{Im}\,(X_u)]^2$	
Cross-spectrum	$CC_u^{\langle x,z \rangle} = \overline{X_u} \cdot Z_u$	
Cross-correlation	$\underline{cc}^{<x,z>} = \mathrm{IDFT}\,(\underline{CC}^{<x,z>})$	
Frequency response	$H_u = \dfrac{Y_u}{X_u}$	$H_u = \dfrac{(CC_u^{<x,z>})_{avr}}{(AC_u^{<x>})_{avr}}$
Phase shift	$\varphi_u = \tan^{-1}\left[\dfrac{\mathrm{Im}(H_u)}{\mathrm{Re}(H_u)}\right]$	
Amplitude	$\lvert H_u \rvert = \sqrt{[\mathrm{Re}(H_u)]^2 + [\mathrm{Im}(H_u)]^2}$	
Coherence function	Not applicable	$\gamma_u^2 = \dfrac{\lvert (CC_u^{<x,z>})_{avr} \rvert^2}{(AC_u^{<x>})_{avr} \cdot (AC_u^{<x>})_{avr}}$
Noise-to-signal ratio	Not applicable	$SNR_u = \dfrac{\gamma_u^2}{1 - \gamma_u^2}$

The u-th value of a parameter corresponds to frequency: $\omega_u = u \frac{2\pi}{N \cdot \Delta t}$

6.8 SUMMARY

- Sinusoids and complex exponentials are eigenfunctions for LTI systems: the output is a scaled and shifted sinusoid or complex exponential of the same frequency.
- The frequency response $\underline{H}$ fully characterizes the LTI system. It is equal to the DFT of the impulse response $\underline{H} = \mathrm{DFT}(\underline{h})$.

- The convolution sum in the time domain $\underline{y} = \underline{x} * \underline{h}$ becomes a point-by-point multiplication in the frequency domain $Y_u = X_u \cdot H_u$. Cross-correlation becomes a multiplication in the Fourier domain as well.
- Circular convolution and cross-correlation reflect the inherent periodicity in the DFT. Zero-padding helps reduce the effects of circularity.
- A window is applied as a point-by-point multiplication with a signal. Windows are used in the time domain to reduce leakage in truncated signals. In the frequency domain, windows are used to implement filters.
- Windowing in one domain is equivalent to convolution with a kernel in the other domain. The kernel and the window are related by the DFT.
- The frequency response $\underline{H}$ can be computed in the frequency domain using any generic broadband input signal, including random noise. This is a salient advantage of the frequency domain.
- The presence of noise requires adequate signal processing procedures.
- Low coherence indicates noise, unmeasured inputs, inadequate resolution, or nonlinear system behavior.
- The simpler, albeit slower, approach of determining each value H_u by exciting the system with steady-state single-frequency sinusoids can render high signal-to-noise ratios even when the signal is buried in noise.

FURTHER READING AND REFERENCES

Bendat, J. S. and Piersol, A. G. (1993). Engineering Applications of Correlation and Spectral Analysis. John Wiley & Sons, New York. 458 pages.

Gonzalez, R. C. and Woods, R. E. (2002). Digital Image Processing, 2nd edn. Prentice-Hall, Upper Saddle River. 793 pages.

Kearey, P. and Brooks, M. (1991). An Introduction to Geophysical Exploration, 2nd edn. Blackwell Scientific Publications, Oxford UK. 254 pages.

Libbey, R. B. (1994). Signal and Image Processing Sourcebook. Van Nostrand Reinhold, New York. 456 pages.

Resenfeld, A. and Kak, A. C. (1982). Digital Picture Processing. Academic Press, New York. 457 pages.

Telford, W. M., Geldart, L. P., and Sheriff, R. E. (1990). Applied Geophysics, 2nd edn. Cambridge University Press, Cambridge, UK. 770 pages.

SOLVED PROBLEMS

P6.1 *Properties of the Fourier transform.* Demonstrate that if $\underline{X} = \text{DFT}(\underline{x})$, then the DFT of the tail-reversed rev($\underline{x}$) is the complex conjugate of $\underline{X}$.

Solution: Let us consider a real-valued single-frequency sinusoid computed as the sum of a cosine and a sine. Its tail-reversed rev($\underline{x}$) is obtained by changing the sign of the sine component:

$x_i = \cos\left(\frac{2\pi}{N} i\right) + 1.3 \sin\left(\frac{2\pi}{N} i\right)$ then $\text{rev}(x_i) = \cos\left(1\frac{2\pi}{N} i\right) - 1.3 \sin\left(\frac{2\pi}{N} i\right)$

The signal is evaluated for $N = 8$ points. The real part of the double-sided DFT of $\underline{x}$ has positive amplitude at frequency counters $u = 1$ and $u = N - 1$. The imaginary part Im($\underline{X}$) has negative amplitude at $u = 2$ and positive at $u = N - 2$ in agreement with the symmetry property of the DFT (Section 5.3). All other terms are zero. The DFT of the tail-reversed signal has the same real components but the imaginary components have opposite sign. Indeed, the DFT(rev($\underline{x}$)) is the conjugate of $\underline{X}$.

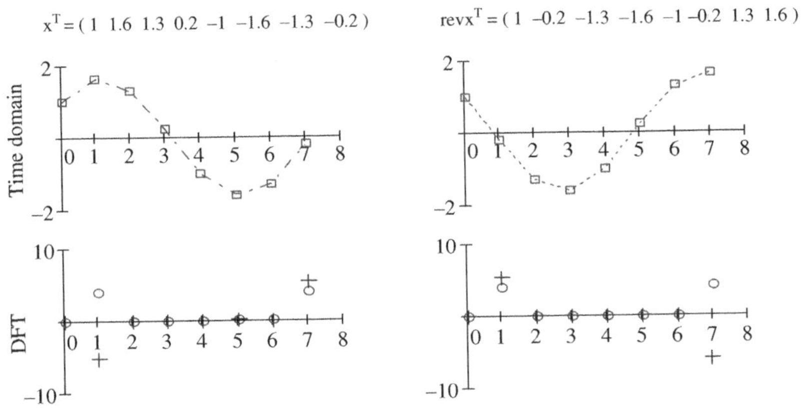

Important: notice that the array is not obtained by reversing the array! In fact, $x_0 = \text{rev}(x_0)$.

P6.2 *Convolution in the frequency domain.* Demonstrate that $Y_u = X_u \cdot H_u$ starting with the expression for time-domain convolution and assuming that $\underline{H} = \text{DFT}(\underline{h})$.

Solution: Convolution in the time domain is $y_i = \sum_k x_k \cdot h_{i-k}$

Its DFT is $$Y_u = \sum_i \left(\sum_k x_k \cdot h_{i-k} \right) \cdot e^{-j \cdot \left(u \frac{2\pi}{N} i\right)}$$

But according to the shift property (Equation 5.22): $$h_{i-k} = h_i \cdot e^{-j\left(u \frac{2\pi}{N} k\right)}$$

Replacing $$Y_u = \sum_k \sum_i x_k \cdot h_i \cdot e^{-j\left(u \frac{2\pi}{N} k\right)} \cdot e^{-j \cdot \left(u \frac{2\pi}{N} i\right)}$$

Rearranging $Y_u = \left(\sum_k x_k \cdot e^{-j\left(u\frac{2\pi}{N}k\right)}\right)\left(\sum_i h_i \cdot e^{-j\cdot\left(u\frac{2\pi}{N}i\right)}\right)$

The factors in brackets are the u-th components of the DFT of $\underline{x}$ and $\underline{h}$; therefore, $Y_u = X_u \cdot H_u$.

P6.3 *Filters.* All-pass filters are used for phase control, and the magnitude response is $|H_u| = 1.0$ for all u. Typically, all-pass filters are used to correct the phase shift imposed by other filters. Design an all-pass filter that will cause a linear phase shift with frequency and apply the filter to a sine sweep. Conclude on the response of the filter and the behavior of the output signals.

Answer: All-pass filter defined as $|H_u| = 1$ and $\varphi_u = v\frac{2\pi}{N}u$

The filter frequency response is $H_u = |H_u| \cdot [\cos(\varphi_u) + j \cdot \sin(\varphi_u)]$

Consider a frequency sweep $x_i = \sin\left(\frac{2\pi}{N}i^{1.5}\right)$

Its DFT is $\underline{X} = \mathrm{DFT}(\underline{x})$

The filtered signal is $Y_u = H_u \cdot X_u$

The filtered signal in time is $\underline{y} = \mathrm{IDFT}(\underline{Y})$

The original and filtered signals with v = 64, 128 and 192 are presented next. As the rate of phase shift v increases, the signal is shifted to the left, advancing in time. Because a linear phase shift is equivalent to a constant time shift δt, there is no distortion in the signal. The effect of circularity is clearly seen (periodicity assumption in the DFT).

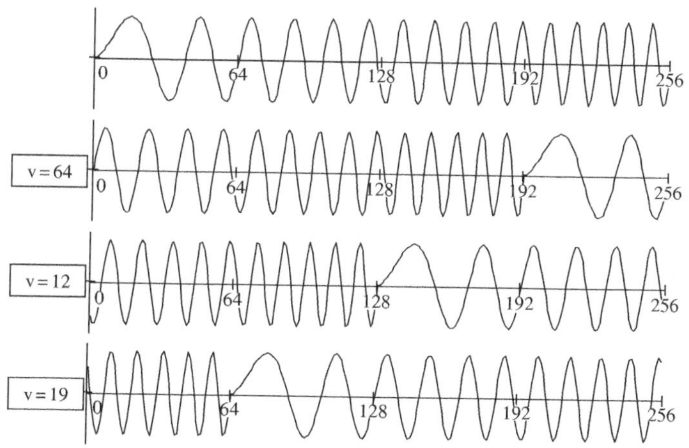

ADDITIONAL PROBLEMS

P6.4 *Power spectra.* Demonstrate that the additivity rule does not apply to either the cross- or the autospectra: $\underline{AC}^{<x1+x2>} \neq \underline{AC}^{<x1>} + \underline{AC}^{<x2>}$.

P6.5 *Filters – Hanning window.* Consider arrays of length N·Δt. A Hanning window of width E·Δt is used to extract signal segments. Compute and plot the DFT of Hanning windows with width E = N, E = N/2 and E = N/4. Realizing that these transforms act as kernels in the frequency domain, what is the effect of windowing a single-frequency sinusoid with a Hanning window? Repeat for a Gaussian window. Move the windows off-center; how do real and imaginary components react?

P6.6 *Filters: windows and kernels.* Band-pass filtering in the frequency domain is equivalent to the convolution of the kernel with the signal in the time domain. Study the kernel characteristics for band-pass filters of different width and transition rates at boundaries. What is the kernel of a Gaussian window? Explore the effects of these filters on a sawtooth signal. Draw conclusions.

P6.7 *Noise and frequency response.* Study the effect of noise in the determination of $\underline{H}$ when (a) the input is measured with noise but noise does not go through the system; and (b) the input is measured without noise, but noise gets introduced into the system at the input and manifests in the output. (Note that these two cases are different from the one covered in the chapter.)

P6.8 *Frequency response determination.* What is the DFT of random noise (consider spectral averages)? What is the DFT of an impulse signal? What can you conclude about the application of these signals to determine the frequency response of a system? What are the differences from the point of view of a system with a limited linear range?

P6.9 *Coherence.* Expand the definition of coherence and show that coherence $\gamma_u^2 = 1.0$ for all frequencies when a single measurement is used.

P6.10 *Coherence and signal-to-noise ratio.* Compare the theoretical and practical definitions of the coherence and signal-to-ratio functions. Vary the noise level and the number of signals in the ensemble. Conclude. (Hint: define the input signal $\underline{x}$, noiseless output $\underline{y}$, noise $\underline{r}$ and noisy output $z_i = y_i + r_i$ for each signal in the ensemble.)

P6.11 *Application: echo testing* (e.g. ultrasound). An exponentially increasing amplification is sometimes applied to the received signal to compensate for the effects of geometric and material attenuation. This amplification is a window in the time domain. What is the corresponding kernel in the frequency domain? What are the implications of this method?

P6.12 *Application: transducers and peripheral electronics.* Find the manual or specifications for standard laboratory devices that you use (transducers, amplifiers, signal generators). Identify the information provided by the manufacturer about the device frequency response $H(\omega)$. Is the information sufficient to completely define $H(\omega)$? Implement a numerical simulation of the effect of the device on a known measurement $\underline{x}$ as a convolution between $\underline{x}$ and $\underline{H}$. Draw conclusions on the effect of transducers and peripheral electronics on your system.

P6.13 *Application: system characterization.* Design a step-by-step procedure to determine the frequency response of a system of your interest (e.g. transducer, image analyzer, bridge, city traffic). Consider both the experimental setup and the numerical processing of signals. Make sure you include guidelines to reduce noise and experimental and computational details to correct for the frequency response of transducers and peripherals used in the measurements.

P6.14 *Application: system characterization with random noise (Part 1).* Systems with low damping readily respond in their resonant frequency, and the measured response Z for any broadband signal will resemble the system frequency response H. Consider a single DoF oscillator. Prepare an ensemble of M input signals $\underline{x}$ of length N generated with a random number generator. For each input signal $\underline{x}$, compute the output $\underline{y}$ as a convolution with the system response, and add random noise to obtain the ensemble of "measured" output signals $\underline{z}$. Then (1) compute the frequency response with average spectra, and (2) consider the possible use of ambient noise to explore $\underline{H}$ without measuring the input. Repeat these studies for different number of signals M, duration N, and system damping. Draw conclusions. Can the system be characterized using ambient noise as excitation without measuring the input?

P6.15 *Application: system characterization with random noise (Part 2).* Background noise is omnipresent and may be used as a source to study systems without additional excitation. Design a detailed procedure – both experiment and data-reduction components – to determine the frequency response of a system of your interest using background noise. Include detailed information about the transducer, the sampling interval, the signal duration, the number of records, and the data processing procedure.

P6.16 *Application: cepstrum analysis.*The cepstrum (from spectrum) of a signal is defined as IDFT[log($\underline{AC}^{<x>}$)]. Standard signal processing terminology is changed when using cepstrum analysis in order to denote this type of transformation, for example: gamnitude (magnitude), quefrency (frequency), rahmonics (harmonics), and liftering (filtering). Cepstrum analysis may facilitate detecting changes in the system such as the formation of cracks (or the effect of repairs) and recovering the signal at the source without the multiple reflections. Explore the viability and benefits of cepstrum analysis in your system of interest.

P6.17 *Application: spectrum of velocity and attenuation.* Consider 1D wave propagation in an infinite rod. A traveling wavelet is measured at two locations at a distance L apart. There is only material attenuation in this system $e^{-\alpha \ell}$ where α is the attenuation coefficient and ℓ the travel length. Detail the algorithm to determine the velocity and attenuation spectrum given the two measurements (include phase unwrapping). Consider (a) noiseless signals, (b) noisy signals, and (c) known transducer transfer functions.

7 Time Variation and Nonlinearity

The operations presented in previous chapters are versatile and effective and facilitate the interpretation of signals and the characterization of systems in a wide range of problems in engineering and science. However, there are some restrictions. For example, consider a musical score: it simultaneously tells us the timing and the frequency of each note; yet, the frequency domain representation of sound would convey no information about timing. On the other hand, the efficient algorithms for system analysis described in previous chapters were developed on the bases of linear, time-invariant system behavior; yet many systems do not satisfy either or both assumptions. Alternatives to analyze nonstationary signals, and time-varying nonlinear systems, are explored in this chapter.

7.1 NONSTATIONARY SIGNALS: IMPLICATIONS

The discrete Fourier transform (DFT) perfectly fits the N-points of a discrete signal with a finite series of harmonically related sinusoids. Each nonzero Fourier coefficient indicates the existence of a sinusoid that is present at all times, not only in the time interval of the signal $[0, T[$ but from $t = -\infty$ to $t = +\infty$. The lack of timing-related information in the Fourier transform would suggest a stationary signal with constant statistics across broad time segments. By contrast, speech, earthquakes, music, topography and color pictures consist of different frequencies or "notes" that take place at different times and for a fixed duration!

For example, Figure 7.1a shows successive wave trains of different single-frequency sinusoids. The autospectrum of the complete signal is shown in Figure 7.1b. The spectral peaks reflect the frequency of the wave trains, but there is no information in the frequency domain about the timing of the events.

Discrete Signals and Inverse Problems J. C. Santamarina and D. Fratta

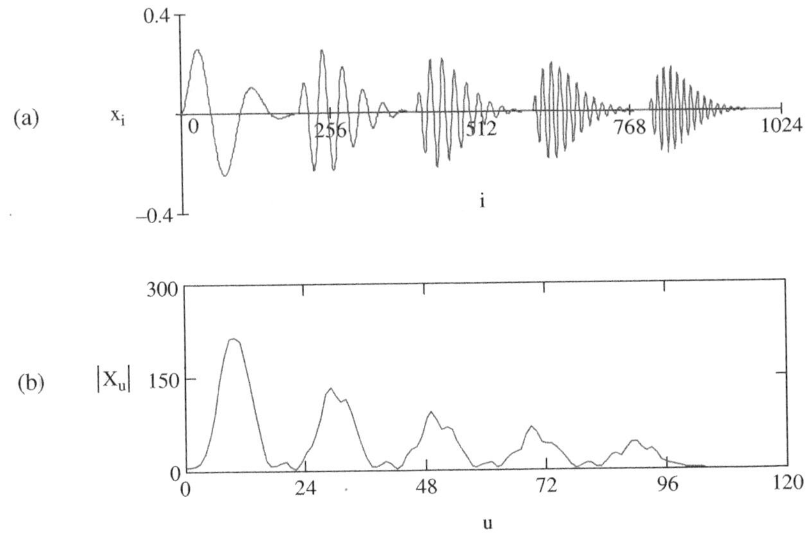

Figure 7.1 The frequency domain representation of the signal yields information about frequency content, but fails to define the time location of each frequency: (a) a nonstationary signal with different frequencies at different times; (b) autospectral density

There are important, yet often subtle, effects related to the interpretation of signal processing results in the case of nonstationary signals. Consider the multiple transmission paths an emitted sound experiences (Figure 7.2). The signature recorded with the microphone includes the direct wave, signals reflected at walls, floor and ceiling, and signals diffracted around or traveling across any anomaly within the medium. Each arriving component will have experienced a travel-length and frequency-dependent phase shift and attenuation (geometric, backscatter, and material loss). Assume the complete record obtained with the geophone is discrete Fourier transformed. What do amplitude $|Y_u|$ and phase φ_u indicate?

The implications of this situation are analyzed with the help of Figure 7.3. Figure 7.3a consists of a sinusoidal of 8 cycles in 512 points. The DFT is an impulse at the corresponding frequency ($u = 8$) and phase $\varphi_8 = -\pi/2$. By contrast, Figure 7.3b shows a windowed version of the same single-frequency sinusoid ($u = 8$), showing only two cycles followed by zero entries from $i = 128$ to $i = 511$ points. In spite of having the same frequency, the DFT of the windowed signal is fairly broadband. The phase is correct, $\varphi_8 = -\pi/2$. *Even though a single*

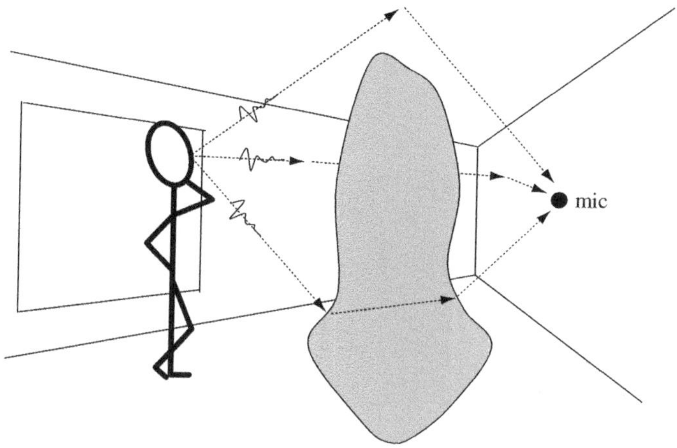

Figure 7.2 A visit to the Museum of Modern Art. Emitted sound: "Oh..!?". The signal detected at the microphone includes the multiple arrivals along different transmission paths, with different arrival times and frequency content partially filtered by the medium and reflections

frequency acts for a short time, the DFT implies the presence of multiple sinusoids at all times. There are two ways to interpret this result:

1. The DFT is equivalent to fitting the array $\underline{x}$ with the Fourier series. In this case, it is clear that several nonzero Fourier coefficients are required to fit not only the two cycles of the sinusoid but the full signal including the zero amplitude region as well (see Figure 7.3b). All sinusoids are present at all times; yet their amplitude and phases are such that their contributions in the synthesized signal render the correct values of x_i at all discrete times t_i, including the $x_i = 0$ values.
2. The alternative view is to consider the signal $\underline{x}$ in Figure 7.3b as a sinusoid 512 points long but multiplied point by point with a square window $\underline{w}$ in which the first 128 points are ones, and the rest are zeros. The point-by-point multiplication in the time domain implies a convolution in the frequency domain between the DFT of the sinusoid (which is an impulse) and the DFT of the square window. Therefore, the DFT in Figure 7.3b is the DFT of the window shifted to the frequency of the sinusoid $u = 8$.

Figure 7.3c shows the original signal plus a "reflected" signal with no attenuation. The reflection arrives at $i = 255$. Given the phase of the reflection, this signal can be considered as the original sinusoid in Figure 7.3a, but windowed with a square

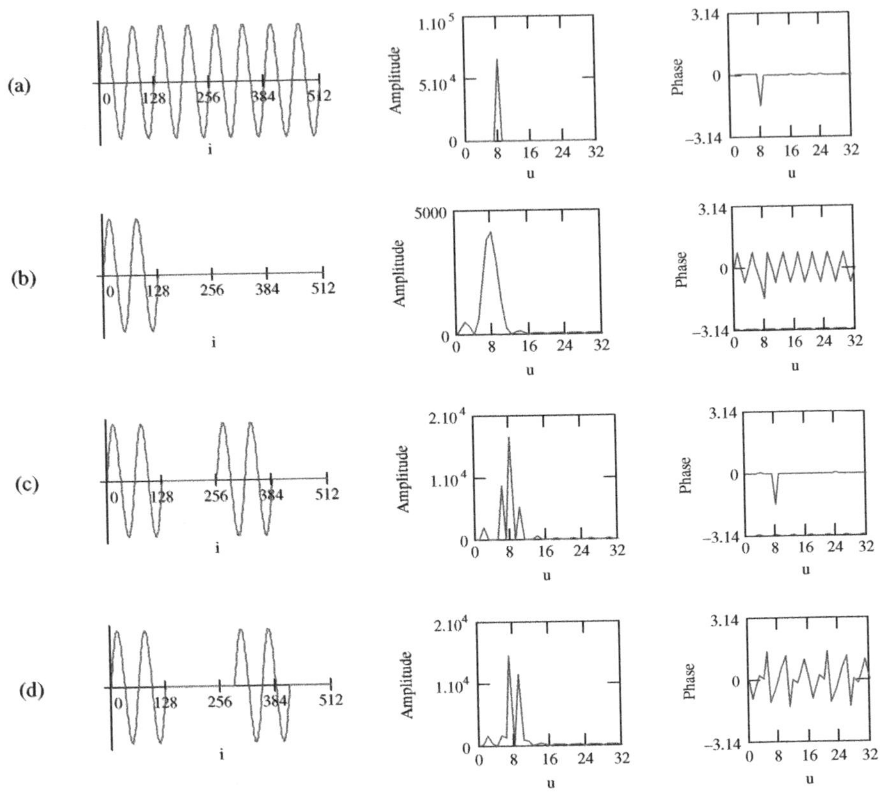

Figure 7.3 Some effects of nonstationarity: (a) a single-frequency sinusoid transforms to a single peak in amplitude and phase; (b) two cycles of a single-frequency sinusoid with zero-padding. The amplitude spectrum is broadband; the phase remains as $-\pi/2$ in the dominant frequency; (c) two wave trains of the same frequency with time shift equal to twice the period; (d) two wave trains of the same frequency but with time shift equal to 2.5 times the period. The computed amplitude and phase are zero at the otherwise "dominant" frequency

wave with two nonzero regions. Then, the DFT of the signal in Figure 7.3c is the DFT of this new window with two nonzero regions shifted to $u = 8$. The phase computed for the frequency of the sinusoid is still $\varphi_8 = -\pi/2$.

Finally, the case analyzed in Figure 7.3d consists of the initial two-cycle signal followed by a "reflection" that arrives at $i = 288$, that is, half a period after the reflection in Figure 7.3c. This signal cannot be obtained as a window of the sinusoid in the first frame. In fact, the computed energy is $AC_8^{<x>} = 0$ and the phase is $\varphi_8 = 0$ at the frequency of the sinusoid $u = 8$. In general, the computed phase corresponding to the frequency of the wave train cannot be associated with any of the two arrivals.

7.2 NONSTATIONARY SIGNALS: INSTANTANEOUS PARAMETERS

Let us reconsider the fundamental signals used in frequency domain analyses: cosine, sine, and complex exponentials $e^{j\omega t}$ and $e^{-j\omega t}$. The signals are plotted in Figure 7.4, including both the real and the imaginary components of the complex exponentials. Their corresponding double-sided DFTs are shown from $u = 0$ to $u = N - 1$, where the Nyquist frequency corresponds to $u = N/2$.

The DFT(cos) is real and symmetric, whereas the DFT(sin) is imaginary and antisymmetric (Chapter 5: periodicity and symmetry properties). However, the DFT of complex exponentials are single-sided. Furthermore, visual inspection allows the confirmation of Euler's identities in the frequency domain:

$$\text{DFT}(e^{j\omega t}) = \text{DFT}(\cos) + j \cdot \text{DFT}(\sin) \tag{7.1}$$

$$\text{and} \quad \text{DFT}(e^{-j\omega t}) = \text{DFT}(\cos) - j \cdot \text{DFT}(\sin) \tag{7.2}$$

7.2.1 The Hilbert Transform

The Hilbert transform $\underline{x}^{<ht>}$ is a new signal, orthogonal to the original signal $\underline{x}$, obtained by imposing $-\pi/2$ phase shift, and of the same spectral density. By definition, the following is a chain of interrelated Hilbert transforms:

$$\cos(\omega t) \xrightarrow{ht} \sin(\omega t) \xrightarrow{ht} -\cos(\omega t) \xrightarrow{ht} -\sin(\omega t) \xrightarrow{ht} \cos(\omega t) \tag{7.3}$$

These transforms are readily confirmed by visual inspection of results in Figure 7.4. Moreover, the detailed analysis of these figures allows us to identify a procedure to implement the Hilbert transform:

- Given a signal $\underline{x}$, compute its $\underline{X} = \text{DFT}(\underline{x})$.
- For $0 \leq u < N/2$, set $X_u^{<ht>} = -j \cdot X_u$.
- For $N/2 \leq u \leq N-1$, set $X_u^{<ht>} = j \cdot X_u$.
- The array $\underline{X}^{<ht>}$ is the Hilbert transform of the signal in the frequency domain.
- The Hilbert transform in the time domain is $\underline{x}^{<ht>} = \text{IDFT}(\underline{X}^{<ht>})$.

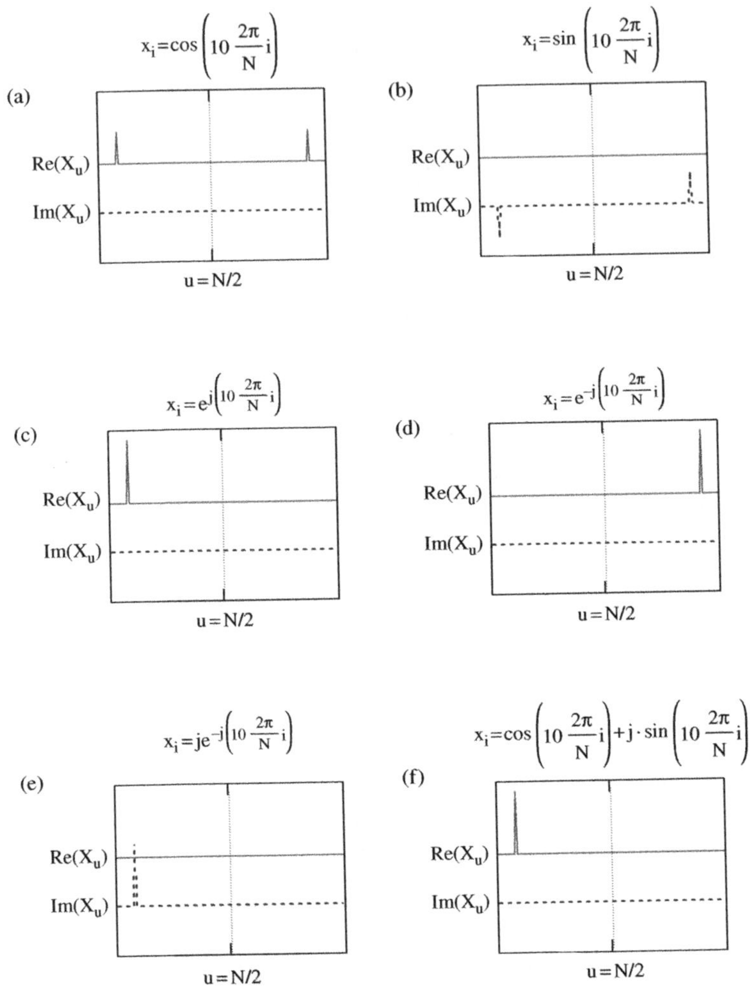

Figure 7.4 The double-sided DFT of elemental signals. For reference, the signal length is N = 128 and includes 10 periodic cycles in all cases. (a and b) Sinusoids are real-valued in the time domain and double-sided in the frequency domain. (c and d) Complex exponential are complex-valued in the time domain and single-sided in the frequency domain. (e) Multiplying the signal by j converts real into imaginary components of the same sign, and imaginary into real components of opposite sign as $j \cdot j = j^2 = -1$. (f) Verification of Euler's identity

7.2.2 The Analytic Signal

Let us define the "analytic signal" as the array of complex numbers formed with the original signal $\underline{x}$ as the real part and its Hilbert transform $\underline{x}^{<ht>}$ as the imaginary component:

$$x_i^{<A>} = x_i + j \cdot x_i^{<ht>} \qquad \textit{analytic signal} \tag{7.4}$$

The following analytic signals are computed by visual inspection in relation to Figure 7.4, or by invoking the results in Equation 7.3 and Euler's identities:

$$\text{if } x = \cos(\omega t) \quad \text{then} \quad x^{<A>} = \cos(\omega t) + j \cdot \sin(\omega t) = e^{j\omega t} \tag{7.5}$$

$$\text{if } x = \sin(\omega t) \quad \text{then} \quad x^{<A>} = \sin(\omega t) - j \cdot \cos(\omega t) = -j \cdot e^{j\omega t} \tag{7.6}$$

Notice that the DFT of these two analytic signals is an impulse between 0 and the Nyquist frequency, $0 < u < N/2$. This is always the case: it follows from the definition of the analytic signal (Equation 7.4) that its Fourier transform is $DFT(\underline{x}^{<A>}) = DFT(\underline{x}) + j \cdot DFT(\underline{x}^{<ht>})$. Then, recalling the procedure for computing the Hilbert transform, the values of the analytic signal at the u-th frequency become

$$X_u^{<A>} = X_u + j(-j \cdot X_u) = 2X_u \qquad \text{for } 0 \le u < N/2 \tag{7.7}$$

$$\text{and} \quad X_u^{<A>} = X_u + j(j \cdot X_u) = 0 \qquad \text{for } N/2 \le u \le N. \tag{7.8}$$

These observations lead to an effective procedure to compute the analytic signal $\underline{x}^{<A>}$ associated with a signal $\underline{x}$:

- Compute the DFT of the signal: $\underline{X} = DFT(\underline{x})$.
- Set all values above the Nyquist frequency to zero: $X_u^{<A>} = 0$ for $N/2 \le u \le N-1$.
- Multiply values times 2 for $0 \le u < N/2$: $X_u^{<A>} = 2X_u$.
- This is the analytic signal in the frequency domain $\underline{X}^{<A>}$.
- The analytic signal in the time domain is $\underline{x}^{<A>} = IDFT[\underline{X}^{<A>}]$.

By definition (Equation 7.4), the real part of the analytic signal is the signal itself, $Re(x_i^{<A>}) = x_i$.

7.2.3 Instantaneous Parameters

The analytic signal can be processed to extract "instantaneous amplitude" and "instantaneous frequency" information at each i-th position in time (see exercise at the end of the chapter). The instantaneous amplitude is

$$\text{amp}_i = \sqrt{\text{Re}\left(x_i^{\langle A\rangle}\right)^2 + \text{Im}\left(x_i^{\langle A\rangle}\right)^2} \qquad \textit{instantaneous amplitude} \qquad (7.9)$$

The instantaneous frequency requires an intermediate computation of "instantaneous phase":

$$\phi_i = \tan^{-1}\left[\frac{\text{Im}\left(x_i^{\langle A\rangle}\right)}{\text{Re}\left(x_i^{\langle A\rangle}\right)}\right] \qquad (7.10)$$

Finally, the instantaneous frequency is computed as the time derivative of the instantaneous phase. Using the first order-finite difference approximation

$$\omega_i = \frac{\phi_i - \phi_{i+1}}{\Delta t} \qquad \textit{instantaneous frequency} \qquad (7.11)$$

The methodology is summarized in Implementation Procedure 7.1 and demonstrated in Figure 7.5. The instantaneous frequency and amplitude are plotted versus time to resemble a musical score (see solved problem at the end of this Chapter).

Implementation Procedure 7.1 Analytic signal and instantaneous parameters

Determination of the analytic signal

1. Detrend the signal.
2. Compute the DFT of the signal $\underline{X} = \text{DFT}(\underline{x})$.
3. Create a single-sided array:

$$X_u^{<A>} = 0 \quad \text{for} \quad N/2 \le u \le N-1 \text{ (above Nyquist frequency)}$$

$$X_u^{<A>} = 2 \cdot X_u \quad \text{for } 0 \le u < N/2 \quad \text{(below Nyquist frequency)}$$

4. Calculate the analytic signal as $\underline{x}^{<A>} = \text{IDFT}[\underline{X}^{<A>}]$.

Evaluation of instantaneous parameters

Instantaneous amplitude: $\text{amp}_i = \sqrt{\text{Re}\left(x_i^{\langle A\rangle}\right)^2 + \text{Im}\left(x_i^{\langle A\rangle}\right)^2}$

Instantaneous phase: $\phi_i = \tan^{-1}\left[\frac{\text{Im}\left(x_i^{\langle A\rangle}\right)}{\text{Re}\left(x_i^{\langle A\rangle}\right)}\right]$

Instantaneous frequency: $\omega_i = \frac{\phi_i - \phi_{i+1}}{\Delta t}$

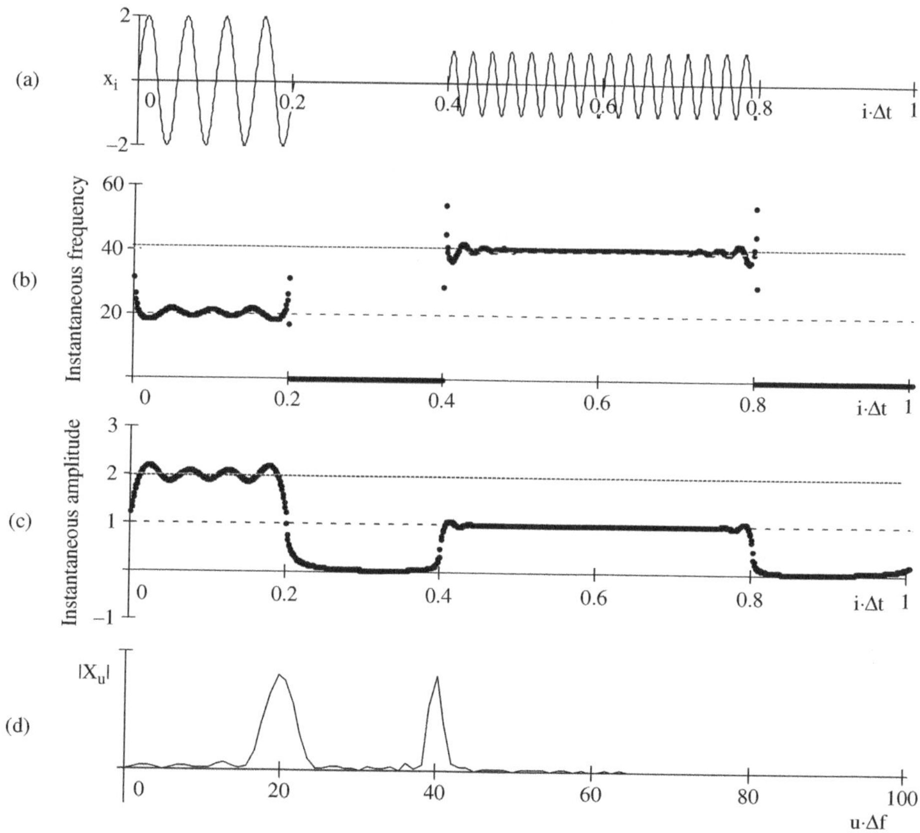

Figure 7.5 Analytic signal: (a) signal composed of two wave trains of different frequency; (b) instantaneous frequency versus time; (c) instantaneous amplitude versus time; (d) autospectral density

Comments

In most cases, results are not as clear as in the example in Figure 7.5, and require careful interpretation. For example:

- The instantaneous frequency of a beat function $x = A\cos(\omega_1 t) + B\sin(\omega_2 t)$ oscillates from $\omega_i = 0$ to $\omega_i = (A\omega_1 + B\omega_2)/(A + B)$ with periodicity $\omega_1 - \omega_2$.
- Whereas the instantaneous amplitude is quite stable, the instantaneous frequency is very sensitive to noise, which becomes magnified by a factor ω when the derivative of the instantaneous phase is computed. Thus, it is recommended that at least a low-pass filter be applied in the frequency domain.
- The instantaneous frequency may be outside the bandwidth of the signal observed in the autospectral density.

An alternative analysis of nonstationary signals involves its transformation into the time-frequency space, where the "momentary" signal characteristics are determined at different times. In this case, the one-dimensional (1D) signal in time is transformed into a two-dimensional (2D) signal in time-frequency. Three time-frequency signal processing methods are introduced in the following sections.

7.3 NONSTATIONARY SIGNALS: TIME WINDOWS

Drawbacks in the global DFT can be lessened by extracting time windows of the original nonstationary signal $\underline{x}$ and analyzing each windowed in the frequency domain. Then frequency content is plotted versus the time position of each window. Implementation details follow.

The k-th windowed signal is obtained as a point-by-point multiplication $y_i^{<k>} = w_i^{<k>} \cdot x_i$. The Fourier transforms of the extracted subsignals are assembled into a matrix $\underline{\underline{Y}}$ that defines the short-time Fourier transform (STFT) of the original signal $\underline{x}$

$$\underline{\underline{Y}} = \text{STFT}(\underline{x}) \qquad \textit{short-time Fourier transform} \tag{7.12}$$

where the k-th column of $\underline{\underline{Y}}$ is the DFT of the k-th windowed signal $\underline{y}^{<k>}$. Therefore, while the DFT of a signal converts the 1D array $\underline{x}$ in time into the 1D array $\underline{X}$ in frequency, the STFT converts the 1D array $\underline{x}$ in time into the 2D array $\underline{\underline{Y}}$ in the time-frequency space.

If the window width is $M \cdot \Delta t$, only M entries are kept in the windowed signals (where M is the number of points in the window, and Δt is the sampling rate of the signal), the maximum period of the window is $M \cdot \Delta t$ and the u-th element in the k-th column $Y_{u,k}$ is the Fourier coefficient that corresponds to the frequency

$$\omega_u = u\frac{2\pi}{M \cdot \Delta t} \qquad \text{(for window width } M \cdot \Delta t\text{)} \tag{7.13}$$

The timing assigned to the k-th window is the time at the center of the window.

The presentation of STFT results is more cumbersome than a simple DFT. Typically, the graphical display of $\underline{\underline{Y}}$ involves the amplitude $|Y_{u,k}|$ on the time-frequency information plane. This is the "spectrogram" of the signal. If contour plots or collapsed three-dimensional (3D) graphs are used, the value $|Y_{u,k}|$ is mapped onto a color scale or shades of gray. Figure 7.6 displays the DFT of individual windowed records and the STFT of the signal.

7.3.1 Time and Frequency Resolutions

The separation between two adjacent window positions $\delta t = q \cdot \Delta t$ and the window width $M \cdot \Delta t$ define the overlap between windows, and affect the STFT and its interpretation. The analyst must select the integers q and M, and the shape of the window.

Window width, and to a lesser extent its form, determine time and frequency resolutions. The longest discernible period is obtained with a square window; however, it causes spurious frequencies. For any other window, $T_{max} < M \cdot \Delta t$. Thus, the lowest resolvable frequency in the windowed signal is $>1/(M \cdot \Delta t)$ and this is frequency resolution Δf_{res} between successive harmonics:

$$\Delta f_{res} >\approx \frac{1}{M \cdot \Delta t} \qquad \textit{frequency resolution} \tag{7.14}$$

The maximum frequency remains the Nyquist frequency, which is determined by the sampling rate Δt used in digitizing the signal $\underline{x}$, and is independent of the characteristics of the window.

While a wide window enhances frequency resolution, it also leads to the analysis of longer time segments, and the timing of a certain frequency component in $\underline{x}$ loses precision. For windows that are square at the center with smooth edge transitions, time resolution is worse than half the window width $M \cdot \Delta t$,

$$\Delta t_{res} > \frac{M \cdot \Delta t}{2} \qquad \textit{time resolution} \tag{7.15}$$

Optimal coverage of the time-frequency plane takes place when two neighbor windows just touch. In practice, higher overlap is used, yet, the separation δt between windows need not be less than the time resolution Δt_{res}.

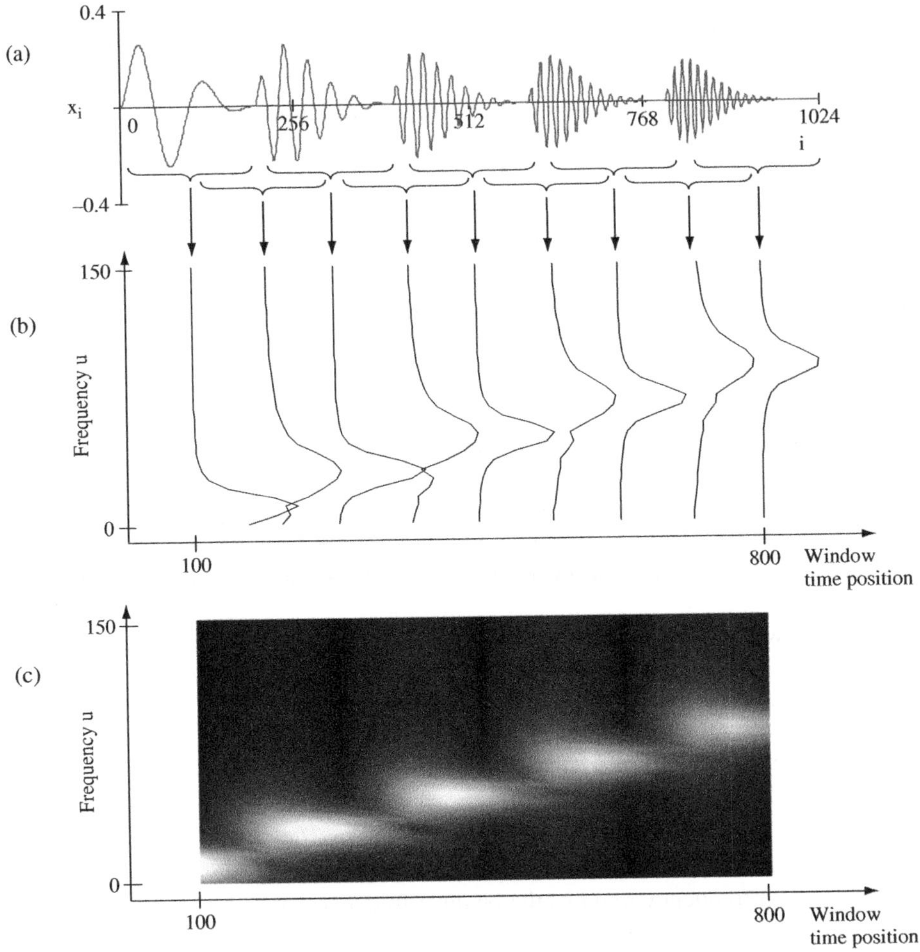

Figure 7.6 Short time Fourier transform: (a) original signal; (b) amplitude-frequency plots for windowed signals indicate the time location corresponding to each wave train – only 9 of the 16 windows are shown; (c) contour plot – amplitude normal to the page

Equations 7.14 and 7.15 can be combined as

$$\Delta f_{res} \cdot \Delta t_{res} > \frac{1}{2} \tag{7.16}$$

This relation summarizes the trade-off between time and frequency resolution, and determines the rate of scanning of the time-frequency information plane.

The successful implementation of the STFT depends on balancing the trade-off between frequency and time resolutions.

7.3.2 Procedure

The STFT is intuitively appealing and enhances signal analysis. Implementation Procedure 7.2 outlines the steps to compute the short-time Fourier transform $\underline{\underline{Y}} = \text{STFT}(\underline{x})$. The technique can be readily extended to image processing; the display in this case involves frequency-specific plots.

Implementation Procedure 7.2 The short time Fourier transform (nonstationary signals)

1. Digitize and store the N-point signal $\underline{x}$.
2. Define the form and length of the window $\underline{w}$. Square windows with smooth transitions are adequate. Consider the following criteria when selecting the width M:
   - Each signal segment, length $M \cdot \Delta t$, will be considered stationary.
   - The time and frequency resolutions are $\Delta t_{res} \geq 0.5 \cdot M \cdot \Delta t$ and $\Delta f_{res} \geq 1/(M \cdot \Delta t)$.
   - The longest period that can be discerned is $T_{max} \leq M \cdot \Delta t$.
   - A wider window will improve frequency resolution but decrease time resolution (windowed segments may be zero padded).
3. Select the time distance between two successive windows $\delta t = q \cdot \Delta t$ where q is an integer. The value of δt need not exceed the time resolution, $\delta t < 0.5 \cdot M \cdot \Delta t$. The separation δt and the width of time windows $M \cdot \Delta t$ define the overlap between windows.
4. For the k-th window position, compute the windowed signal $y^{<k\text{-th}>}$ consisting of M points,

$$y_i^{<k\text{-th}>} = w_{i+k \cdot q} \cdot X_i$$

5. Compute the DFT of each windowed signal $y^{<k\text{-th}>}$

$$\underline{Y}^{<k\text{-th}>} = \text{DFT}\left(\underline{y}^{<k\text{-th}>}\right)$$

6. Ensemble these arrays into a matrix $\underline{\underline{Y}}$, so that the k-th column of $\underline{\underline{Y}}$ is the DFT of the windowed signal $\underline{y}^{<k\text{-th}>}$. This is the STFT of $\underline{x}$:

$$\underline{\underline{Y}} = \text{STFT}(\underline{x})$$

   The u-th element in the k-th column $Y_{u,k}$ is the Fourier coefficient that corresponds to frequency $\omega_u = u \cdot 2\pi/(M \cdot \Delta t)$ and central time $t_k = k \cdot \delta t$.

7. The STFT may be presented as a collapsed 3D plot of magnitude $|Y_{u,k}|$.

The trade-off between time and frequency resolution is explored in Figure 7.7. The two wave packets in the simulated signal are of different frequency. STFTs computed with two window widths are presented in Figures 7.7b and c as contour plots of amplitude. Results confirm the dependency the STFT has on the selected window width, and the lower time resolution attained with wider windows.

7.4 NONSTATIONARY SIGNALS: FREQUENCY WINDOWS

The STFT seeks to identify the frequency content at selected time segments. One could also wonder about the time when selected frequency bands take place. In this case, the DFT is computed for the whole signal, $\underline{X} = \text{DFT}(\underline{x})$, and frequency windows of $\underline{X}$ are extracted and inverse-transformed to time. Once again, windowing is a point-by-point multiplication, in this case in the frequency domain. For the s-th window $\underline{W}^{<s>}$

$$Y_u^{<s>} = W_u^{<s>} \cdot X_u \tag{7.17}$$

The band-pass filtered spectrum $\underline{Y}^{<s>}$ is inverse transformed to the time domain and placed in the s-th column of the matrix $\underline{\underline{y}}$. Thus, this is also a transformation from a 1D array $\underline{x}$ in time into a 2D array in time-frequency $\underline{\underline{y}}$. The plotting strategies for the spectrogram resemble those used in STFT.

A window in frequency is a band-pass filter. Therefore, each column of $\underline{\underline{y}}$ is a band-pass filtered version of $\underline{x}$. Yet, why does this procedure work for nonstationary signals? The time-frequency duality predicts that the point-by-

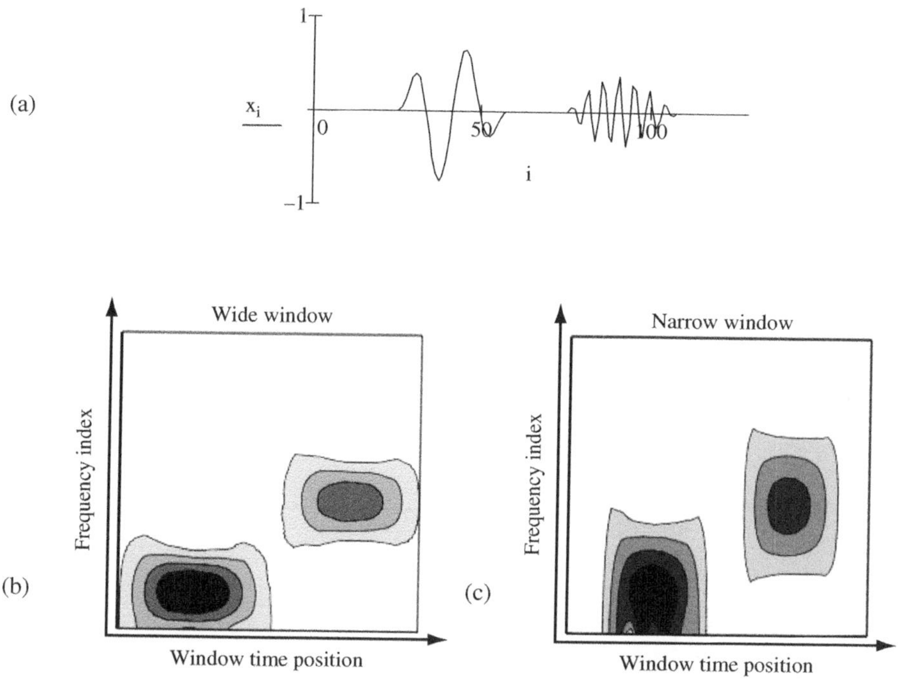

Figure 7.7 The time window size has important consequences on the results of the STFT: (a) signal made of two wave trains of different frequencies; (b) STFT performed with a 64-point wide window presents high resolution in frequency, but does not discriminate well in time; (c) STFT performed with a 16-point narrow window presents higher resolution in time, but does not discriminate the fast-varying frequencies very well

point multiplication in the frequency domain (Equation 7.17) is equivalent to the convolution between the signal and the kernel $\underline{\kappa}$ that contains the inverse transform of the frequency window, $\underline{x} * \underline{\kappa}$. But convolution is a tail-reversed cross-correlation (see Sections 4.2, 4.4, 6.3 and 6.4). Therefore, the procedure can be viewed as the identification of similarities between the original signal $\underline{x}$ and the IDFT of the frequency window.

7.4.1 Resolution

The trade-off between the resolution in time and in frequency persists: a narrow filter enhances frequency resolution but it corresponds to a wide kernel in time, decreasing time resolution. Furthermore, a narrow frequency band deforms the

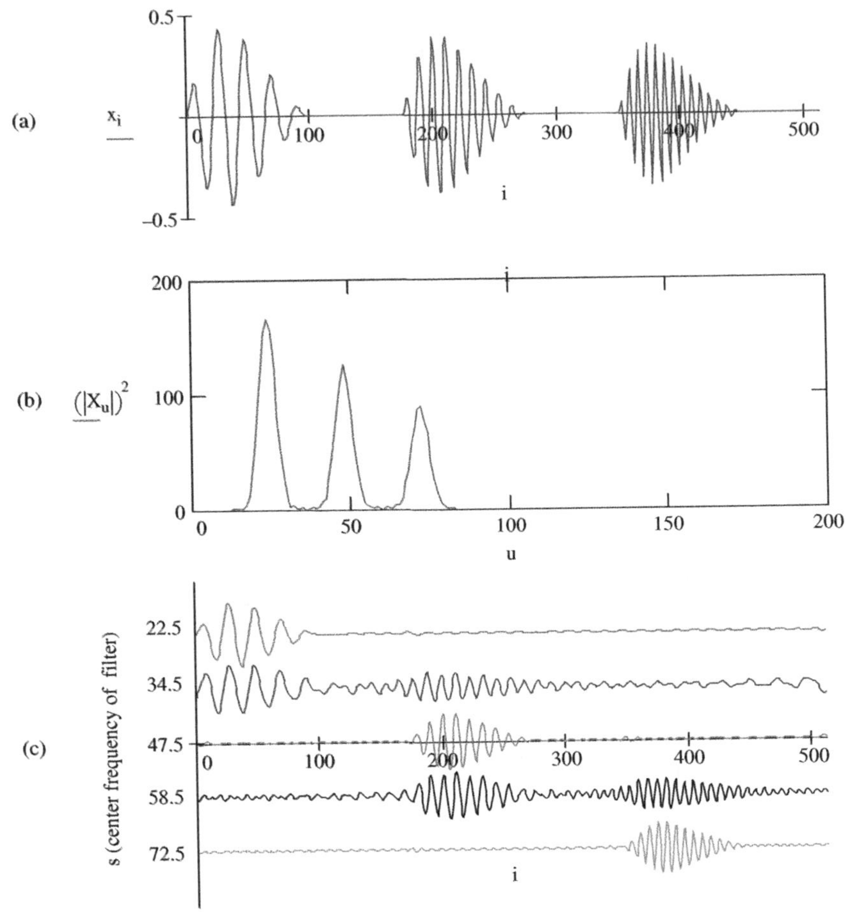

Figure 7.8 Windows in the frequency domain – Band-pass filters: (a) the signal consists of three wave trains of different frequency; (b) the autospectrum of the signal; (c) the DFT of the original signal is windowed to extract the frequency bands that are inverse transformed

signal, creating phantoms of the signature before the true signal appears in time (typically fish-shaped; Figure 7.8).

7.4.2 Procedure – Example

Implementation Procedure 7.3 presents the step-by-step algorithm to compute the band-filtered time-frequency analysis of a signal $\underline{x}$. This type of algorithm can be used to unmask events that are hidden within other frequency components, including noise.

The procedure is demonstrated in Figure 7.8. The signal $\underline{x}$ in Figure 7.8a is transformed to the frequency domain $\underline{X}$ (Figure 7.8b) and analyzed by extracting successive frequency windows. Figure 7.8c presents the spectrogram assembled with the inverse transforms of the filtered signals.

Implementation Procedure 7.3 Band-pass filters and nonstationary signals (windows in the frequency domain)

1. Digitize and store the N-point signal $\underline{x}$.
2. Compute the DFT of the signal: $\underline{X} = \text{DFT}(\underline{x})$.
3. Define the form and width of the filter $\underline{W}$ and the frequency separation between adjacent windows $\delta\omega$. Consider the trade-off in the frequency-time resolution.
4. For each position of the window, multiply $\underline{x}$ by the window $\underline{W}^{<s\text{-th}>}$ centered at frequency $\omega_s = s \cdot \delta\omega$ (a symmetric window must be applied above the Nyquist frequency when double-sided Fourier transforms are used – see Section 6.5): $Y_u^{<s\text{-th}>} = W_u^{<s\text{-th}>} \cdot X_u$. This array has the same length as $\underline{x}$.
5. Compute the IDFT of the filtered signal: $\underline{y}^{<s\text{-th}>} = \text{IDFT}\left(\underline{Y}^{<s\text{-th}>}\right)$.
6. Ensemble these arrays into a matrix $\underline{\underline{y}}$, so that the s-th column of $\underline{\underline{y}}$ is the IDFT of the filtered signal $\underline{y}^{<s>}$. The i-th element in the s-th column $y_{i,s}$ is the value of the band-pass filtered signal at time $t_i = i \cdot \Delta t$.
7. Display the results.

7.5 NONSTATIONARY SIGNALS: WAVELET ANALYSIS

The STFT highlights time resolution in the spectrogram; on the other hand, band filtering enhances frequency resolution. It would be advantageous to improve the frequency resolution of low-frequency events while enhancing the time resolution of high-frequency events. This could be achieved by increasing the width of the band-pass filter as it is moved along the frequency axis. Let us analyze this suggestion:

- A band-pass filter is a window that is multiplied point by point with the DFT of the signal. This is equivalent to the convolution of the signal with a kernel $\underline{\kappa}$ in the time domain.
- The IDFT of the band-pass filter is a wavelet-type kernel. The duration of the wavelet is inversely related to the width of the band-pass filter.
- The convolution of the kernel $\underline{\kappa}$ with the signal $\underline{x}$ is equivalent to cross-correlation (Chapter 4); that is, it identifies similarities between the signal and the kernel.

These observations are the foundations for the wavelet transform.

7.5.1 Wavelet Transform

The wavelet transform of a signal $\underline{x}$ consists of identifying similarities between the signal and the wavelet kernel $\underline{\kappa}$. The wavelet is translated by imposing time shifts $\tau = b \cdot \Delta t$, and its frequency is varied by successive time contractions α:

$$\alpha^{-\frac{1}{2}} \cdot \kappa_{\frac{i-b}{\alpha}} \qquad \textit{controlling time shift and frequency} \tag{7.18}$$

Mathematically, the wavelet transform is the cross-correlation of the signal with wavelets of increasing central frequency, and it converts a 1D signal $\underline{x}$ onto the 2D wavelet transform $\underline{\underline{G}}$:[1]

$$G_{\alpha,b} = \alpha^{-\frac{1}{2}} \cdot \sum_{i=0}^{N-1} x_i \cdot \kappa_{\frac{i-b}{\alpha}} \tag{7.19}$$

Some wavelet functions form a base, and the inverse wavelet transform exists. This is important if the intention is to conduct signal processing in the time-frequency domain, followed by a reconstruction of the signal back to the time domain, for example, in the processing of noisy signals. In other cases, wavelet analysis may be restricted to the detailed study of the signal within some frequency range, for example, to analyze dispersion. In this case, wavelet analysis involves a finer scanning of the time-frequency information space, within the region of interest and the constraints imposed by the time-frequency resolution.

---

[1] The wavelet transform in continuous time is (the bar indicates complex conjugate):

$$G(a,b) = \frac{1}{\sqrt{a}} \cdot \int x(t) \cdot \overline{g\left(\frac{t-b}{a}\right)} \cdot dt$$

The wavelet transform depends on the selected wavelet. Therefore, the wavelet used in the analysis must be explicitly stated. There are several well-known wavelets (see exercises at the end of the chapter). The Morlet wavelet is reviewed next.

7.5.2 The Morlet Wavelet

The Morlet wavelet consists of a single-frequency sine and cosine in quadrature (complex, with 90° phase shift). The amplitude is modulated with a Gaussian function,

$$\kappa_i = e^{j\cdot(v\cdot i)} e^{-4\cdot\ln(2)\cdot\left(\frac{i}{M}\right)^2} \tag{7.20}$$

The first complex exponential represents the sinusoid and the second exponential captures the Gaussian amplitude modulation. The wavelet central frequency is indicated in the exponent $v \cdot i = (v/\Delta t) \cdot (i \cdot \Delta t)$; thus $\omega = v/\Delta t$. The value of v must be $v < \pi$ to satisfy the Nyquist criterion $\omega_{Nyq} = \pi/\Delta t$. The wavelet width $M{\cdot}\Delta t$ is measured at half the peak amplitude of the wavelet. Figure 7.9 shows a Morlet wavelet in time and frequency domains. The DFT is single-sided (refer to Figure 7.4) and its spectral density is a Gaussian curve (not a single frequency).

The wavelet transform of a signal $\underline{x}$ in terms of the Morlet wavelet is

$$G_{a,b} = 2^{-\frac{a}{2}} \cdot \sum_{i=0}^{N-1} \left[e^{j\frac{\pi}{2^a}(i-b)} \cdot e^{-\left(\frac{i-b}{2^a}\right)^2} \cdot x_i \right] \tag{7.21}$$

where the central frequency is $\omega_a = \pi/(2^a \cdot \Delta t)$, and the Nyquist frequency corresponds to $a = 0$, that is $\omega = \pi/\Delta t$. The time shift for each value of b is $\tau = b \cdot \Delta t$. If the signal $\underline{x}$ has N points, then $2^a \leq N/2$. Finally, the width of the wavelet is $M = 2^a \cdot \sqrt{4 \cdot \ln(2)}$. The parameters "a" and "b" are indices in the frequency-time information space.

7.5.3 Resolution

The trade-off in time-frequency resolution is also manifest in wavelet analysis. The time resolution attained in the wavelet transform using the Morlet wavelet is related to its width $M{\cdot}\Delta t$, and the frequency resolution Δf_{res} is related to the frequency band of the transformed wavelet. If the time and frequency widths are determined on the Gaussian envelopes at half the peak, the uncertainty principle becomes

$$\Delta f_{res} \cdot \Delta t_{res} \approx 0.9 \tag{7.22}$$

(a)

$$g_i = e^{j \cdot v \cdot (i-b)} \cdot e^{-4 \cdot \ln(2) \cdot \left(\frac{i-b}{M}\right)^2}$$

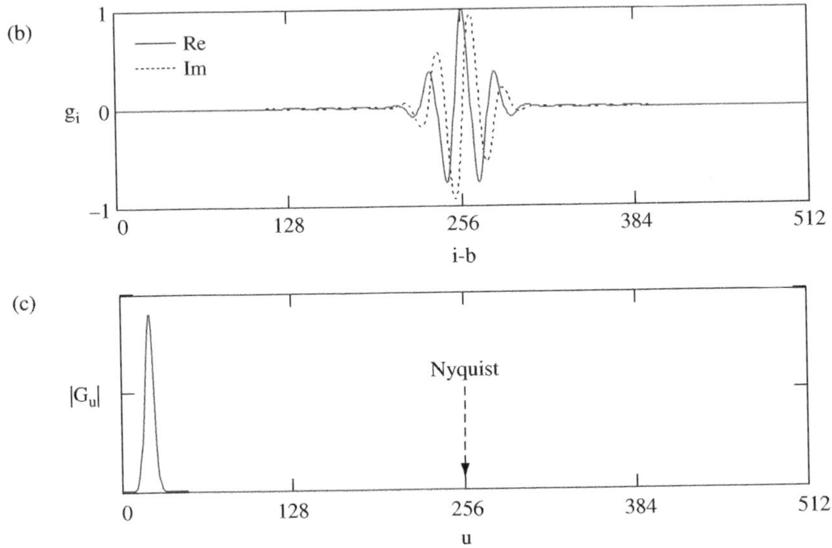

Figure 7.9 Morlet wavelet: (a) mathematical definition of the complex series; (b) Morlet wavelet with parameters $v = 0.15 \cdot \pi$ and $M = 40$ – real and imaginary components; (c) autospectral density – note that it is single-sided – refer to Figure 7.4

7.5.4 Procedure – Example

Implementation Procedure 7.4 outlines the computation of the wavelet transform of a signal in terms of the Morlet wavelet. A numerical example is shown in Figure 7.10. Small values of "a" give information about high-frequency content details in the signal, whereas high values of "a" show the low-frequency global trends.

Implementation Procedure 7.4 Wavelet transform of nonstationary signals (Morlet wavelet)

1. Select a wavelet $\kappa(t)$. Copies in discrete time are obtained by time shifts b and contractions α: $\alpha^{-\frac{1}{2}} \cdot \kappa_{\frac{i-b}{\alpha}}$

2. Decide the scanning rate for the time-frequency space, keeping in mind the restrictions imposed by the uncertainty principle. This rate will determine shift b and contraction α parameters.

3. Calculate the wavelet transform as a cross-correlation of the signal $\underline{x}$ and the wavelet for each degree of time contraction

$$G_{\alpha,b} = \alpha^{-\frac{1}{2}} \cdot \sum_{i=0}^{N-1} x_i \cdot \kappa_{\frac{i-b}{\alpha}}$$

4. If the Morlet wavelet is used, the wavelet transform of $\underline{x}$ is computed as

$$G_{a,b} = 2^{-\frac{a}{2}} \cdot \sum_{i=0}^{N-1} \left[e^{j\frac{\pi}{2^a}(i-b)} \cdot e^{-\left(\frac{i-b}{2^a}\right)^2} \cdot x_i \right]$$

   - The Nyquist frequency corresponds to $a = 0$.
   - If the signal $\underline{x}$ has N points, $2^a \leq N/2$.
   - The width of the wavelet is $M \cdot \Delta t = 2^a \sqrt{4 \cdot \ln(2)}$.

5. The values of the wavelet transform for each combination of contraction "a" and shift "b" are plotted versus a and b, or versus time shift $\tau_b = b \cdot \Delta t$ and frequency $\omega_a = \pi/(2^a \cdot \Delta t)$.

Example

A numerical example is presented in Figure 7.10.

Note: Efficient algorithms are implemented with filter banks and decimation or down-sampling. (Recall the time-scaling properties of the DFT in Chapter 5.) The wavelet transform can be designed to minimize oversampling the time-frequency information space, while assuring invertibility.

In practice, the scanning of the time-frequency space is planned to avoid redundancy, in agreement with the superposition of windows in the STFT, and the scaling parameter α is varied in powers of 2, $\alpha = 2^a$ (compare Equations 7.19 and 7.20). Likewise, it is not necessary to compute the cross-correlation for time shifts that differ in only one sampling interval ($\tau = b \cdot \Delta t$); in fact, the time shifts can be related to the central period of the wavelet.

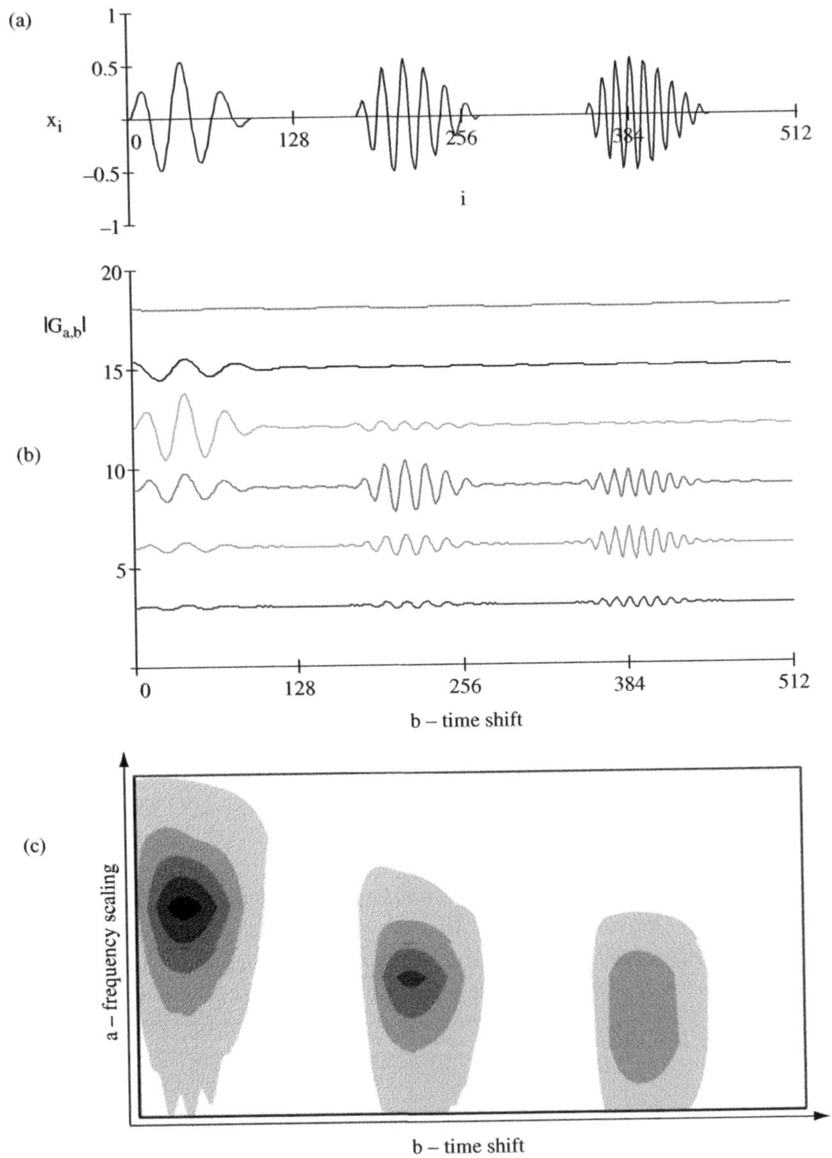

Figure 7.10 The wavelet transform: (a) the signal in discrete time; (b) the wavelet transform presented as amplitude versus time shift b for different frequencies; (c) contour plot: wavelet transform presented in the frequency-time space denoted by parameters a and b

7.6 NONLINEAR SYSTEMS: DETECTING NONLINEARITY

The application of the convolution operator is restricted to linear time-invariant (LTI) systems where the generalized superposition principle applies: "the input is expressed as a sum of scaled and shifted elemental signals and the output is computed as a sum of equally scaled and shifted system responses". The superposition principle loses validity in nonlinear or time-varying systems.

7.6.1 Nonlinear Oscillator

The single DoF oscillator in Figure 7.11a is the archetypal LTI system applicable to wide range of engineering and science applications, ranging from atomic phenomena to mechanical and electrical engineering systems (see also Figure 4.10). The frequency response $\underline{H}$ is independent of the amplitude of the forcing function $\underline{x}$.

By contrast, the other four systems displayed in Figures 7.11b–e are nonlinear. The systems in frames b and c include frictional elements: nonrecoverable slip

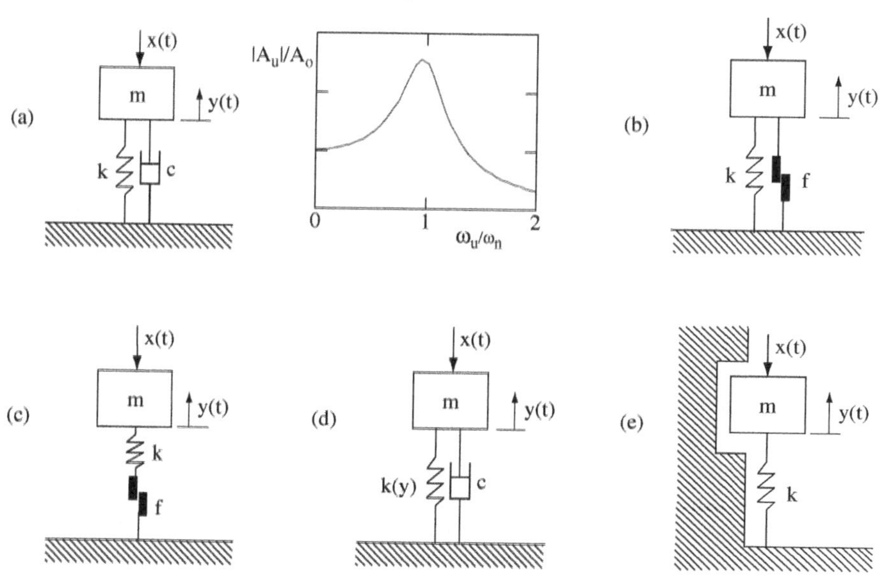

Figure 7.11 Linear and nonlinear single DoF systems: (a) linear system; (b, c) nonlinear frictional systems; (d) Duffing nonlinear system (nonlinear spring); (e) nonlinearity caused by physical constraints

takes place whenever the force transmitted to the frictional element exceeds its capacity. Nonlinearity is first exhibited near resonance and its effect spreads to a wider frequency range as the amplitude of the imposed excitation increases. At low frequencies – below resonance – the inertial response of the mass is very small and most of the applied force is transmitted to the support; this is not the case at high frequencies – above resonance – owing to the inertial resistance of the mass. Hence, the quasi-symmetry of $|\underline{H}|$ in a linear viscoelastic system is gradually lost as the system becomes nonlinear.

Figure 7.11d presents another simple yet revealing nonlinear system in which the restoring force is nonlinear with the displacement. This is called the Duffing system. The equation of motion is

$$m \cdot \ddot{y} + c \cdot \dot{y} + \left(k \cdot y + \alpha \cdot y^3\right) = F_0 \cdot \cos(\omega t) \tag{7.23}$$

When $\alpha = 0$, the equation of motion becomes the equation of motion of a linear system. If $\alpha > 0$, the restoring force increases with amplitude, and $|H|$ is skewed to the right. If $\alpha < 0$, the restoring force decreases with amplitude, and $|H|$ is skewed to the left (Figure 7.12a). These examples show that a shift in the peak value of $|H|$ with increasing input amplitude is another indicator of nonlinearity.

7.6.2 Multiples

Consider the nonlinear system in Figure 7.11e subjected to a single-frequency sinusoidal input $\underline{x}$. As the excitation amplitude is increased, the mass oscillation eventually reaches the boundaries, the displacement is stopped, and the rest of the motion is distorted. Without further analysis, assume that the mass displacement history resembles the signal $\underline{y}$ shown in Figure 7.13a. The response repeats with periodicity $T = 2\pi/\omega_0$ where ω_0 is the frequency of the input sinusoid; however, it is not a sinusoid.

How is this periodic response $\underline{y}$ fitted with a Fourier series? Clearly, the sinusoid corresponding to frequency ω_0 remains an important component of the response. But other frequency components are needed to fit the response, and their contribution to the synthesis of $\underline{y}$ must take place at locations that are repeatable with periodicity $T = 2\pi/\omega_0$. Therefore, the other components must be harmonics of ω_0. The DFT of the response $\underline{y}$ is shown in Figure 7.13b where harmonics or "multiples" are readily seen (see problems at the end of this Chapter).

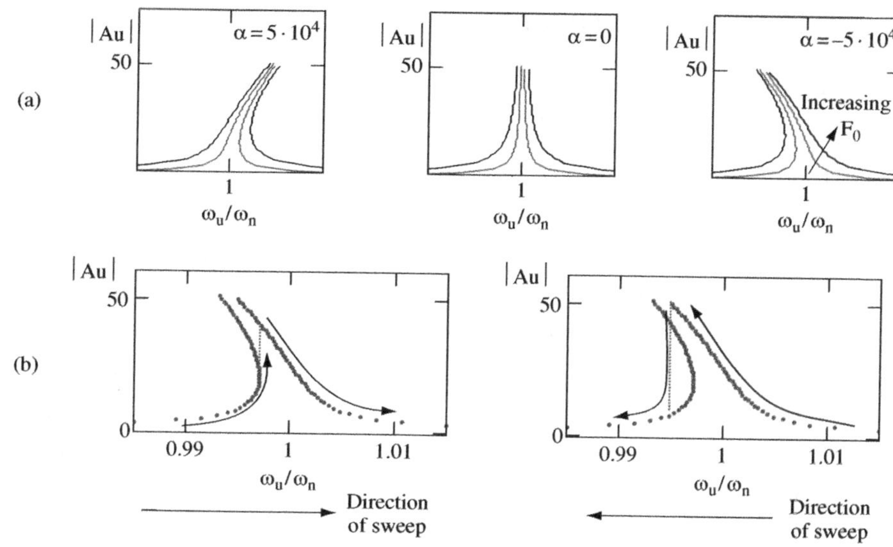

Figure 7.12 Duffing nonlinear system: (a) spectral response as a function of α and the amplitude of excitation F_0; (b) measured response varies with the sweep direction (the case shown corresponds to a soft spring, $\alpha < 0$; the same applies for $\alpha > 0$

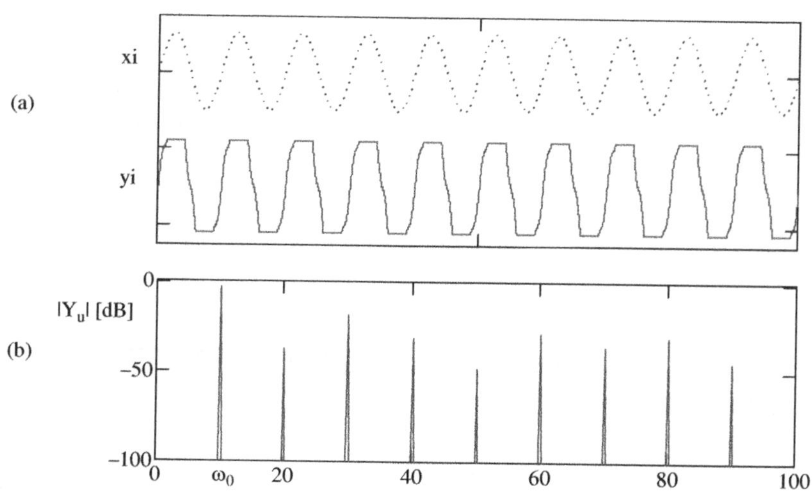

Figure 7.13 Nonlinear response of the nonlinear system presented in Figure 7.11e: (a) input $\underline{x}$ and output $\underline{y}$ signals; (b) the amplitude of the discrete Fourier transform $\underline{Y}$ in dB. The multiple peaks are harmonics of the excitation frequency (only the first 10 harmonics are shown)

7.6.3 Detecting Nonlinearity

Because the application of classical signal processing and system analysis methods presumes linearity, it is important to assess whether the system under consideration exhibits nonlinear response. Let us list here those previously discussed methods that can be used for this purpose:

- *Scaling and additive rules.* Test whether the scaling or additive rules are fulfilled by exciting the system with the same signal at different amplitudes, or with two different signals and their sum (Section 3.5).
- *Preserving statistics.* Compare the statistics of input and output signals (Section 3.5; also Chapter 9).
- *Input-independent frequency response.* Compute the frequency response $\underline{H}$ for different levels of excitation. $\underline{H}$ does not change in shape, amplitude, or position if the system is linear (Sections 6.6 and 6.7).
- *Loss in coherence near peak.* Check the value of coherence, particularly at frequencies near the peak of $\underline{H}$. (Recall: loss in coherence near the peak is also an indicator of poor frequency resolution, Section 6.7.)
- *Multiples.* Check higher harmonics or multiples in the DFT of the response to narrow-band input.
- *Compatible spectral variation.* The spectral variation of the real and imaginary parts of $\underline{H}$ are related through the Hilbert transform (known as Kramers–Kronig relations in materials research, Section 7.2.1).

Other aspects in the detection of nonlinearity are described next.

7.7 NONLINEAR SYSTEMS: RESPONSE TO DIFFERENT EXCITATIONS

The frequency response $\underline{H}$ of linear systems is independent of the excitation used. (Techniques based on frequency sweep and broadband signals were discussed in Implementation Procedures 6.1, 6.5, and 6.6.) This is not the case in nonlinear systems, as is demonstrated in this section. For clarity, a step-by-step description of each experiment is summarized in Implementation Procedure 7.5.

Implementation Procedure 7.5 Nonlinear systems – Different excitations

Frequency sweep at constant amplitude input

1. Select the amplitude of the input $Amp(\underline{x})$. This is the excitation force for the case of a single DoF oscillator.
2. For a given frequency ω, apply the forcing function with amplitude $Amp(\underline{x})$ and determine the amplitude of the response $Amp(\underline{y})$.
3. Compute the magnitude of the frequency response $|H_\omega| = Amp(\underline{y})_\omega / Amp(\underline{y})_\omega$.
4. Repeat for other frequencies ω.
5. Repeat for other selected input amplitudes $Amp(\underline{x})$.

Frequency sweep at constant amplitude output

1. Select the amplitude of the response $\Re$ for a given frequency ω.
   - Apply the input with frequency ω and amplitude $Amp(\underline{x})$.
   - Determine the amplitude of the response $Amp(\underline{y})$. Modify the amplitude of the input until $Amp(\underline{y}) = \Re$ (feedback loop).
   - Compute the magnitude of the frequency response $|H_\omega| = Amp(\underline{y})_\omega / Amp(\underline{x})_\omega$.
2. Repeat for other frequencies ω.
3. Repeat for other selected magnitudes of the amplitude of the response $\Re$.

Random input signal

1. Select the amplitude of the random input signal.
2. Apply the signal, compute the coherence, and determine the number of signals M to be stacked.
3. Compute the frequency response using the average cross- and autospectra (Implementation Procedure 6.6):

$$H_u = \frac{(CC_u^{<x,z>})_{avr}}{(AC_u^{<x>})_{avr}} \quad \text{for frequency } \omega_u = u \cdot 2\pi/(N \cdot \Delta t)$$

The computed frequency response $\underline{H}$ is the *equivalent linear model that is least squares fitted to the data*, within the extent of the input random signal.

4. Repeat for other selected amplitudes of the random signal.

Example

Figure 7.14 compares data gathered with these three methods for the shear stiffness of a soil column.

7.7.1 Input: Single-frequency, Constant-amplitude Sinusoid

In this method, the frequency of the forcing function is gradually stepped while keeping the input amplitude constant, so that $|X_u| =$ constant for all frequencies ω_u. The system response is measured at each frequency step ω_u. The frequency response H_u is the measured response Y_u divided by a constant. This is a robust method to study the system response, including conditions with high background noise.

Analytical results are presented in Figure 7.12b for a "soft spring" Duffing system. There is a "jump" from the low-frequency jump to the high frequency branch in the response. The frequency at which the jump occurs changes with the direction of the frequency sweep. This phenomenon, also known as "galloping", indicates that the frequency response is not only dependent on the amplitude of the input signal but also on the evolution of the experiment.

Consider the following experiment: sand is poured inside a thin cylindrical latex balloon, and it is then subjected to vacuum to form a stiff sand specimen. The sand column is then subjected to torsional excitation to study its response. (This is a fairly standard device known as torsional-resonant column.) Results obtained for a frequency-increasing sweep are presented in Figure 7.14a. Note the gradually increasing asymmetry of the frequency response, the shift of the peak response to lower frequencies, and the increase in attenuation (lower peak and wider band) with increasing excitation amplitude.

7.7.2 Input: Random Signal

Frequency-domain analysis is required to determine the frequency response $\underline{H}$ when a system is excited with wide-band signals. However, assumptions in frequency domain analyses are violated when the system is nonlinear and the computed frequency response is inadequate or misleading.

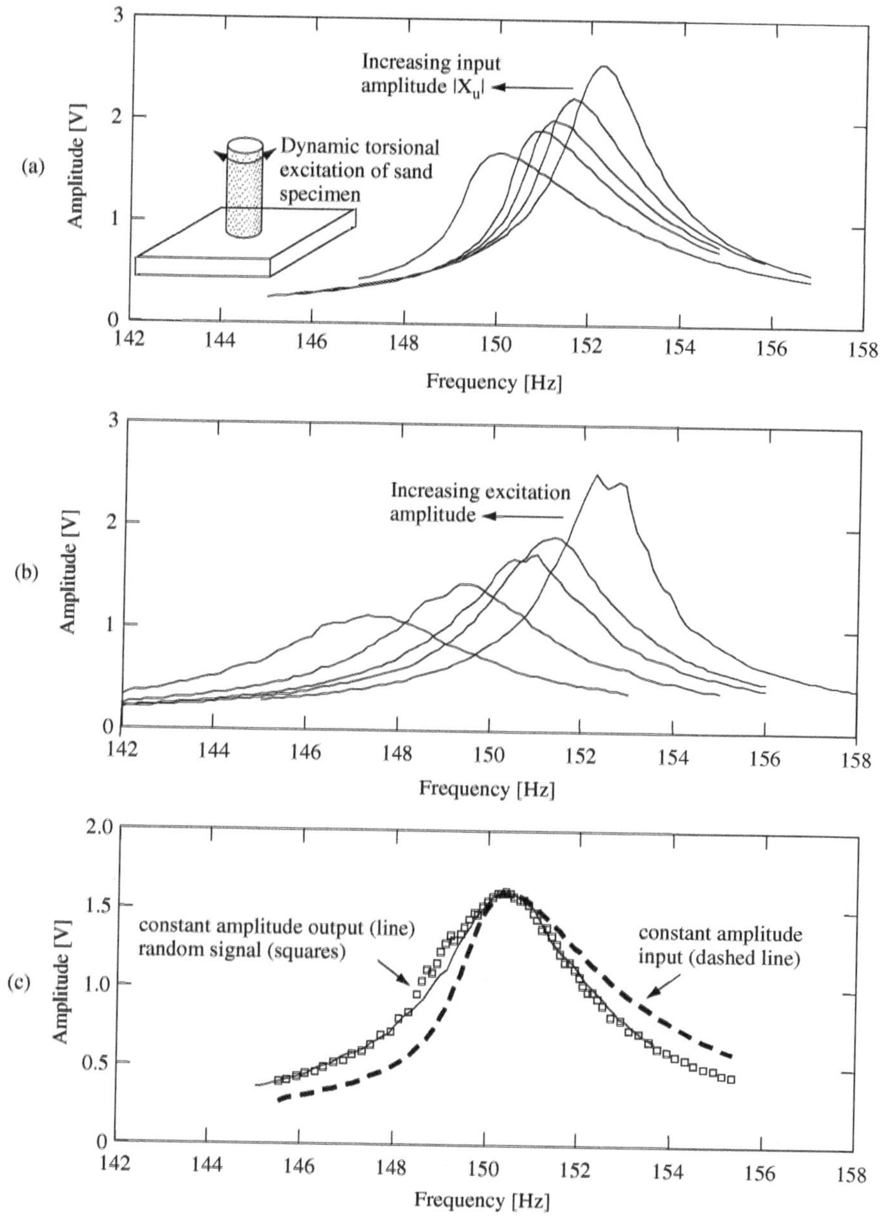

Figure 7.14 Determination of frequency response in nonlinear systems: (a) constant amplitude input – frequency sweep; (b) random input signal; (c) comparison for similar amplitude (Cascante and Santamarina, 1997)

Let us explore the nonlinear system response to random noise using the same sand column tested previously. A random signal is of particular interest: like the previous method, a random signal contains the same energy in all frequencies as in the previous method; however, all frequencies are present at all times in a random signal.

Figure 7.14b shows the frequency response obtained with random noise using cross- and autospectral densities (see Implementation Procedure 7.5). The computed frequency response curves shift to lower frequencies and exhibit higher attenuation with increasing excitation amplitude. But there is an important distinction with the results in Figure 7.14a: the responses measured with random noise are quasi-symmetric and resemble the response of linear systems. In fact, the system parameters inverted by fitting a linear viscoelastic model to any of these curves would be the parameters of an equivalent linear system for the corresponding strain level. *It can be concluded that the frequency response* $\underline{H}$ *computed from cross- and autospectra is the best-fit linear model to the data, within the extent of the input.*

7.7.3 Input: Single-frequency Sinusoid – Output: Constant Amplitude

Consider the same experimental device and a single-frequency sinusoid, but in this case the input amplitude is modified to produce the same amplitude output for all frequencies; that is, $|Y_u| =$ constant at all ω_u. The methodology requires a feedback loop (see Implementation Procedure 7.5).

Results obtained at similar peak strains using constant amplitude output and random signal are almost identical (Figure 7.14e).

Why are measured responses obtained with the two single-frequency sweep methods so different? The degree of nonlinearity and the associated frictional energy consumed per cycle are strain-dependent in sands. In the constant input method, the amplitude of the displacement varies with frequency; hence, the level of nonlinearity also varies across the spectrum. However, in the constant output procedure, the displacement and the strain are constant at all frequencies causing the same degree of nonlinearity and energy loss across the spectrum.

7.8 TIME-VARYING SYSTEMS

The response of a nonlinear system inherently varies in time according to the imposed excitation history. Furthermore, there are linear systems that experience time-varying system parameters. Both cases fail the time-invariant assumption that underlies frequency domain system analysis: the system response must not change during the measurement.

If slowly changing time-varying systems are assumed time-invariant within short-time windows, data processing is based on spectral ratios, as discussed in Chapter 6. The selection of the window width $M \cdot \Delta t$ depends on the system rate of change. In turn, the window width affects the ability to identify the response to low-frequency components, the determination of "instantaneous" rather than time-averaged system parameters, and time resolution.

A versatile time domain methodology for the analysis of time-varying systems is introduced next.

7.8.1 ARMA Model

Systems have "inertia" or "memory"; therefore, the current output y_i can be *forecast* on the basis of the prior output values. On the other hand, the convolution equation *in the time domain* shows that the current output y_i is a moving average of the current and prior inputs according to the entries in the impulse response $\underline{h}$. In general these two approaches are valid and can be used in combination,

ARMA		***Auto-Regressive***		***Moving-Average***
Current output y_i	=	linear combination of prior output values $y_{i-1}, y_{i-2}, y_{i-3} \ldots$	+	linear combination of current and prior input values $x_i, x_{i-1}, x_{i-2}, \ldots.$

Formally, the predictive equation is written as

$$\begin{aligned} y_i &= (a_1 \cdot y_{i-1} + a_2 \cdot y_{i-2} + \ldots..) + (b_0 \cdot x_i + b_1 \cdot x_{i-1} + \ldots..) \\ &= \sum_{h=1}^{na} a_h \cdot y_{i-h} + \sum_{k=0}^{nb} b_k \cdot x_{i-k} \end{aligned} \tag{7.24}$$

where the output y_i at discrete time t_i is computed as an Auto-Regressive linear combination of the "na" prior output values, and a causal Moving Average of the "nb" current and prior input values. The values na and nb define the order of the auto-regressive and the moving average components of the ARMA model; proper implementation requires adequate a priori selection of orders na and nb. The coefficients a_h and b_k capture all the information relevant to the system response.

7.8.2 A Physically Based Example

Let us develop a physically based ARMA model for a single DoF oscillator. The equation of motion is (details in Section 4.3)

$$\begin{aligned} m \cdot \ddot{y} + c \cdot \dot{y} + k \cdot y &= x \\ \ddot{y} + 2 \cdot D \cdot \omega_n \cdot \dot{y} + \omega_n^2\, y &= \frac{1}{m} \cdot x \end{aligned} \tag{7.25}$$

where the driving force x and the displacement y are functions of time. In discrete time, the values of velocity and acceleration can be replaced by forward finite difference approximations in terms of the displacements at time $t_i \leq i \cdot \Delta t$:

$$\dot{y}_i = \frac{y_i - y_{i-1}}{\Delta t} \qquad \textit{velocity} \tag{7.26}$$

$$\ddot{y}_i = \frac{y_i - 2 \cdot y_{i-1} + y_{i-2}}{\Delta t^2} \qquad \textit{acceleration} \tag{7.27}$$

Substituting into Equation 7.25,

$$y_i = \underbrace{\left(\frac{2\,(1 + D \cdot \omega_n \cdot \Delta t)}{C}\right) \cdot y_{i-1} + \left(\frac{-1}{C}\right) y_{i-2}}_{\text{Auto-Regressive}} + \underbrace{\left(\frac{\Delta t^2}{m}\right) x_i}_{\text{Moving-Average}} \tag{7.28}$$

where $C = (1 + 2 \cdot D \cdot \omega \cdot \Delta t + \omega^2 \cdot \Delta t^2)$. Therefore, the physical meaning of all ARMA parameters is readily apparent (the factors of y_{i-1}, y_{i-z} and X_i). The methodology can be extended to other linear and nonlinear systems.

7.8.3 Time-Varying System Analysis

An equation similar to Equation 7.28 can be written for each value of the measured output. This leads to a system of equations of the following form:

$$\underline{[y]} = \left[\underline{\underline{[y]}}\ \underline{\underline{[x]}}\right] \cdot \left[\begin{matrix} \underline{[a]} \\ \underline{[b]} \end{matrix}\right] \tag{7.29}$$

where the values of $\underline{y}$ and $\underline{x}$ are known. The total number of unknowns is $N = na + nb$. *For example, for* $na = 3$ and $nb = 2$, the system of Equations 7.29 becomes

$$\begin{bmatrix} y_1 \\ y_2 \\ y_3 \\ y_4 \\ y_5 \\ \cdots \end{bmatrix} = \begin{bmatrix} 0 & 0 & 0 & x_1 & 0 & 0 \\ y_1 & 0 & 0 & x_2 & x_1 & 0 \\ y_2 & y_1 & 0 & x_3 & x_2 & x_1 \\ y_3 & y_2 & y_1 & x_4 & x_3 & x_2 \\ y_4 & y_3 & y_2 & x_5 & x_4 & x_3 \\ \cdots & \cdots & \cdots & \cdots & \cdots & \cdots \end{bmatrix} \cdot \begin{bmatrix} a_1 \\ a_2 \\ a_3 \\ b_0 \\ b_1 \\ b_2 \end{bmatrix} \tag{7.30}$$

The goal is to extract the values $\underline{a}$ and $\underline{b}$ that characterize the system parameters. This is an inverse problem. If the system is time-invariant, the inverse problem is solved using all equations. If the system is time-variant or nonlinear, a limited number of equations around time t_i is used to obtain the equivalent time-invariant behavior that corresponds to time t_i. The calculation is repeated at all times of interest. This will render momentary system properties within the time covered by the model.

The selected number of rows M corresponding to known values y_i must be equal or greater than the number of unknown model parameters $na + nb$. If $M > (na + nb)$, a least squares approach is used to determine the unknowns $[\underline{a}, \underline{b}]$, the inferred values are less sensitive to noise, and the system parameters average over the time interval $M \cdot \Delta t$. The least squares solution to the inverse problem is presented in Chapter 9.

If MA models are used instead of ARMA models, Equations 7.24 and 7.29 become the convolution operation, and inverted MA model parameters are the system impulse response $\underline{h} = \underline{b}$. However, the convolutional nature of MA does not accommodate systems with feedback, which are common from mechanics, to biology and medicine; however, this is readily considered in the AR component of ARMA models. Furthermore, more complex models can be developed; for example, forecasting does not need to be based on linear combinations but may involve polynomial auto-regressive models.

7.9 SUMMARY

7.9.1 Nonstationary Signals

- The DFT of a signal converts a 1D array in time into a 1D array in frequency by decomposing the signal into a series of harmonically related, scaled and phase shifted, infinitely long sinusoids. The signal is presumed periodic. It is always possible to compute the DFT of nonstationary signals; however, the unequivocal interpretation of the DFT requires stationary signals.

- Techniques for the analysis of nonstationary signals include short-time Fourier transform, band-pass filters and wavelet analysis. They convert the 1D array in the time domain into a 2D array in the time-frequency space. Alternatively, the analytic signal presents instantaneous amplitude and frequency versus time. Like a musical score, these methods capture the time-varying frequency content of the signal.
- The uncertainty principle is inherent to all forms of signal analysis: an increase in frequency resolution (for a given number of digitized points) can only take place at the expense of a loss in time resolution.
- All the available information is encoded in the signal. Hence, *transformations do not generate new information, but facilitate the interpretation* of the information encoded within the signal.

7.9.2 Nonlinear Time-Varying Systems

- The analysis of signals and systems in the frequency domain presumes linear time invariance; thus, the generalized superposition principle applies. Under these conditions, there are equivalent operations in the time and frequency domains for all linear or convolutional operators. The choice between time or frequency domain reflects computational demands, enhanced interpretation of information, or the nature of the application at hand. Given the efficiency of FFT algorithms, frequency domain operations are often preferred.
- Several procedures permit the detection of system nonlinearity: verification of scaling or additive rules in the superposition principle, determination of the frequency response $\underline{H}$ for different excitation levels, similitude between input and output statistics, presence of multiples, verification of coherence, and compatible spectral variation between real and imaginary components of the frequency response.
- Both test procedures and data analysis methods may hide the nonlinear system response. Thus, the experimenter must remain skeptical and alert to the selected methodology.
- Slowly changing time-varying systems can be studied in the frequency domain by extracting short-time windows. Alternatively, locally fitted auto-regressive moving-average ARMA models extract momentary system properties in the time domain.

FURTHER READING AND REFERENCES

Bendat, J. S. (1990). Non-linear System Analysis and Identification from Random Data. Wiley-Interscience, New York. 267 pages.

Cascante, G. C. and Santamarina, J. C. (1997). Low Strain Measurements Using Random Excitation. Geotechnical Testing Journal. Vol. 20, No. 1, pp. 29–39.

Cohen, L. (1995). Time-Frequency Analysis. Prentice-Hall, Englewood Cliffs, NJ. 299 pages.

Dimarogonas, A. D. and Haddad, A. (1992). Vibration for Engineers. Prentice-Hall, Inc., Englewood Cliffs. 749 pages.

Ewins, D. J. (1988). Modal Analysis. Theory and Practice. England: Research Study Press Ltd. 269 pages.

Saskar, T. K., Salazar-Palma, M., and Wicks, M. C. (2002). Wavelets Applications in Engineering Electromagnetics. Artech House, Boston. 347 pages.

Strang, G. and Nguyen, T. (1996). Wavelets and Filter Banks. Wellesley-Cambridge, Wellesley, Mass. 490 pages.

SOLVED PROBLEMS

P7.1 *Analytic signal.* Consider the signal $x = A \cdot \cos(\omega \cdot t) + B \cdot \sin(\omega \cdot t)$. Combine the results in Equations 7.5 and 7.6 to compute the analytic signal $\underline{x}^{<A>}$. What is the instantaneous amplitude?

Solution: The analytic signals for sine and cosine functions are (Equations 7.5 and 7.6):

$$x = \cos(\omega t) \quad \Rightarrow \quad x^{<A>} = \cos(\omega t) + j \cdot \sin(\omega t)$$
$$x = \sin(\omega t) \quad \Rightarrow \quad x^{<A>} = \sin(\omega t) - j \cdot \cos(\omega t)$$

Invoking the linearity property:

$$\begin{aligned} x^{<A>} &= A \cdot \cos(\omega \cdot t) + j \cdot A \cdot \sin(\omega \cdot t) + B \cdot \sin(\omega \cdot t) - j \cdot B \cdot \cos(\omega \cdot t) \\ &= A \cdot \cos(\omega \cdot t) + B \cdot \sin(\omega \cdot t) + j \cdot [A \cdot \sin(\omega \cdot t) - B \cdot \cos(\omega \cdot t)] \end{aligned}$$

The instantaneous amplitude is $A_i = \sqrt{[\mathrm{Re}(x_i^{<A>})]^2 + [\mathrm{Im}(x_i^{<A>})]^2}$, therefore

$$\begin{aligned} A_i &= \sqrt{[A \cdot \cos(\omega \cdot t) + B \cdot \sin(\omega \cdot t)]^2 + [A \cdot \sin(\omega \cdot t) - B \cdot \cos(\omega \cdot t)]^2} \\ &= \sqrt{A^2 \cdot \cos^2(\omega \cdot t) + B^2 \cdot \sin^2(\omega \cdot t) + A^2 \cdot \sin^2(\omega \cdot t) + B^2 \cdot \cos^2(\omega \cdot t)} \\ &= \sqrt{(A^2 + B^2) \cdot [\cos^2(\omega \cdot t) + \sin^2(\omega \cdot t)]} \\ &= \sqrt{(A^2 + B^2)} \end{aligned}$$

P7.2 *Analytic signal*. Does the instantaneous amplitude follow the exponential decay observed just with the peaks of an attenuating sinusoid?
Solution: To explore this question, let us plot

$$x_i = A \cdot e^{\alpha \cdot i} \cdot \sin\left(50\frac{2\pi}{N} i\right)$$

for $N = 512$, $A = 1$, and attenuation $\alpha = -0.005$. The signal $\underline{x}$ and $\underline{X} =$ DFT(x) are

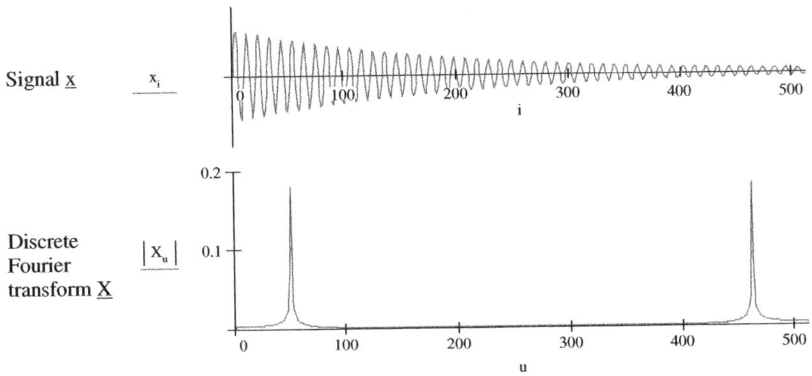

Follow the step-by-step approach in Implementation Procedure 7.1 to compute the analytic signal. The real and imaginary parts of $\underline{x}^{<A>}$ are

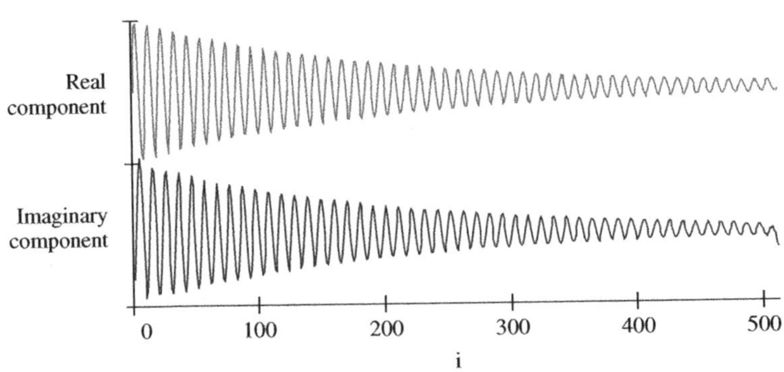

Let us plot the instantaneous amplitude $A_i = \sqrt{[\mathrm{Re}(x_i^{<A>})]^2 + [\mathrm{Im}(x_i^{<A>})]^2}$ in semilogarithmic scale together with $A_i = A \cdot e^{\alpha \cdot i}$

Semilogarithm plot: instantaneous amplitude versus time

$$\alpha = \frac{\Delta\left[\ln\left(\frac{A}{A_i}\right)\right]}{\Delta i}$$

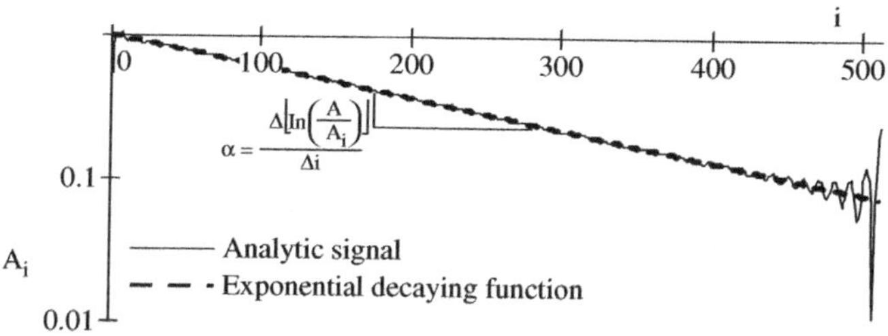

Results show that the instantaneous amplitude properly follows the attenuation law. Add high frequency and random noise to the signal $\underline{x}$ and repeat this analysis. Furthermore, test this approach with real data. Evaluate potential applications and limitations of the analytical signal.

P7.3 *Nonlinear systems.* Explore the multiples in a thresholded beat function. Consider the following input $\underline{x}$ and output $\underline{y}$ signals:

$$x_i = 2 \cdot \sin\left(10\frac{2\pi}{N}i\right) + 2 \cdot \sin\left(25\frac{2\pi}{N}i\right) \text{ and } y_i = \begin{cases} th \cdot \left(\frac{x_i}{|x_i|}\right) & \text{for } |x_i| > th \\ x_i & \text{otherwise} \end{cases}$$

Set the threshold at th $= 2.5$. Plot the amplitude of the discrete Fourier transforms $\underline{X}$ and $\underline{Y}$. Analyze the frequency where multiples are observed.
Solution: The input and output signals $\underline{x}$ and $\underline{y}$ are

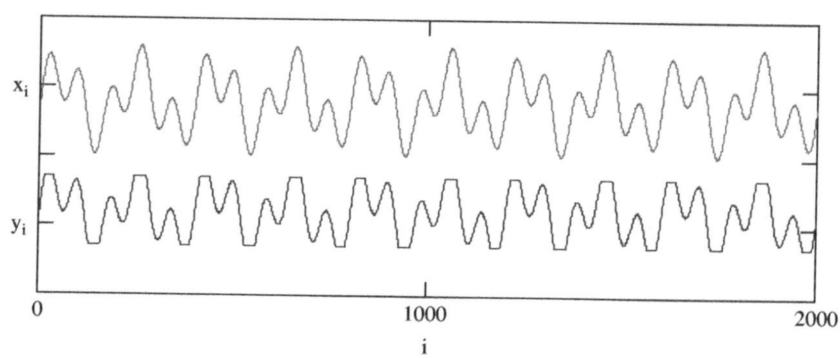

The magnitude of discrete Fourier transforms $\underline{X}$ and $\underline{Y}$ is plotted in dB scale to enhance the identification of multiples:

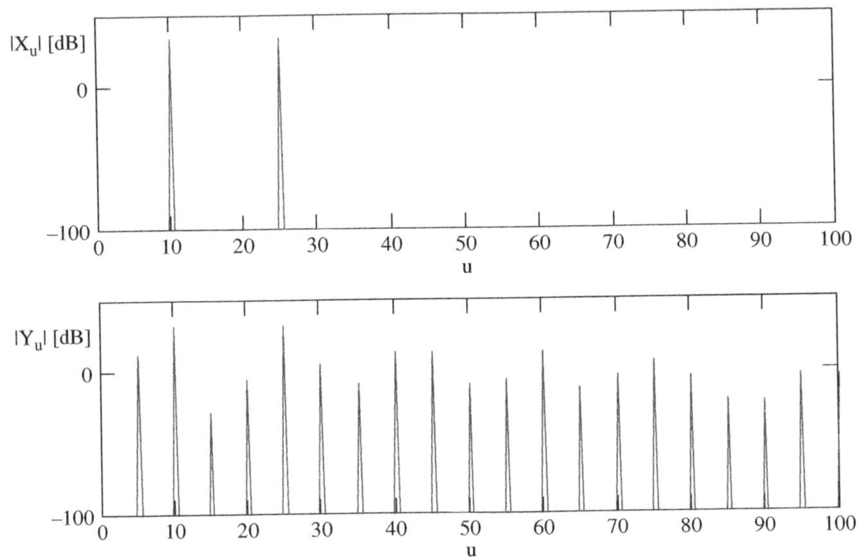

The two peaks in $|\underline{X}|$ correspond to the two frequencies f_1 and f_2 that make the beat function $\underline{x}$ (frequency counters $u = 10$ and $u = 25$). However, the nonlinear thresholding transformation causes multiple peaks in $|\underline{Y}|$. Some develop in the harmonics of the primary frequencies f_1 ($u = 20, 30, 40, \ldots$) and f_2, ($u = 50, 75, 100, \ldots$), while others are manifest in the beat frequency $f_2 - f_1$ and its harmonics. Therefore, expect peaks at frequencies $[pf_1 + qf_2 \pm r(f_2 - f_1)]$ where p, q and r are integers 0, 1, 2, ...

ADDITIONAL PROBLEMS

P7.4 *Nonstationary signals*. Express the transformation of a nonstationary signal into the time-frequency space in matrix form for: (a) time windows (STFT), (b) frequency windows (band-pass filtering), and (c) wavelet transform.

P7.5 *Wavelets.* Consider the Morlet, Mexican hat and Sinc wavelets. Plot these wavelets for different control parameters (v, M). Compute their DFT.

$$\text{Mexican hat wavelet} \quad \kappa_i = \left[1 - \left(\frac{i}{M}\right)^2\right] \cdot e^{-\frac{1}{2}\cdot\left(\frac{i}{M}\right)^2}$$

$$\text{“Sinc” wavelet} \quad \kappa_i = \frac{\sin\left(2\pi \cdot \frac{i}{M}\right)}{2\pi \cdot \frac{i}{M}} \quad \text{with } \kappa_0 = 1$$

P7.6 *Nonlinearity.* Knowing the linear response $\underline{y}$ of a single DoF oscillator ($f_n = 100\,\text{Hz}$, $D = 0.4$), assume that the response of a quasi-linear oscillator is $y_i^{<quasi>} = |y_i|^{1.3} (y_i/|y_i|)$. Simulate the output $\underline{y}^{<quasi>}$ for the following input $\underline{x}$ signals: (1) a random signal, (2) an ensemble of single-frequency sinusoids of the same amplitude, that spans across the resonant frequency, and (3) an ensemble of single-frequency sinusoids with variable amplitude to render the same output amplitude. Compute the frequency response $\underline{\underline{H}}$ in each case. Compare results, analyze and draw conclusions.

P7.7 *Time varying system – ARMA.* Reconsider the stock market problem in Chapter 1 using the ARMA approach. Download the data from the Internet. Fit ARMA models of order 2, 4, and 8 to 10-year data until 12 months ago. Then use the fitted models to extrapolate the Dow Jones values into the present time. Analyze the results and discuss.

P7.8 *Application in your area of interest: nonstationary signals.* Obtain a long signal of your interest – either run experiments or download similar signals from the Internet. (1) Test whether the signal is stationary. (2) Analyze the signal with techniques described in this chapter: analytical signal, STFT, band-pass filtering, and wavelet transform. Modify the control parameters to optimize the information extracted in each case. Compare results and draw conclusions.

P7.9 *Application in your area of interest: linearity and time invariance.* Identify a system in your area of interest. (1) Develop and implement a procedure to test whether the system remains time-invariant within the timescale of interest. (2) Run the different tests to explore nonlinearity, as in Section 7.6.3. (3) Analyze and discuss the results and their physical relevance.

8

Concepts in Discrete Inverse Problems

Many engineering and science tasks are inverse problems (Tables 1.1 and 1.3). The goal of inverse problem solving is to determine unknown parameters from measured quantities by assuming a model that relates the two. This chapter begins with a few examples of inverse problems, introduces the general concept of data-driven solutions, and identifies some of the difficulties involved in inverse problem solving. Solution methods for inverse problems are presented in subsequent chapters.

8.1 Inverse Problems – Discrete Formulation

Forward problems start from the known input. Conversely, inverse problems start from the known output and attempt to determine either the input or the properties of the system. Inverse problems appear in all engineering applications and scientific tasks.

Possible forward and inverse problems are identified for simple examples in Figure 8.1. These examples underlie more complex problems: shadow inversion underscores tomographic imaging; water flowing out of the vase is analogous to rain falling in a river basin and causing surface runoff and flooding downstream; the moving weight on the beam is a simple model of a bridge structure; and the source of heat within a body is common to conduction phenomena of all kinds and it is directly relevant to geothermal resources as well as to infrared detection systems.

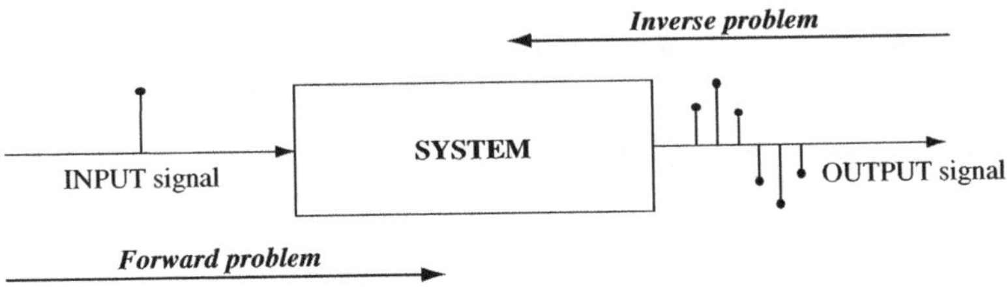

Description: The tube-lamp illuminates the medium and a shadow is created on the wall.

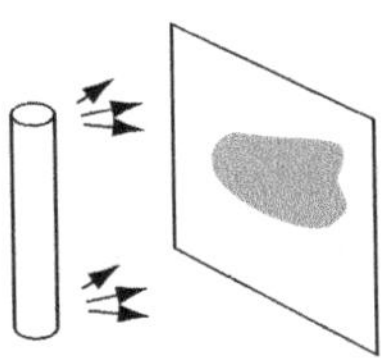

Forward problem: Given a semi-opaque object and a known source, compute the shadow on the wall.

Inverse problems: for a known medium and light intensity on the wall,

Input: determine the tube position, orientation and intensity.

System: infer the characteristics of the semi-opaque object.

Description: At time t, the height of water in the vessel is y(t), the surface of the water has a radius r(t), and the stream strikes at distance x(t).

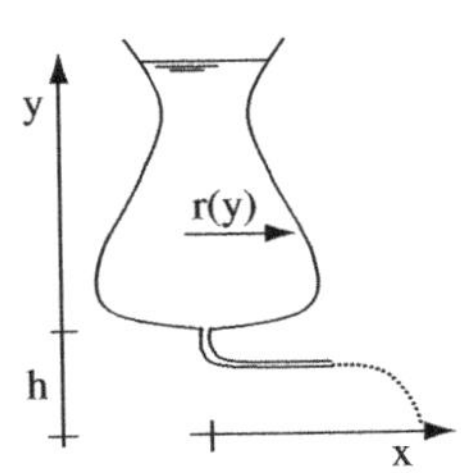

Forward problems: Determine the striking distance x when the vessel is filled to height y, or the time required to drain the vessel.

Inverse problems:

Input: knowing x at a certain time, what is the height of the water in the vessel at that moment?

System: knowing the time history x(t), what is the vessel's shape r(y)?

Figure 8.1 Inverse problems in engineering and science – simple examples

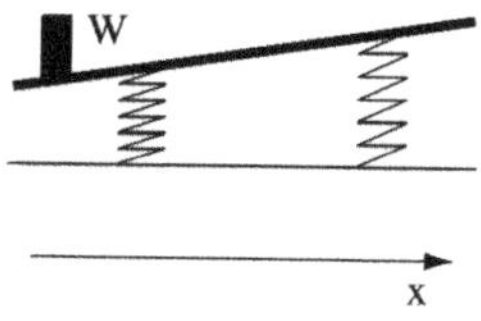

Description: The rigid beam supported on two springs is loaded with a known body weight W at position x.

Forward problem: Compute the deflection of the beam given the known position and stiffness of both springs.

Inverse problems:

Input: knowing the characteristics of the system and its deflection, infer the body's location and weight.

System: knowing the deflection of the beam for different positions of the weight, infer each spring position and stiffness.

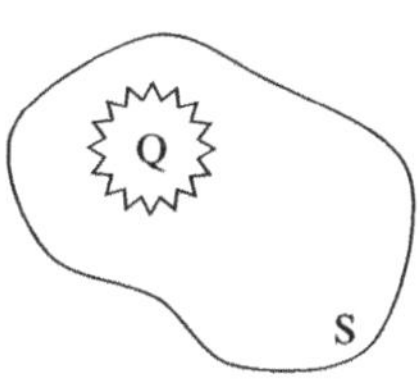

Description: There is a source of heat Q within the body. Temperature can be measured anywhere on the surface S.

Forward problems: Compute the surface temperature knowing the source Q, location and size, and medium properties.

Inverse Problems:

Input: knowing the medium properties and the spatial distribution of surface temperature, infer the source position and size.

System: for a given source position and size, and surface temperature, determine the volumetric distribution for thermal conductivity.

Figure 8.1 (*Continued*)

8.1.1 Continuous vs. Discrete

Examples in Figure 8.1 can be formulated using continuous or discrete mathematics. Light attenuation from the tube source to the screen is an integral of the absorption that takes place in each differential ray length ds, along the ray path. On the other hand, the cumulative output from the vessel at time t depends on the corresponding elevation z of the water inside the vessel at time t and the geometry of the vessel above it. These two examples can be captured in equations of the following form:

$$y(p) = \int_a^b h(p, s) \cdot x(s) \cdot ds \qquad \textit{Fredholm equation} \tag{8.1}$$

$$\text{or} \quad y(p) = \int_a^p h(p, s) \cdot x(s) \cdot ds \qquad \textit{Volterra equation} \tag{8.2}$$

where the function h(p, s) is the kernel. When the kernel h(p, s) describes the system response at location p owing to a unit input at location s, the function h(p, s) is the *Green's function.*

In inverse problems, y(p) and the kernel h(p, s) are known, but the function x(s) is unknown. When the unknown function appears inside the integral, the expression is known as an "integral equation". There are two main types of integral equations: *Fredholm equations* when both integration limits are fixed (Equation 8.1), *Volterra equations* when one integration limit is variable (Equation 8.2). Either integral equation is of the *first kind* when the unknown function appears only inside the integral, and it is of the *second kind* when the unknown function appears both inside and outside the integral; therefore, both Equations 8.1 and 8.2 are of the first kind. Note that convolution (Chapter 4) and even the Fourier transform (Chapter 5) are integrals of the product of two functions, and their inverse operations are integral equations.

The discrete form of integral equations is a summation:

$$y_i = \sum_k h_{i,k}\ x_k \tag{8.3}$$

When many measurements are available, the system of equations can be expressed as matrix multiplication

$$\underline{y} = \underline{\underline{h}} \cdot \underline{x} \tag{8.4}$$

where the array $\underline{x}$ captures the unknown values. (Note that the matrix $\underline{\underline{h}}$ is lower triangular in Volterra-type problems.) If the matrix $\underline{\underline{h}}$ is invertible, its inverse is computed $\underline{\underline{h}}^{-1}$, and the solution of the inverse problem becomes

$$\underline{x} = \underline{\underline{h}}^{-1} \cdot \underline{y} \qquad \textit{inverse problem} \tag{8.5}$$

However, the matrix $\underline{\underline{h}}$ is noninvertible in most cases, and a "pseudoinverse" is computed instead.

Vectors and matrices are the natural data structure for discrete signals and linear transformations that operate on discrete data values. Therefore, in accordance with the scope of this book, we seek to express inverse problems in discrete form, like Equation 8.5. (Note: not all problems are amenable to this representation.) Once the forward problem is encoded in matrix form, we can implement simple yet powerful and versatile algebraic procedures to compute a pseudoinverse and solve the inverse problem. Matrix algebra also facilitates the analysis and diagnosis of inherent difficulties in inverse problems (Chapter 9).

Selected examples are explored in the following sections. As you read these examples, consider simple problems of your own interest, identify the governing physical laws, express them in mathematical form and convert them to a discrete formulation like Equation 8.4 that would be compatible with some possible measurement scheme. This association with a specific problem will facilitate understanding this and subsequent chapters, and enhance the interpretation of underlying implications and limitations. (See exercises at the end of this chapter.)

8.1.2 Revisiting Signal Processing: Inverse Operations

Many signal processing operations have an inverse or involve the solution of an inverse problem, for example: deconvolution in the time domain, inverse Fourier transform, system identification including the case of time-varying systems using ARMA models, and adaptive filters (Chapters 4–7).

Convolution is the forward problem of determining the output signal $\underline{y}$ knowing the input $\underline{x}$ and the impulse response $\underline{h}$. In terms of discrete mathematics, convolution is a sum of pairwise multiplications and it can be readily expressed in matrix form (Section 4.5)

$$\underline{y} = \underline{\underline{h}} \cdot \underline{x} \qquad \textit{forward problem: convolution} \tag{8.6}$$

where the columns of matrix $\underline{\underline{h}}$ are shifted versions of the impulse response $\underline{h}$. The inverse problem of *deconvolution* is to determine the input $\underline{x}$ knowing the output $\underline{y}$ and the impulse response $\underline{h}$. If the matrix $\underline{\underline{h}}$ were invertible,

$$\underline{x} = \underline{\underline{h}}^{-1} \cdot \underline{y} \qquad \textit{inverse problem: deconvolution} \tag{8.7}$$

The other type of inverse problem is *system identification*. Because convolution is commutative, the convolution sum in matrix form can be expressed as the multiplication of a matrix $\underline{\underline{x}}$ whose columns are shifted versions of the input signal $\underline{x}$ times the vector of the impulse response $\underline{h}$, $\underline{y} = \underline{\underline{x}} \cdot \underline{h}$. Following a similar reasoning as before, if the matrix $\underline{\underline{x}}$ were invertible, the impulse response could be extracted as

$$\underline{h} = \underline{\underline{x}}^{-1} \cdot \underline{y} \qquad \textit{inverse problem: system identification} \tag{8.8}$$

The inverse Fourier transform in matrix form was developed in Section 5.4.

8.1.3 Regression Analysis

System identification problems can be seen as fitting a *hypothesized model* to the *measurements* in order to extract the *unknown model parameters*. The solution procedure starts by selecting a plausible model or physical law. This defines the function to be fitted.

Consider fitting a polynomial of order N – 1, with constants $(c_0, c_1, \ldots, c_{N-1})$ to measurements of distance z_i traveled by a free-falling object at times t_i. A polynomial equation can be written for each i-th measurement:

$$\begin{aligned}
t_1 &= c_0 \cdot z_1^0 + c_1 \cdot z_1^1 + c_2 \cdot z_1^2 + \ldots + c_j \cdot z_1^j + \ldots + c_{N-1} \cdot z_1^{N-1} \\
&\ldots \quad \ldots \quad \ldots \quad \ldots \quad \ldots \quad \ldots \\
t_i &= c_0 \cdot z_i^0 + c_1 \cdot z_i^1 + c_2 \cdot z_i^2 + \ldots + c_j \cdot z_i^j + \ldots + c_{N-1} \cdot z_i^{N-1} \\
&\ldots \quad \ldots \quad \ldots \quad \ldots \quad \ldots \quad \ldots \\
t_M &= c_0 \cdot z_M^0 + c_1 \cdot z_M^1 + c_2 \cdot z_M^2 + \ldots + c_j \cdot z_M^j + \ldots + c_{N-1} \cdot z_M^{N-1}
\end{aligned} \tag{8.9}$$

This set of equations can be rewritten in matrix form as

$$\begin{bmatrix} t_1 \\ \ldots \\ t_i \\ \ldots \\ t_M \end{bmatrix} = \begin{bmatrix} 1 & z_1 & \ldots & z_1^k & \ldots & z_1^{N-1} \\ \ldots & \ldots & \ldots & \ldots & \ldots & \ldots \\ 1 & z_i & \ldots & z_i^k & \ldots & z_i^{N-1} \\ \ldots & \ldots & \ldots & \ldots & \ldots & \ldots \\ 1 & z_M & \ldots & z_M^k & \ldots & z_M^{N-1} \end{bmatrix} \cdot \begin{bmatrix} c_0 \\ c_1 \\ \ldots \\ c_k \\ \ldots \\ c_{N-1} \end{bmatrix} \tag{8.10}$$

or

$$\underline{t} = \underline{\underline{z}} \cdot \underline{c}$$

The N model parameters $\underline{c} = (c_0, \ldots, c_k, \ldots, c_{N-1})$ are *unknown*. In general, there are more measurements than unknowns (M > N, overdetermined) and measurements are noisy (inconsistent set of equations).

Note that setting the problem in matrix form does not require a linear functional relation $t = f(z)$, but a linear combination of basis functions of z. The Fourier series is a good example: $t = c_0 + c_1 \cdot \cos(\omega \cdot z) + c_2 \cdot \sin(\omega \cdot z) + \ldots$, where sines and cosines are the basis functions. A hyperbolic model is fitted to experimental data in Figure 8.2.

When several competing models are available, the goodness of the fit helps identify the most plausible one. However, this is a necessary but not sufficient selection criterion. Why is it not sufficient? The answer becomes apparent later in this chapter.

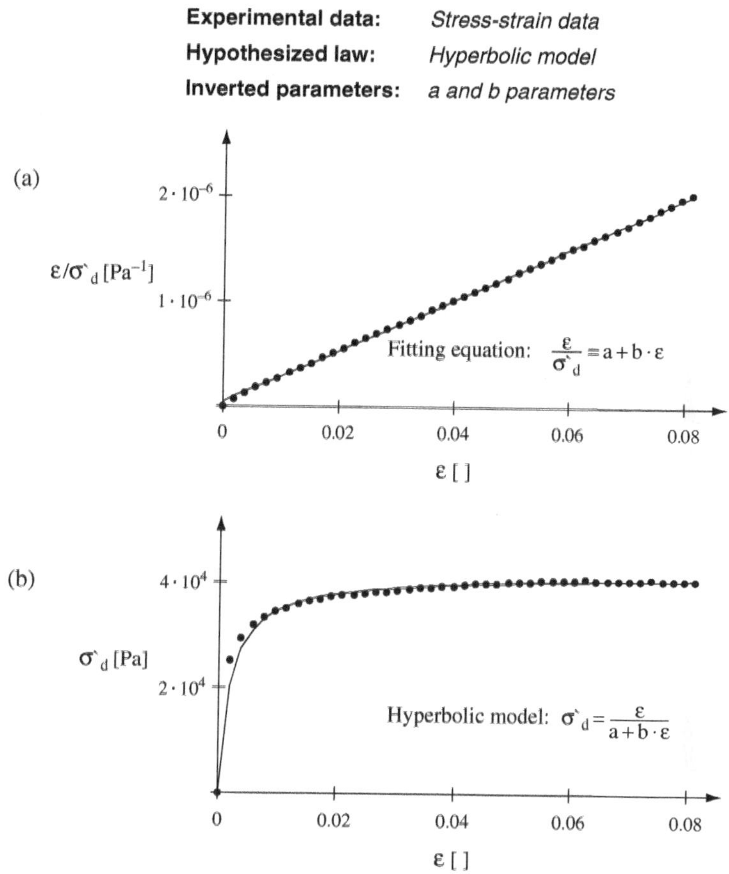

Figure 8.2 Calibration of constitutive models. Experimental load–deformation data for a kaolinite specimen. The hypothesized stress–strain relationship is the hyperbolic model. Inverted parameters: initial Young's modulus $E = a^{-1} = 20.8\,MPa$, material strength $b^{-1} = 41.5\,kPa$. (a) Transformed coordinates; (b) data in standard stress–strain space (data courtesy of E. J. Macari)

8.1.4 Travel Time Tomographic Imaging

Tomographic inversion attempts to infer material parameters and their spatial variability within a body by mathematically processing measurements obtained at the boundary. The technique applies to chemical, electrical, thermal, or mechanical parameters. Hence, this is a powerful approach in the study of many systems in engineering and science. In all cases, a *physical model must be presumed.*

Consider an ultrasound diagnostics tool where travel time measurements are inverted to render a tomographic image of the spatial variability of velocity V within the body. Both transmission data and reflection data may be used (Figures 8.3a and b). The signal emitted at the source travels through the medium and is detected at the receiver (Figure 8.3c). The travel time from the source to the receiver is the integral of differential times spent in traveling differential lengths "dh" along the ray path,

$$t = \int_{\text{source}}^{\text{receiver}} \frac{1}{V(p, q)} dh \tag{8.11}$$

where V(p, q) is the velocity of propagation within the medium at location (p, q).

The problem can be set in discrete form by dividing the region of interest into pixels. For example, the unknown region shown in Figure 8.4 has been discretized into four subregions or "pixels", $N = 4$, such that each pixel k has a constant velocity V_k. For simplicity, straight ray propagation is assumed as the governing physical model. Then, the travel time t_1 between source S_1 and receiver R_1 can be computed as

$$t_1 = \frac{h_{1,1}}{V_1} + \frac{h_{1,2}}{V_2} = \sum_{k=1}^{4} \frac{h_{1,k}}{V_k} \tag{8.12}$$

This is the discrete form of Equation 8.11. The value $h_{i,k}$ is the distance traveled by ray i in pixel k. Defining "slowness" s as the inverse of velocity $s_k = 1/V_k$, the travel times for the four measurements in Figure 8.4 are (rays 1, 2, 3, and 4; $M = 4$):

$$\begin{aligned} t_1 &= h_{1,1} \cdot s_1 + h_{1,2} \cdot s_2 \\ t_2 &= h_{2,3} \cdot s_3 + h_{2,4} \cdot s_4 \\ t_3 &= h_{3,1} \cdot s_1 + h_{3,3} \cdot s_3 \\ t_4 &= h_{4,2} \cdot s_2 + h_{4,4} \cdot s_4 \end{aligned} \tag{8.13}$$

These equations can be arranged in matrix form as

$$\begin{bmatrix} t_1 \\ t_2 \\ t_3 \\ t_4 \end{bmatrix} = \begin{bmatrix} h_{1,1} & h_{1,2} & 0 & 0 \\ 0 & 0 & h_{2,3} & h_{2,4} \\ h_{3,1} & 0 & h_{3,3} & 0 \\ 0 & h_{4,2} & 0 & h_{4,4} \end{bmatrix} \cdot \begin{bmatrix} s_1 \\ s_2 \\ s_3 \\ s_4 \end{bmatrix} \tag{8.14}$$

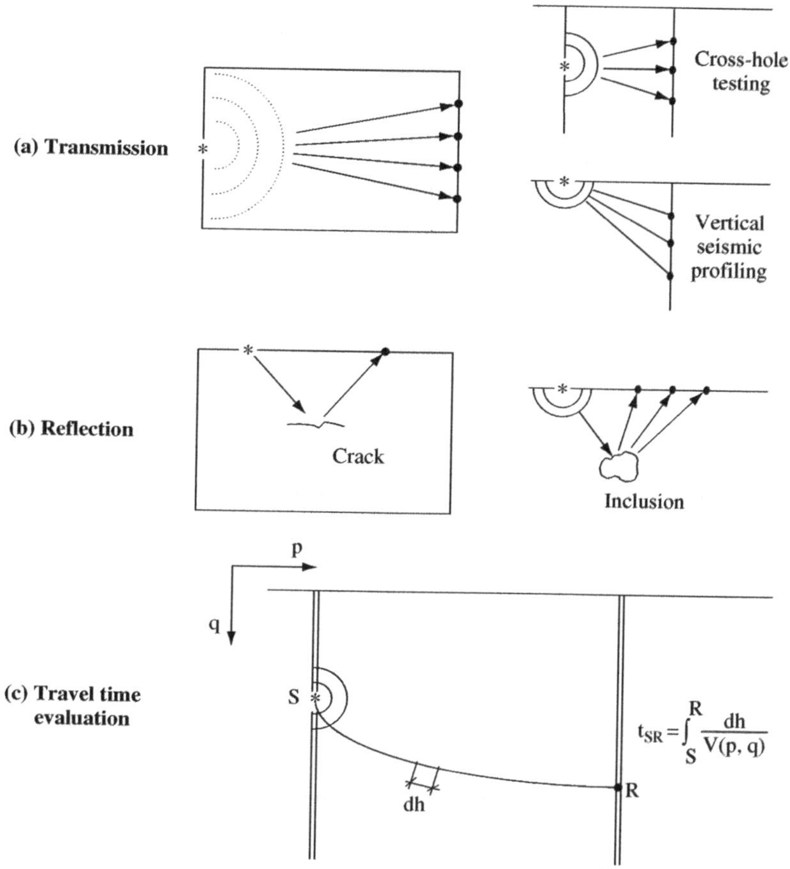

Figure 8.3 Travel time techniques. Data gathered with (a) transmission or (b) reflection techniques can be analyzed using inverse problem-solving techniques. (c) Travel time is the line integral of slowness (the inverse of velocity) along the ray path. Note: asterisks indicate the location of sources; dots show the location of receivers

In the general case of M-measurements of travel time and N-unknown pixel values,

$$\begin{bmatrix} t_1 \\ \cdots \\ t_i \\ \cdots \\ t_M \end{bmatrix} = \begin{bmatrix} h_{1,1} & \cdots & h_{1,k} & \cdots & h_{1,N} \\ \cdots & \cdots & \cdots & \cdots & \cdots \\ h_{i,1} & \cdots & h_{i,k} & \cdots & h_{i,N} \\ \cdots & \cdots & \cdots & \cdots & \cdots \\ h_{M,1} & \cdots & h_{M,k} & \cdots & h_{M,N} \end{bmatrix} \cdot \begin{bmatrix} s_1 \\ \cdots \\ s_k \\ \cdots \\ s_N \end{bmatrix} \tag{8.15}$$

or

$$\underline{t} = \underline{\underline{h}} \cdot \underline{s}$$

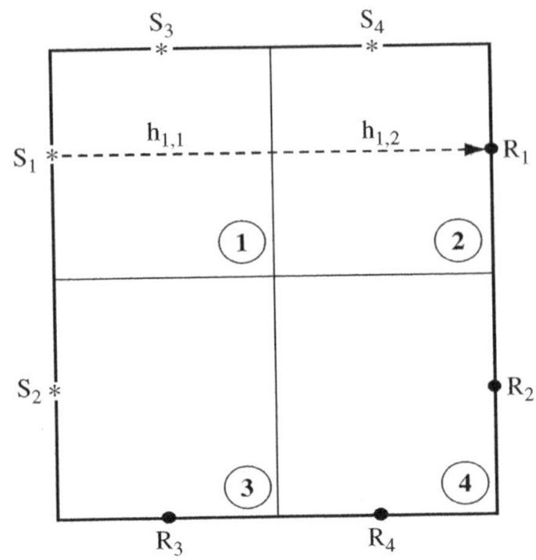

Figure 8.4 Tomography. The unknown region is digitized in four subregions or pixels. The region is "illuminated" with straight rays. Sources S and receivers R are placed on the boundary. Only one ray is shown $S_1 \rightarrow R_1$. The other three rays in this example are $S_2 \rightarrow R_2$, $S_3 \rightarrow R_3$, and $S_4 \rightarrow R_4$

where

i refers to ray number;
k refers to pixel number;
$\underline{t}$ is the $[M \times 1]$ vector of measured travel times;
$h_{i,k}$ is the length traveled by the i-th ray in the k-th pixel;
$\underline{\underline{h}}$ is the $[M \times N]$ matrix of travel lengths; and
$\underline{s}$ is the $[N \times 1]$ vector of unknown pixel slowness.

Equation 8.15 is the forward problem: travel times $\underline{t}$ are computed knowing the travel lengths $\underline{\underline{h}}$ and pixel slowness $\underline{s}$. The aim of the inverse problem is to determine the pixel values $\underline{s}$ by measuring travel times $\underline{t}$. Note that a physical wave propagation model is presumed to estimate the travel lengths in $\underline{\underline{h}}$. Once pixel values $\underline{s}$ are computed, the image is rendered by coloring pixels according to their slowness using a selected color scheme.

8.1.5 Determination of Source Location

Many tasks require precise information regarding the location of sources. Proper source location is used to (a) determine the evolution of material failure using either electromagnetic or acoustic emissions; (b) locate brain lesion from electroencephalograms; (c) assess fault rupture evolution on the bases of successive hypocenter locations during earthquakes; (d) identify the instantaneous position of a transmitting antenna; or (e) define the trajectory of a pill transmitter that is swallowed and travels inside the body.

The travel time t_i from the source to the i-th receiver in a homogeneous medium is (Figure 8.5)

$$t_i = \frac{1}{V} \cdot \sqrt{(p_s - p_i)^2 + (q_s - q_i)^2 + (r_s - r_i)^2} \quad \textit{homogeneous medium} \tag{8.16}$$

where p_s, q_s, and r_s are the unknown coordinates of the source, and p_i, q_i, and r_i are the known coordinates of the i-th receiver. The travel time t_i can be expressed as the addition of (1) the time that the wave takes to arrive at a reference receiver t_0 (unknown), and (2) the time difference between the arrival at the reference and the i-th transducers Δt_i

$$t_0 + \Delta t_i = \frac{1}{V} \cdot \sqrt{(p_s - p_i)^2 + (q_s - q_i)^2 + (r_s - r_i)^2} \tag{8.17}$$

A similar equation can be written for each monitoring transducer. Each equation is a *nonlinear* combination of spatial coordinates. The problem can be linearized

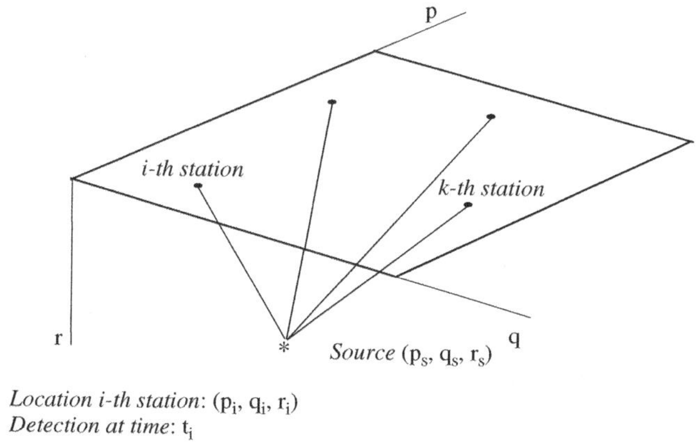

Figure 8.5 Passive emissions. The unknowns are source location p_s, q_s, r_s, and time of emission t_0. Note: the asterisk indicates the location of the source; dots show the location of receivers

if equations for two different receivers are subtracted. Consider two receivers i and k:

$$V^2 \cdot \left[(t_0 + \Delta t_i)^2 - (t_0 + \Delta t_k)^2\right] = (p_s - p_i)^2 + (q_s - q_i)^2 + (r_s - r_i)^2 - (p_s - p_k)^2 - (q_s - q_k)^2 - (r_s - r_k)^2 \quad (8.18)$$

Expanding and simplifying,

$$V^2 \cdot \left[2 \cdot t_0 \cdot (\Delta t_i - \Delta t_k) + \Delta t_i^2 - \Delta t_k^2\right] = 2 \cdot p_s \cdot (p_k - p_i) + 2 \cdot q_s \cdot (q_k - q_i) + 2 \cdot r_s \cdot (r_k - r_i) + p_i^2 + q_i^2 + r_i^2 - p_k^2 - q_k^2 - r_k^2 \quad (8.19)$$

Finally,

$$\underbrace{t_0 \cdot \overbrace{V^2 \cdot (\Delta t_i - \Delta t_k)}^{\text{known } a_{i,k}} - p_s \cdot \overbrace{(p_k - p_i)}^{\text{known } b_{i,k}} - q_s \cdot \overbrace{(q_k - q_i)}^{\text{known } c_{i,k}} - r_s \cdot \overbrace{(r_k - r_i)}^{\text{known } d_{i,k}}}_{} = \underbrace{\frac{1}{2} \cdot \left[p_i^2 + q_i^2 + r_i^2 - p_k^2 - q_k^2 - r_k^2 - V^2\left(\Delta t_i^2 - \Delta t_k^2\right)\right]}_{\text{known } e_{i,k}} \quad (8.20)$$

where a, b, c, d, and e are auxiliary parameters that depend on known values related to receivers i and k and the reference receiver. Equation 8.20 is a linear equation in terms of the reference time t_0 and the coordinates of the source p_s, q_s, and r_s. It can be written for each pair of receivers i and k (in relation to the reference transducer),

$$\underbrace{\begin{bmatrix} e_{1,2} \\ \cdots \\ e_{i,k} \\ \cdots \\ \cdots \end{bmatrix}}_{\underline{e}} = \underbrace{\begin{bmatrix} a_{1,2} & b_{1,2} & c_{1,2} & d_{1,2} \\ \cdots & \cdots & \cdots & \cdots \\ a_{i,k} & b_{i,k} & c_{i,k} & d_{i,k} \\ \cdots & \cdots & \cdots & \cdots \\ \cdots & \cdots & \cdots & \cdots \end{bmatrix}}_{\underline{\underline{A}}} \cdot \underbrace{\begin{bmatrix} t_0 \\ p_s \\ q_s \\ r_s \end{bmatrix}}_{\underline{u}} \quad (8.21)$$

The goal of the inverse problem is to solve for the time of the event (with respect to its arrival at the reference receiver) and the coordinates of the source. This is the vector of unknowns $\underline{u} = (t_0, p_s, q_s, r_s)$.

If a large number of monitoring stations are available, Equation 8.21 can be extended to take into consideration the spatial variability in the medium. In this case, inversion will not only render the timing of the event and the location of the source but the characteristics of the medium as well.

8.2 LINEARIZATION OF NONLINEAR PROBLEMS

The expression $\underline{y} = \underline{\underline{h}} \cdot \underline{x}$ implies a linear relation. However, linearity is only a tangential approximation to any physical process. For example, the trend between conductivity and electrolyte concentration deviates from linearity as the salt concentration approaches saturation, and Ohm's linear law between current and voltage fails at high current densities. In nonlinear cases, the analyst must decide how far the linear model can be used within acceptable deviations.

Still, a nonlinear relation can be linearized about a point using a first-order Taylor expansion. (See solved problems at the end of the chapter.) Consider the nonlinear function $z = f(x)$ shown in Figure 8.6. The first-order Taylor expansion about $x^{<0>}$ permits the estimation of the value of the function z at $x^{<1>}$ from the value of z at $x^{<0>}$

$$z\left(x^{<1>}\right) \simeq z(x^{<0>}) + \left.\frac{dz}{dx}\right|_{x^{<0>}} \cdot (x^{<1>} - x^{<0>}) \tag{8.22}$$

where the slope $dz/dx|_{x^{<0>}}$ is the derivative of the function evaluated at $x^{<0>}$. If z is a function of two variables x_1 and x_2, the value of $z\left(x_1^{<1>}, x_2^{<1>}\right)$ at a point $x_1^{<1>} = x_1^{<0>} + \Delta x_1$ and $x_2^{<1>} = x_2^{<0>} + \Delta x_2$ can be estimated as

$$z\left(x_1^{<1>}, x_2^{<1>}\right) \simeq z\left(x_1^{<0>}, x_2^{<0>}\right) + \left.\frac{\partial z}{\partial x_1}\right|_{\left(x_1^{<0>}, x_2^{<0>}\right)} \cdot \Delta x_1 + \left.\frac{\partial z}{\partial x_2}\right|_{\left(x_1^{<0>}, x_2^{<0>}\right)} \cdot \Delta x_2 \tag{8.23}$$

In general, if z is a function of N variables $\underline{x} = (x_1 \ldots x_N)$,

$$z\left(\underline{x}^{\langle 1\rangle}\right) \simeq z\left(\underline{x}^{\langle 0\rangle}\right) + \sum\left(\left.\frac{\partial z}{\partial x_k}\right|_{\underline{x}^{\langle 0\rangle}} \cdot \Delta x_k\right) \tag{8.24}$$

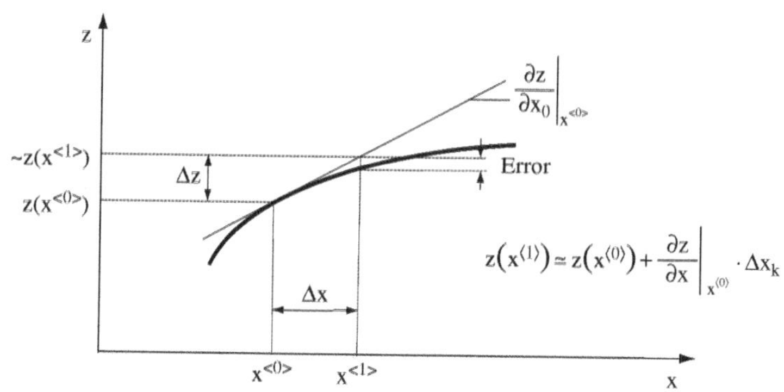

Figure 8.6 Linearization of nonlinear problems: first-order Taylor expansion

If there are M nonlinear equations, the linearized approximations are assembled in matrix form as

$$\underline{\Delta z} = \underline{\underline{J}} \cdot \underline{\Delta x} \tag{8.25}$$

where

$$\Delta z_i = z_i^{\langle 1 \rangle} - z_i^{<0>} \quad \text{for} \quad i = 1 \text{ to M measurements}$$

$$\Delta x_k = x_k^{\langle 1 \rangle} - x_k^{<0>} \quad \text{for } k = 1 \text{ to N variables}$$

$$J_{i,k} = \left. \frac{\partial z_i}{\partial x_k} \right|_{\underline{x}^{\langle 0 \rangle}} \quad [\text{M} \times \text{N}] \ \textit{Jacobian} \text{ matrix of partial derivatives}$$

Equation 8.25 has the same mathematical form as other linear problems discussed in the previous section. Therefore, inverse problems that involve nonlinear systems can be solved by successive linearizations. Difficulties associated with convergence and uniqueness are often exacerbated in nonlinear problems.

8.3 DATA-DRIVEN SOLUTION – ERROR NORMS

The solution of inverse problems is guided by the data and physical requirements about the model. Concepts related to data-driven inverse problem solution are discussed in this section.

8.3.1 Errors

Let us consider the case of curve fitting a second-order polynomial to the data shown in Figure 8.7. The error or residual for the i-th measurement is established between *measured value* $y_i^{<meas>}$ and *predicted value* $y_i^{<pred>}$

$$e_i = y_i^{<meas>} - y_i^{<pred>} \tag{8.26}$$

where the predicted value $y_i^{<pred>} = a + b \cdot t_i + c \cdot t_i^2$ is computed for a given estimate of the unknown coefficients (a, b, c). The goal is to identify the set of coefficients that minimizes the residual, taking all measurements into consideration.

Many physical parameters vary across orders of magnitude. For example, this is typically the case in conductivity of all kinds (fluid, thermal, electrical). Furthermore, complementary manifestations of the same physical phenomenon may take place in very different scaling conditions (see an example in Figures 8.8a and b). When very small and very large values coexist in the data, the error

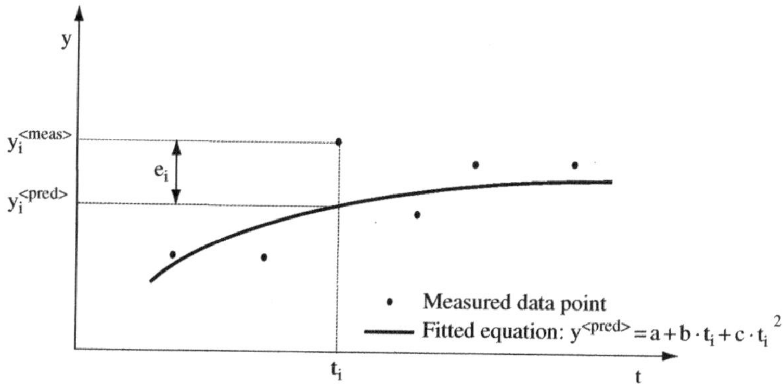

Figure 8.7 Error between the measured and the predicted values for the i-th measurement. (Measured values $y_i^{<meas>}$ and predicted values $y_i^{<pred>} = a + b \cdot t_i + c \cdot t_i^2$)

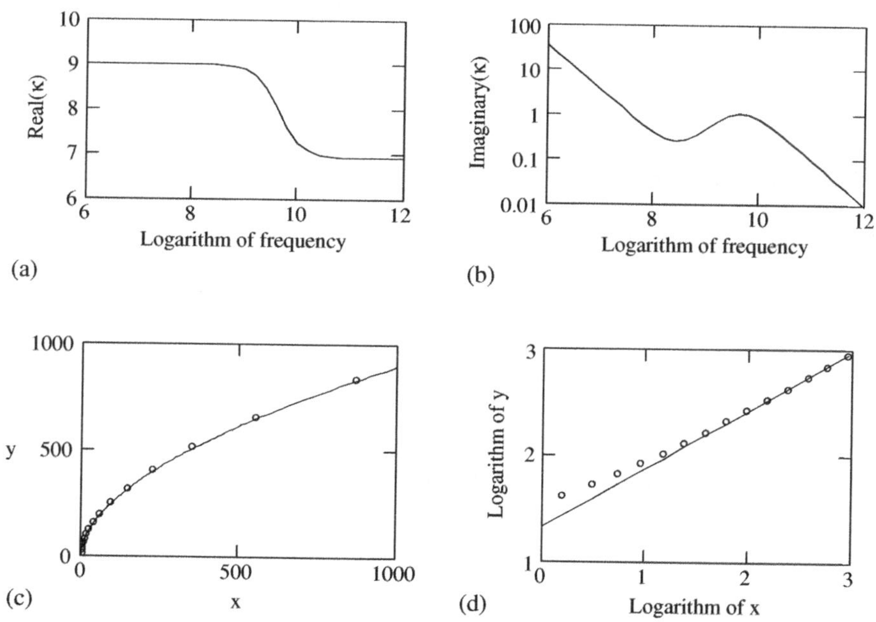

Figure 8.8 Error definition: (a and b) real and imaginary parts of the complex permittivity κ; (c and d) data points with a bias: the fitted line follows the data in linear scale, yet the bias is clearly seen in log-log scale

definition in Equation 8.26 will bias the inversion towards the largest measured values, and alternative error definitions should be considered:

$$e_i = \log\left(y_i^{<meas>}\right) - \log\left(y_i^{<pred>}\right) \qquad \textit{log difference} \tag{8.27}$$

$$e_i = \frac{y_i^{<meas>} - y_i^{<pred>}}{y_i^{<pred>}} \qquad \textit{proportional error} \tag{8.28}$$

$$e_i = \frac{y_i^{<meas>} - y_i^{<pred>}}{\sigma_i} \qquad \textit{standard error} \tag{8.29}$$

where σ_i is the standard deviation for the i-th measurement. It is also possible to define the *perpendicular error* as the distance normal to the trend. While this definition has advantages (for example when inverting very steep trends), the implementation is more involved.

Be aware that the selected error definition affects the inverted parameters. This is demonstrated in Figures 8.8c and d where the same data and trend are plotted in linear and logarithmic scales. While the trend fits the data well in linear scale, it is clearly biased in logarithmic scale. (See problems at the end of this chapter to explore these issues further.)

8.3.2 Error norms

The global "goodness of the fit" is evaluated by computing the norm of the vector of residuals $\underline{e}$. A useful family of "error norms" is the set of n-norms:

$$L_n = \left(\sum_i |e_i|^n\right)^{\frac{1}{n}} \tag{8.30}$$

Three notable norms are those corresponding to $n = 1$, $n = 2$ and $n = \infty$:

$$L_1 = \sum_i |e_i| \qquad \textit{sum of absolute errors} \tag{8.31}$$

$$L_2 = \left(\sum_i |e_i|^2\right)^{\frac{1}{2}} = \sqrt{\underline{e}^T\underline{e}} \qquad \textit{sum of squared errors} \tag{8.32}$$

$$L_\infty = \max\left(|e_1|, \ldots, |e_i|, \ldots, |e_M|\right) \qquad \textit{maximum absolute error} \tag{8.33}$$

The search for the "best fit" attempts to minimize the selected error norm and leads to three distinct criteria:

1. *min(L_1) or minimum total absolute value criterion.* This norm is not sensitive to a few large errors and it yields "robust solutions".
2. *min(L_2) or "least squares" criterion.* This criterion is compatible with additive Gaussian noise present in the data.
3. *min(L_∞) or "min-max" criterion.* The higher the order of the norm, the higher the weight placed on the larger errors. The L_∞ norm is the extreme case and it considers only the single worst error. This criterion is most sensitive to errors in the data and has a higher probability of yielding nonunique solutions.

The term "robust" describes a procedure or algorithm that is not sensitive to a few large deviations or outliers in the data. For comparison, the term "stable" in this context refers to a procedure where errors are damped rather than magnified.

Figure 8.9 presents the fitting of data points with a linear regression $y = a + b \cdot t$, and the error surfaces computed with the three norms. The minimum in the L_∞ surface is not a point but an area (Figure 8.9b): any combination of parameters a and b in this region gives the same minimum error, and the solution is nonunique. The lines displayed in Figure 8.9c have the same minimum L_∞.

Error norms and associated surfaces assess the goodness of a solution and provide information about convergence towards the optimal solution. A simple trial-and-error method to solve inverse problems based on this observation would start with an initial guess, and continue by perturbing one unknown at a time, trying to minimize the error norm between measured and predicted values.

Let us utilize this procedure to study the implications of selecting different norms. Figure 8.10 presents the inverted parameters for straight lines that were "best fitted" to the measured data points by minimizing the three norms L_1, L_2, and L_∞. Observe the effect of the out-of-trend data point on the parameters inverted with each norm.

It is instructive to study the variation of the norm in the vicinity of the optimal set of inverted parameters. This is done by fixing all inverted parameters except one, and varying it about the optimal value. Such plots are shown in Figure 8.11 for the curve-fitting exercise presented in Figure 8.10. Each of these lines is a 2D slice of the error surface across the optimum, that is, the intersection between the error surface and a plane that contains the variable under consideration and goes across the minimum point in the error surface. Figure 8.11 shows that the different norms result in different inverted parameters and exhibit different convergence rates towards the minimum.

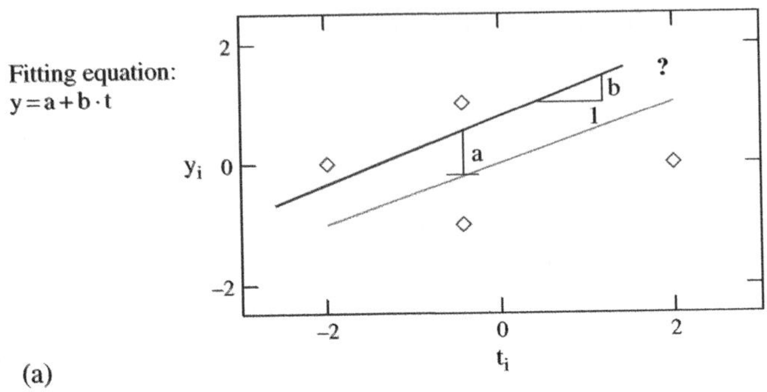

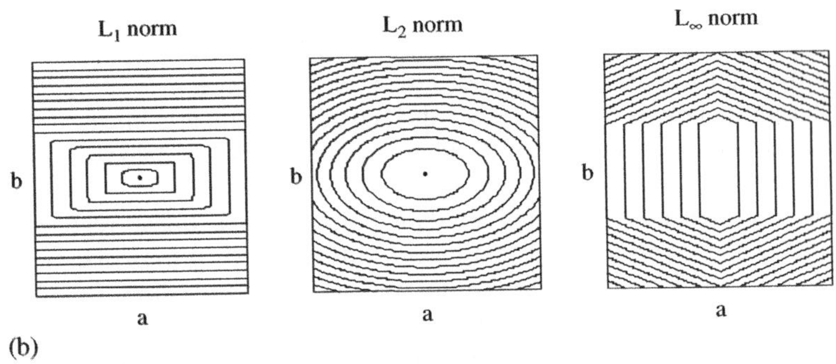

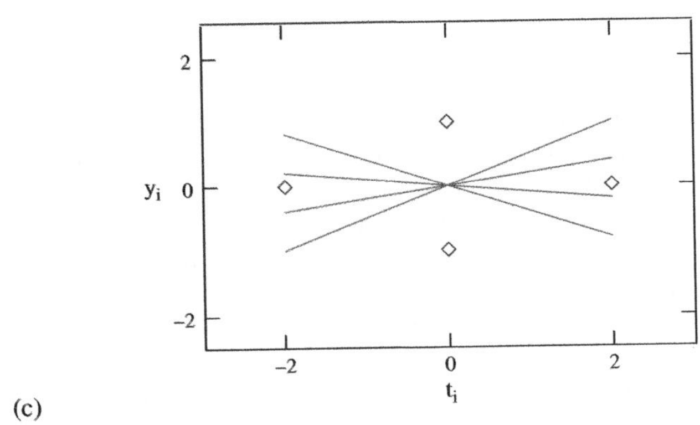

Figure 8.9 Error norms: (a) four data points to be fitted with a straight line $y = a + b \cdot t$. The residual is evaluated using three different error norms: L_1, L_2, and L_∞; (b) contours of equal error. The central plateau for the L_∞ error function suggests that several combinations of a and b parameters yield the same minimum error, as exemplified in (c)

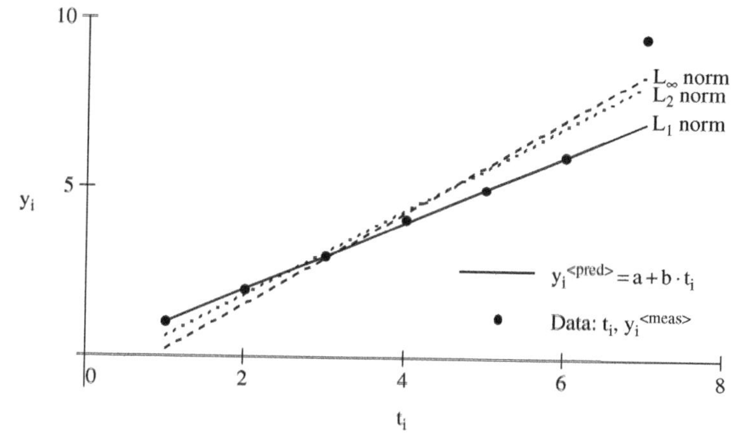

Figure 8.10 Fitted straight lines using different error norms:

L_1:	a = 0.000	b = 1.00
L_2:	a = −0.73	b = 1.27
L_∞:	a = −1.40	b = 1.40

The L_1 norm is least sensitive to the data point that "appears out-of-trend"

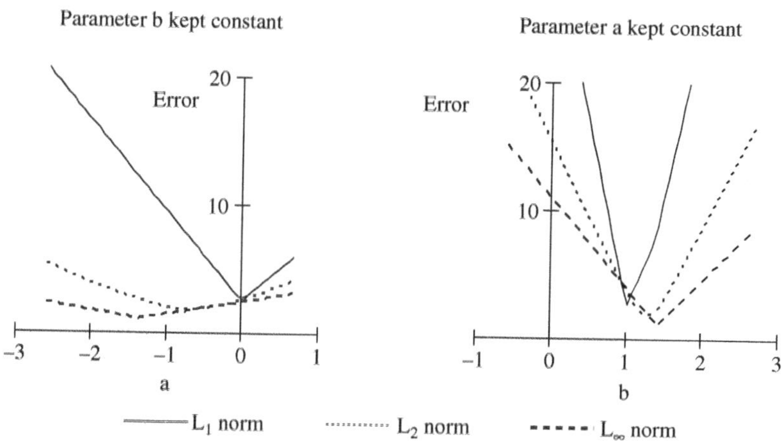

Figure 8.11 Slice of the error surfaces near optimum. Three different error norms (data from Figure 8.10). L_1 and L_2 are divided by the number of measurements to facilitate comparison. Note the differences in convergence gradients

8.4 MODEL SELECTION – OCKHAM'S RAZOR

Error minimization between measured data and predicted values is only a part of inverse problem solving. The other part is selecting a proper model. In the absence of problem-specific criteria, we explore in this section the idea that a "good" model is simple.

8.4.1 Favor Simplicity: Ockham's Razor [1]

While solving inverse problems, the engineer or scientist attempts to extract the most information possible out of the data. Therefore, it is tempting to select models with a large number of unknown parameters. However this may not be a desirable practice.

Let us revisit regression analysis (Section 8.1.3). Given N data points, one may fit increasingly higher-order polynomials to observe that the residual error between measured and predicted values decreases as the number of unknowns increases. In fact, there is a perfect match and zero residual when an $N-1$ order polynomial (N-unknowns) is fitted to the N data points.

But should an $N-1$ polynomial be fitted to the data? Linear and quadratic laws rather than high-order polynomials seem to prevail in the physical sciences. For example, Galileo invoked a second-order polynomial to predict distance d as a function of time t, in terms of velocity V and acceleration g: $d = d_0 + V \cdot t + g \cdot t^2/2$. Why did Galileo not consider higher-order terms to fit the data?

High-order polynomials "fit" data points well, but high-order term coefficients are small and add little information about the physical law. In contrast, lower-order polynomials follow trends, filter data noise, and extract the most meaningful information contained in the measurements. Therefore, new data will most likely fall near the low-order polynomial, particularly when it takes place outside the range of the original data. In terms of Bayesian probabilities, "a hypothesis with fewer parameters automatically has an enhanced posterior probability". Figure 8.12 shows a numerical example.

In summary, *a better fit does not necessarily imply a better model.* A model with many unknowns is preferred over a simpler one only if its predictions are significantly more accurate for multiple data sets. If the predictions are similar, the simpler model should be favored.

---

[1] The philosopher William of Ockham (14th century) is known for his principle: "Plurality must not be posited without necessity". The article by Jefferys and Berger (1992) presents an insightful discussion of this principle, also known as the rule of parsimony.

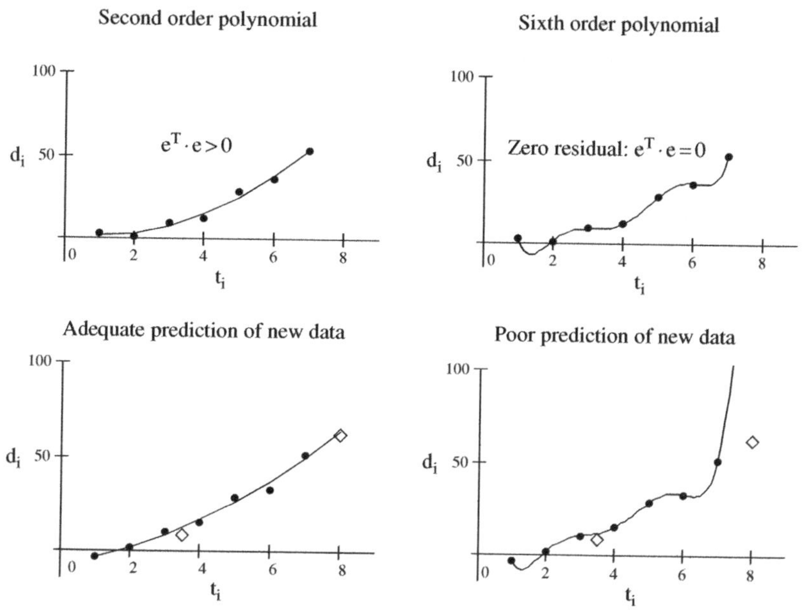

Figure 8.12 Ockham's razor criterion: favor simplicity. Simulated data correspond to $d = \frac{1}{2} \cdot a \cdot t^2$ and includes random noise. The sixth-order polynomial fits the seven data points with zero residual. The higher the order of the polynomial, the higher the probable error will be between the model and new measurements, particularly when new data fall outside the range of the original data set (new data shown as empty diamonds)

8.4.2 Reducing the Number of Unknowns

There are different approaches to implement Ockham's criterion in discrete inverse problems of the form $\underline{y} = \underline{\underline{h}} \cdot \underline{x}$.

Consider tomographic imaging where the vector of known measured travel times $\underline{t}$ [M × 1] is related to the vector of unknown pixel slowness $\underline{s}$ [N × 1] through the matrix of travel lengths $\underline{\underline{h}}$ [M × N] as $\underline{t} = \underline{\underline{h}} \cdot \underline{s}$. It is tempting to seek high-resolution tomograms. Yet physical constraints (e.g. wavelength) and limitations in the implementation of the experiment (e.g. noise, illumination angles) restrict the amount of information in the data. In this case one is well advised to reduce the number of unknowns. (Additional instability reasons are explored in Chapter 9.)

There are several options. The simplest one is to reduce the number of pixels N by increasing their size, but this causes undesirable coarse image granularity. A more effective alternative is to fit a hypothesized slowness function. Let us assume that the field of slowness across the p–q space of the image can be approximated with the following function (Figure 8.13):

$$s(p, q) = a + b \cdot (p) + c \cdot (q) + d \cdot (p \cdot q) + e \cdot (p^2) + f \cdot (q^2) \tag{8.34}$$

with six unknowns [a, b, c, d, e, f]. Then, each pixel value in the equation $\underline{t} = \underline{\underline{h}} \cdot \underline{s}$ can be expressed as a function of its location in the p–q space:

$$\begin{bmatrix} t_1 \\ \cdots \\ t_i \\ \cdots \\ t_M \end{bmatrix} = \begin{bmatrix} h_{1,1} & \cdots & h_{1,k} & \cdots & h_{1,N} \\ \cdots & \cdots & \cdots & \cdots & \cdots \\ h_{i,1} & \cdots & h_{i,k} & \cdots & h_{i,N} \\ \cdots & \cdots & \cdots & \cdots & \cdots \\ h_{M,1} & \cdots & h_{M,k} & \cdots & h_{M,N} \end{bmatrix} \times \begin{bmatrix} a + b \cdot p_1 + c \cdot q_1 + d \cdot p_1 \cdot q_1 + e \cdot p_1^2 + f \cdot q_1^2 \\ \cdots \\ a + b \cdot p_k + c \cdot q_k + d \cdot p_k \cdot q_k + e \cdot p_k^2 + f \cdot q_k^2 \\ \cdots \\ a + b \cdot p_N + c \cdot q_N + d \cdot p_N \cdot q_N + e \cdot p_N^2 + f \cdot q_N^2 \end{bmatrix} \tag{8.35}$$

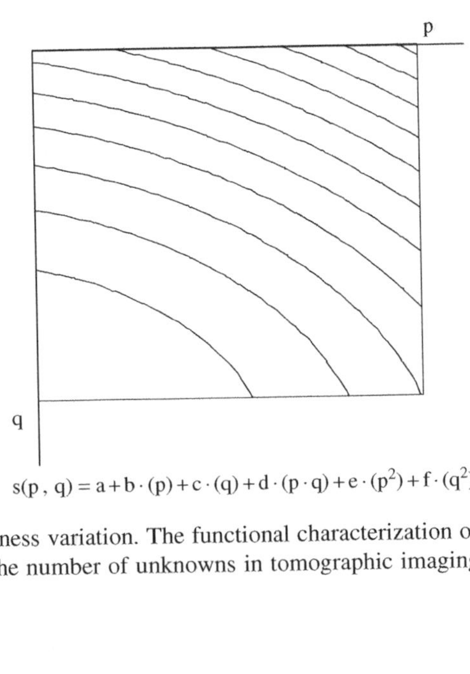

Figure 8.13 Slowness variation. The functional characterization of slowness s(p, q) permits a decrease in the number of unknowns in tomographic imaging

Finally Equation 8.35 can be rearranged as

$$\underbrace{\begin{bmatrix} t_1 \\ \cdots \\ t_i \\ \cdots \\ t_M \end{bmatrix}}_{\underline{t}[M\times 1]} = \underbrace{\begin{bmatrix} h_{1,1} & \cdots & h_{1,k} & \cdots & h_{1,N} \\ \cdots & \cdots & \cdots & \cdots & \cdots \\ h_{i,1} & \cdots & h_{i,k} & \cdots & h_{i,N} \\ \cdots & \cdots & \cdots & \cdots & \cdots \\ h_{M,1} & \cdots & h_{M,k} & \cdots & h_{M,N} \end{bmatrix}}_{\underline{\underline{h}}[M\times N]} \cdot \underbrace{\begin{bmatrix} 1 & p_1 & q_1 & p_1 q_1 & p_1^2 & q_1^2 \\ \cdots & \cdots & \cdots & \cdots & \cdots & \cdots \\ 1 & p_k & q_k & p_k q_k & p_k^2 & q_k^2 \\ \cdots & \cdots & \cdots & \cdots & \cdots & \cdots \\ 1 & p_N & q_N & p_N q_N & p_N^2 & q_N^2 \end{bmatrix}}_{\underline{\underline{O}}[N\times 6]} \cdot \underbrace{\begin{bmatrix} a \\ b \\ c \\ d \\ e \\ f \end{bmatrix}}_{\underline{u}\,[6\times 1]} \tag{8.36}$$

The second matrix $\underline{\underline{O}}$ is a function of the known pixel coordinates p_k and q_k. Likewise, the entries $h_{i,k}$ are known once the geometry is defined. Therefore, the product of the two matrices $\underline{\underline{h}}$ $[M \times N]$ and $\underline{\underline{O}}$ $[N \times 6]$ on the right-hand side can be executed to obtain only one $[M \times 6]$ matrix $\underline{\underline{h}}^* = \underline{\underline{h}} \cdot \underline{\underline{O}}$. The relation between the vector of measurements and the unknowns becomes

$$\underline{t} = \underline{\underline{h}} \cdot \underline{\underline{O}} \cdot \underline{u} = \underline{\underline{h}}^* \cdot \underline{u} \tag{8.37}$$

The result is a system of M equations with only six unknowns $\underline{u}^T = (a, b, c, d, e, f)$, rather than the original system of M equations with N unknown pixel slowness, where typically $N \gg 6$. The form of Equation 8.37 is identical to that of other linear discrete inverse problems analyzed previously in this chapter.

Note: If a slowness function $s(q, p)$ is assumed, then the slowness is not constant within pixels. However a single value of slowness is assigned to each pixel when the problem is discretized as $\underline{t} = \underline{\underline{h}} \cdot \underline{s}$. The error in this approximation can be made as small as desirable by considering smaller pixels. While the number of pixels N increases, the size of the matrix $\underline{\underline{h}}^*$ in Equation 8.37 remains the same: $[M \times N][N \times 6] = [M \times 6]$.

8.4.3 Dimensionless Ratios – Buckingham's π Theorem

Similarity is established in terms of dimensionless ratios. For example, consider the impact that operating a car has on an individual with an annual income I [\$/yr], when the individual drives a vehicle with unit cost C [\$/km] for a total distance D [km/yr]. While the three parameters may vary in a broad range for different individuals, the dimensionless ratio $\pi = D \cdot C/I$ facilitates comparing financial conditions among individuals.

Buckingham's π theorem generalizes this observation: *a physical relation of* N *parameters* $f(x_1, \ldots, x_N)$ *is equivalent to a relation* $F(\pi_1, \ldots, \pi_{N-d})$ *in terms of* $N - d$ *dimensionless parameters* π, *where* d *is the number of dimensions involved.*

The number of dimensions d is small. For example, d = 3 in typical mechanics problems: length [L], mass [M], and time [T]. Therefore, there is a small relative reduction in the number of unknowns when N is large, such as in tomographic imaging. On the other hand, the dimensionless representation will be advantageous when the inverse problem involves a small number of unknowns and it is solved by repeating time-consuming forward simulations.

8.5 INFORMATION

Information is conserved in an invertible transformation: if $\underline{y} = \underline{\underline{h}} \cdot \underline{x}$ and $\underline{\underline{h}}$ is invertible, then $\underline{x} = \underline{\underline{h}}^{-1} \cdot \underline{y}$ without loss of information; in fact, $\underline{y}$ can be fully recovered from $\underline{x}$ as $\underline{y} = \underline{\underline{h}} \cdot \left(\underline{\underline{h}}^{-1} \cdot \underline{y}\right)$. The inverse problem cannot lead to a unique solution when more information is required during inversion (N unknowns) than the amount of available information. In this case, one must either reduce the number of unknowns or provide additional information.

8.5.1 Available Information

A large number of measurements M do not necessarily imply a large amount of available information, as many of the measurements may duplicate the same information. Deciding whether information is duplicated may not be obvious at first glance. For example, the system of equations for the tomographic problem in Figure 8.4 is (assuming square pixels of length 1.0)

$$\underline{t} = \begin{bmatrix} 1 & 1 & 0 & 0 \\ 0 & 0 & 1 & 1 \\ 1 & 0 & 1 & 0 \\ 0 & 1 & 0 & 1 \end{bmatrix} \cdot \underline{s} \tag{8.38}$$

The fourth row can be obtained by adding the first and second rows and then subtracting the third one, and Equation 8.38 becomes

$$\underline{t}' = \begin{bmatrix} 1 & 1 & 0 & 0 \\ 0 & 0 & 1 & 1 \\ 1 & 0 & 1 & 0 \\ 0 & 0 & 0 & 0 \end{bmatrix} \cdot \underline{s} \tag{8.39}$$

Therefore, there are only three independent equations in this system, the rank $r[\underline{\underline{h}}] = 3$, and the information gathered with the M = 4 measurements is insufficient to solve for the N = 4 unknown pixel values. (Note: appropriate diagnostic tools are identified in Chapter 9.)

8.5.2 Information Density - Spatial Distribution

Uneven information density has different effects on the various error norms and on the inverted parameters. In particular, the L_1 and L_2 error norms sum all individual errors e_i; hence, regions with high information density have a stronger effect on the solution than regions with low information density. On the other hand, the L_∞ norm is not an error-averaging function and it is determined by the worst error; hence, it is not affected by information density.

Consider the graphical example in Figure 8.14. In the top frame, the data set includes one out-of-trend measurement; in the bottom frame, three out-of-trend measurements plot on the same point. A straight line $y = a + b \cdot t$ is fitted in each

Data set with one out-of-trend point

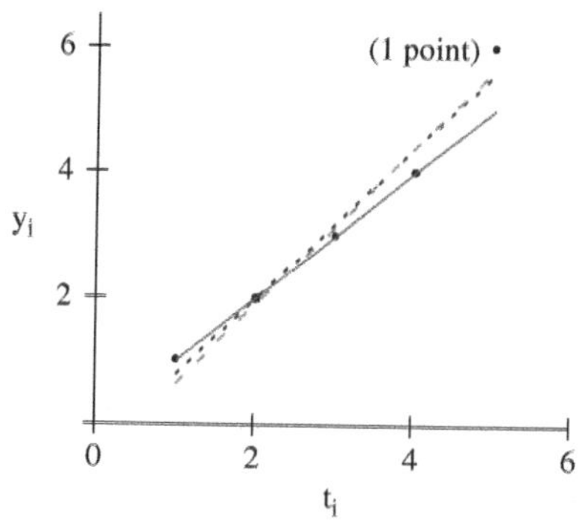

Fitting equation: $y_i = a + b \cdot t_i$

	a	b
L_1	0.00	1.00
L_2	–0.40	1.20
L_∞	–0.60	1.25

Data set with three out-of-trend points located at the same position

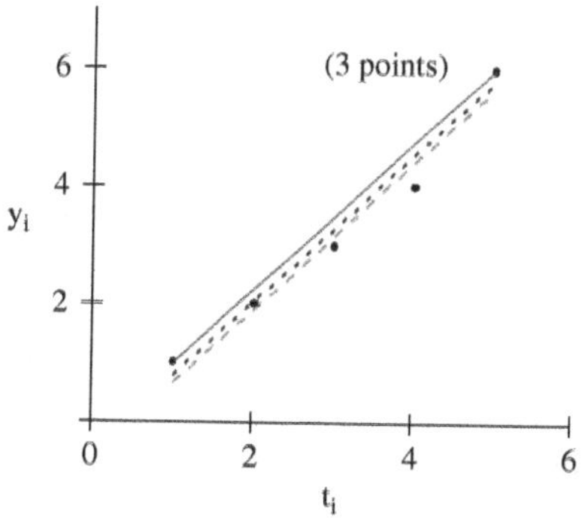

Fitting equation: $y_i = a + b \cdot t_i$

	a	b
L_1	–0.25	1.25
L_2	–0.45	1.25
L_∞	–0.60	1.25

Figure 8.14 Distribution of information. The regression line depends on the selected norm. The L_∞ norm is not sensitive to the number of out-of-trend data points. The L_1 norm is least sensitive to outliers

case; inverted parameters are tabulated on the figure. Results confirm that the L_∞ norm is not sensitive to information density and renders the same solution in both cases.

In summary, high-order norms are most sensitive to outliers and least sensitive to uneven information density. (A persuasive example of the effect of uneven spatial distribution and the benefits of the L_∞ norm is presented in Chapter 11 in the context of tomographic imaging.)

8.6 DATA AND MODEL ERRORS

Errors affect the invertibility of unknown parameters in the inverse problem $\underline{y} = \underline{\underline{h}} \cdot \underline{x}$. There are two main sources of errors. First, *data error* renders the measurements $\underline{y}$ noisy. Second, there is *model error* if the model assumed to compute the matrix $\underline{\underline{h}}$ does not properly reflect the phenomenon being studied; for instance, a linear elastic model is used to analyze a material with nonlinear elastic behavior, or straight rays are used to analyze sound propagation data in heterogeneous media. Data and model errors combine, and are often magnified while inverting the equation $\underline{y} = \underline{\underline{h}} \cdot \underline{x}$.

Implications are explored in the context of least-squares fitting a straight line $y = a + b \cdot t$ to gathered data. Three cases are shown in Figure 8.15: noiseless data along a straight line, noisy data aligned with a straight line, and noiseless data along a nonlinear trend. The error surfaces for the L_2 norm are computed in each case. Once the minimum is identified, the L_2 norm is computed near optimum by perturbing one parameter at the time; these are the 2D cross-sections of the 3D error surface obtained across the point of minimum error (see also Figure 8.9b).

It can be observed that in the absence of model or data errors, data are perfectly fitted with the model and the minimum of the error surface is zero (Figure 8.15a). On the other hand, data or model error widens the error surface, the minimum is above zero, and the curvature around optimum decreases, thus diminishing the ability to resolve the optimal values (Figures 8.15b and c). Furthermore, parameters estimated with an improper model mask real features in the data and bias the interpretation of measurements.

Therefore, the error surface provides information to guide the search for the optimum set of parameters that minimizes the error, and provides an indication of error severity. However, identifying the source of error remains the analyst's task. This forensic exercise requires in-depth understanding of the underlying physical process and detailed knowledge of the measurement procedure.

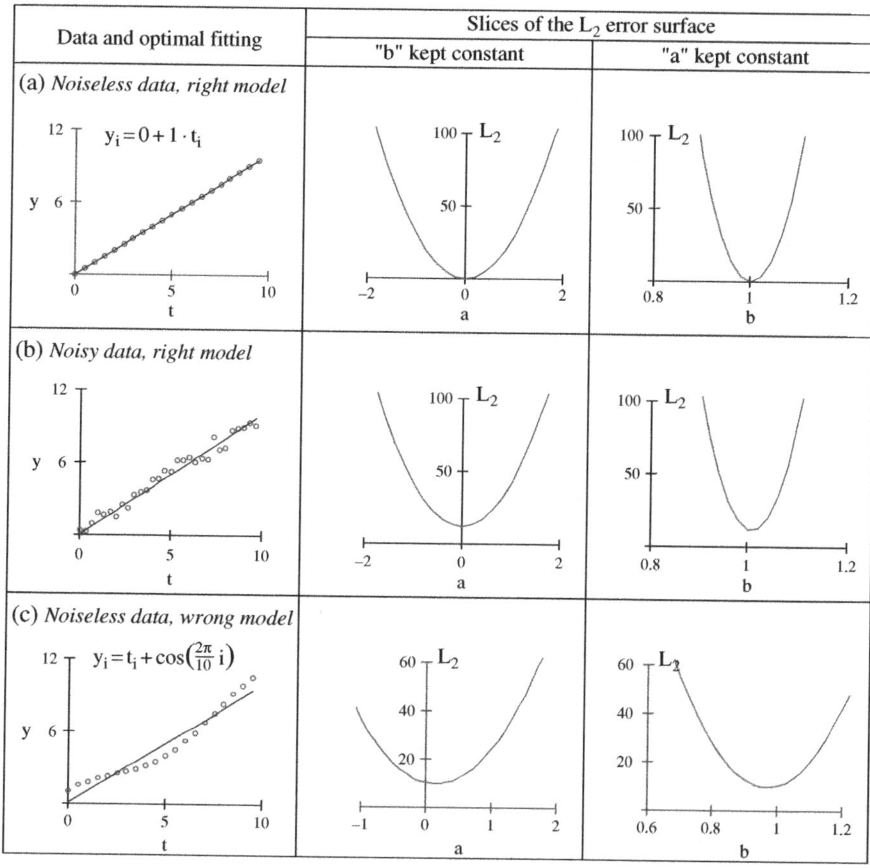

Figure 8.15 Data and model error: (a and b) noise in the data increases the magnitude of the residual at optimum. The minimum in the error function does not necessarily occur at the true values of a and b; (c) If the presumed model is not in agreement with the underlying physical process (within the range of the data), the norm of the residual will not be zero, even for noiseless data

8.7 NONCONVEX ERROR SURFACES

Error surfaces found in previous examples are convex, as in Figures 8.9, 8.11, or 8.15. However, this is not necessarily the case even in simple examples. Figure 8.16 shows a series of data points along a straight line. Data are fitted with a function $y = x \cdot \tan(\alpha)$. The L_2 error norm is computed for angle values between $0° \le \alpha \le 180°$. The error surface – a line in this case – is nonconvex.

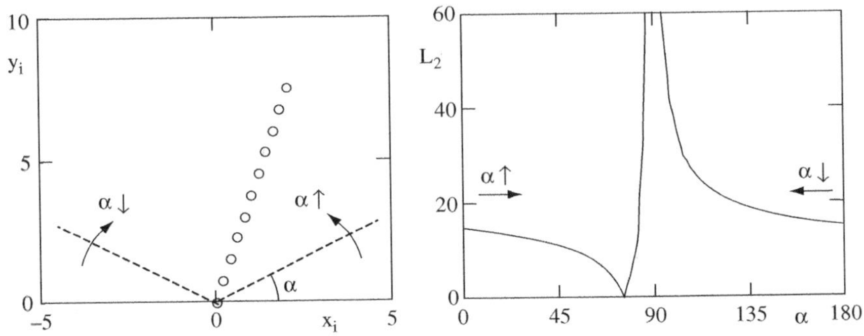

Figure 8.16 Nonconvex objective function. Data points in frame (a) are fitted with a straight line. The inclination of the line α is the only unknown. It can be searched clockwise "α ↓" starting at $\alpha = 180°$, or counterclockwise "α ↑" starting at $\alpha = 0°$. The L_2 norm is plotted in frame (b). Search criteria guided by the gradient in the error surface would not find the minimum when the search starts at $\alpha > 90°$. If the search is extended between 0° and 360°, it would identify either 75° or its symmetric 255°

The minimum in a convex error surface is effectively searched following the gradient. However, gradient-based search algorithms may get trapped in local minima or deviate away from the minimum. This occurs in Figure 8.16 when the search starts anywhere in $90° < \alpha \leq 180°$. Therefore, proper search algorithms must be used when the error surface is suspected to be nonconvex (Chapter 10).

Uneven information density tends to cause nonconvex error functions, particularly when L_1 or L_2 error norms are used (see Chapter 11). But the information density is perfectly even in Figure 8.16: in this example, nonconvexity is caused by the selected error definition $e_i = y_i^{<meas>} - y_i^{<pred>}$: all errors e_i tend to infinity when α approaches 90° and 270°.

8.8 DISCUSSION ON INVERSE PROBLEMS

The goal of inverse problem solving is to obtain physically meaningful values of the unknown parameters $\underline{x}$ *from measured quantities* $\underline{y}^{<meas>}$ *by assuming a proper model or relationship between the two.*

A transformation $x \rightarrow y$ is said to be invertible if there exists another transformation that permits the recovery of x from y, $y \rightarrow x$. Mathematical procedures with inverse operations include multiplication and division, integration and differentiation, DFT and IDFT, among others. But, which of the two operations is the inverse problem in each case? It is difficult to define "inverse problems" a priori, yet we would most likely recognize one when we see it! Inverse problems share one or more of the following characteristics:

- The level of difficulty involved in inverse problem solving is higher than in forward problem solving; in fact, the forward solution may be explicitly used to solve the inverse problem.
- There is no certainty about the physical model that is selected to relate the measurements $\underline{y}^{<meas>}$ to the unknowns $\underline{x}$.
- The available information is effectively much less than the gathered data.
- Data errors are amplified in the solution.
- The solution is ill-posed even if the forward problem is well-posed, where well-posed means that there exists a unique and stable solution that depends continuously on the input.
- The solution is not unique and more than one set of unknown parameters $\underline{x}^{<est>}$ justify the available observations $\underline{y}^{meas}$.
- Additional information is needed to solve the inverse problem; otherwise, the problem must be cast with fewer number of unknowns.
- Complete time history data may be needed to solve the inverse problem, whereas the forward problem is a function of the current state only. (Recall the vase problem in Figure 8.1 and compare the forward computation of the instantaneous seepage velocity versus the inverse computation of the vase shape.)
- Computational demands are higher than for the corresponding forward problem.

8.9 SUMMARY

- Vectors and matrices are the natural data structure to cast forward and inverse problems that operate on discrete data values. The resulting formulation is versatile and facilitates the analysis and diagnosis of inverse problems.
- The goal of inverse problem solving is to determine the value of unmeasured quantities (the unknown parameters) from measured quantities (experimental data).
- A model must be assumed to relate the two. Favor simple models.
- The solution of inverse problems is guided in part by the error between the measured data and model predictions.
- Select an error definition that weights all measurements alike.

- Three salient error norms are identified. The L_1 norm is least sensitive to outliers and supports robust inversion; however, it is most sensitive to uneven information density. The L_2 norm is compatible with additive Gaussian noise in the data and it leads to close-form least squares solutions. The L_∞ norm is most sensitive to outliers but least sensitive to uneven information density; it leads to min-max solution strategies.
- The solution of inverse problems typically faces: a noninvertible transformation matrix, insufficient independent data relative to the number of unknowns, noise in the data, difficulties in selecting a proper theory or a nonlinear model, a nonconvex error surface, uneven distribution of information density, and high computational demands.
- The following are preliminary recommendations for inverse problem solving: (1) plan the experiment carefully to gain evenly distributed information, (2) gather high-quality data, and (3) stay in touch with the *physical reality* of the problem.
- *The solution of the inverse problem must be physically meaningful and adequately justify the data, given an acceptable model.*

FURTHER READING AND REFERENCES

Cho, Z.-H., Jones, J. P., and Singh, M. (1993). Foundation of Medical Imaging. John Wiley & Sons, New York. 586 pages.

Enting, I. G. (2002). Inverse Problems in Atmospheric Constituent Transport. Cambridge University Press, Cambridge.

Groetsch, C. W. (1993). Inverse Problems in the Mathematical Science. Vieweg & Sohn, Braunschweig, Germany. 152 pages.

Groetsch, C. W. (1999). Inverse Problems, Activities for Undergraduates. The Mathematical Society of America. 222 pages.

Gubbins, D. (2004). Time Series Analysis and Inverse Theory for Geophysicists. Cambridge University Press, Cambridge. 255 pages.

Jefferys W. H. and Berger, J. O. (1992). Ockham's Razor and Bayesian Analysis. American Scientist. Vol. 80, No. 1, pp. 64–72.

Plaskowski, A., Beck, M. S., Thorn, R., and Dyamakowski, T. (1995). Imaging Industrial Flow – Applications of Electrical Process Tomography. Institute of Physics Publishing, London. 214 pages.

Rice, R. B. (1962). Inverse Convolution Filters. Geophysics. Vol. 27, pp. 4–18.

SOLVED PROBLEMS

P8.1 *Problem linearization.* Given the function $z = x + 2 \cdot y^2 + 3 \cdot x^2$, estimate z at $x = 1.2$ and $y = 7.2$ using the first-order Taylor expansion around $x_0 = 1.0$ and $y_0 = 7.0$.

Solution: The partial derivates of $z = x + 2 \cdot x \cdot y^2 + 3 \cdot x^2$ with respect to x and y are: $\partial z/\partial x = 1 + 2 \cdot y^2 + 6 \cdot x$ and $\partial z/\partial y = 4 \cdot x \cdot y$. Then, the value of z at $(x, y) = (1.2, 7.2)$ is estimated as:

$$z(x, y) \simeq z(x_0, y_0) + (x - x_0) \cdot \left.\frac{\partial z}{\partial x}\right|_{x_0, y_0} + (y - y_0) \cdot \left.\frac{\partial z}{\partial y}\right|_{x_0, y_0}$$
$$\simeq 102 + (7.2 - 7.0) \cdot 102 + (1.2 - 1.0) \cdot 28 = 128$$

The true value is $z(1.2, 7.2) = 129.2$. Therefore the error is $\sim 1\%$.

P8.2 *Regression analysis.* Data that follow a power trend $y = \alpha \cdot x^\beta$ are fitted with a straight line in *log-log scale*. Show that the inverted parameters are different from those obtained by fitting $y = \alpha \cdot x^\beta$ in linear scale. Write the objective function that is minimized in each case.

Solution: Data are simulated as $y_i = \alpha \cdot (x_i)^\beta + \text{rnd}$ for $\alpha = 100, \beta = 0.25$, where rnd is a uniform random number generator between -10 and $+10$. The objective functions are:

$$\text{log-log} \quad Err(\alpha, \beta) = \sum_{i=0}^{N-1} \{\log(y_i) - [\log(\alpha) + \beta \cdot \log(x_i)]\}^2$$

$$\text{linear} \quad Err(\alpha, \beta) = \sum_{i=0}^{N-1} [y_i - \alpha \cdot (x_i)^\beta]^2$$

Slices of the error surfaces taken across the minimum are shown next for the fitting in log scale:

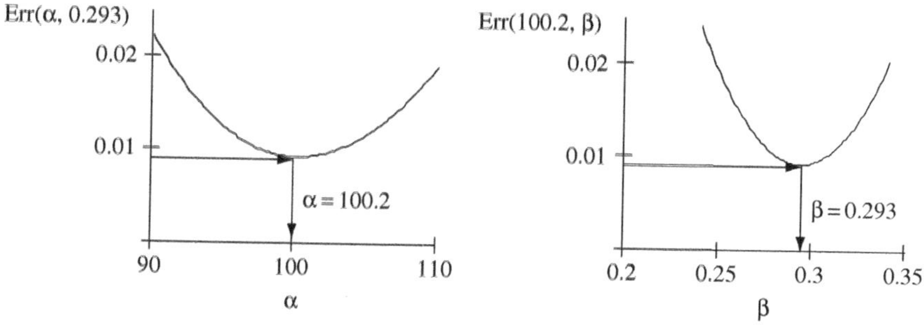

Slices of the error surfaces taken across the minimum are shown next for the fitting in linear scale:

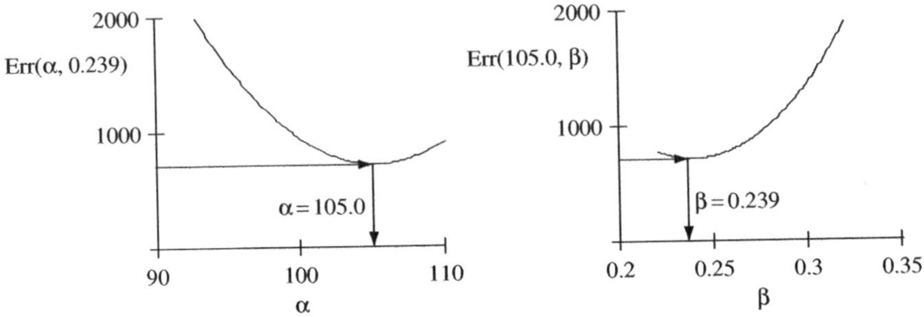

Data and fitted trends follow. Notice that errors in small values gain greater relative weight in log scale:

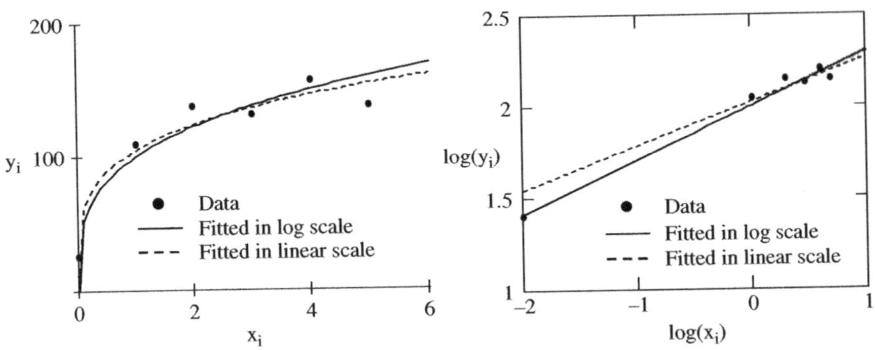

ADDITIONAL PROBLEMS

P8.3 *What is an inverse problem?* Write a short essay to explain what an inverse problem is to a colleague in your own discipline. Identify the main challenges. Highlight guiding principles that prevent pitfalls and facilitate identifying physically meaningful solutions. Provide persuasive examples from your field.

P8.4 *Discrete formulation*. Consider a point load on a beam. Write the Navier's equation for the elastic deformation of the beam in matrix form $\underline{y} = \underline{\underline{h}} \cdot \underline{x}$. Explore the invertibility of $\underline{\underline{h}}$ to infer the position and magnitude of the load from the measured deformation of the beam.

Table P8.1

x	y	z
0	0	9.9
0	5	10.6
1	1	11.2
1	4	11.2
2	2	12.0
3	3	13.3
4	1	14.0
4	4	14.5
5	0	15.2

Table P8.2

x	y
0	10.1
1	11.1
2	12.3
3	13.6
4	15.5
5	18.0
6	19.8
7	22.0
8	24.2

P8.5 *Different norms. Regression analysis.* Use the trial-and-error method to fit a plane $z = a + b \cdot x + c \cdot y$ to the data set given in Table P8.1. Compare the results obtained with the L_1, L_2 and L_∞ error norms. Plot slices of the error surfaces for the three unknowns (a, b, c). Repeat for the three norms. Discuss your results.

P8.6 *Favor simple models. Regression analysis.* Fit a polynominal by trial-and-error to the data presented in Table P8.2. Start by fitting a straight line and repeat the exercise for increasingly higher orders. In each case: plot the data set as points and the polynomial as a continuous line, and extrapolate the polynomial from $x = -5$ to $x = +15$. Plot the residual error versus the order of the polynomial. Discuss your results.

P8.7 *Tomography.* Generate the matrix of travel lengths for the set of sources and receivers and the pixel geometry shown below. (Note: there are 16 possible rays). Assume straight rays.

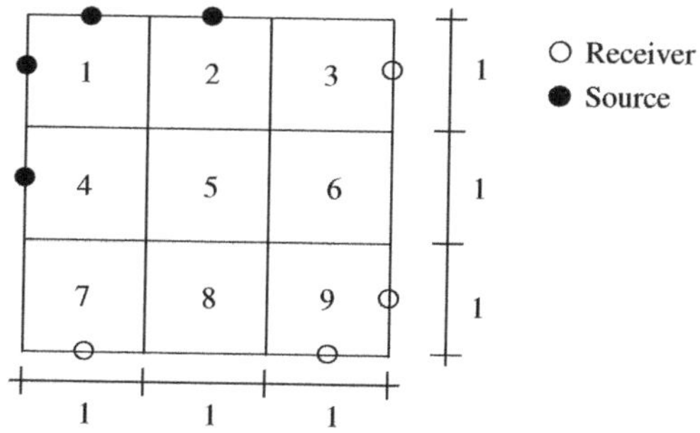

P8.8 *Deconvolution*. Given the output signal $\underline{y} = [0, 4, 2, -1, 0.5, -0.5, 0, 0]^T$ and the impulse response $\underline{h} = [2, -1, 0.5, -0.25, 0, 0, 0, 0]^T$, determine the input signal $\underline{x}$. (Hint: form the matrix $\underline{\underline{h}}$ – refer to Section 4.5.)

P8.9 *Application*. Consider problems in your field of interest. Identify the governing physical relations. Express these relations in discrete form $\underline{y} = \underline{\underline{h}} \cdot \underline{x}$. Rewrite the physical relation in terms of dimensionless π ratios. Explore invertibility.

9

Solution by Matrix Inversion

The inversion of forward problems $\underline{y} = \underline{\underline{h}} \cdot \underline{x}$ is explored in this chapter. The aim is to obtain a physically meaningful solution $\underline{x}^{<est>}$ that can adequately justify the measured data $\underline{y}^{<meas>}$ according to the assumed physical law or model, while taking into consideration all available information.

9.1 PSEUDOINVERSE

The forward solution *predicts* the outcome $\underline{y}^{<pred>}[M \times 1]$ as a function of the vector of known input values $\underline{x}^{<true>}[N \times 1]$ and the transformation matrix $\underline{\underline{h}}\,[M \times N]$, which represents the physical law that connects $\underline{x}$ to $\underline{y}$

$$\underline{y}^{\langle pred \rangle} = \underline{\underline{h}} \cdot \underline{x}^{\langle true \rangle} \qquad \textit{forward problem} \tag{9.1}$$

In the inverse problem, the M-values y_i are *measured*, and the aim is to *estimate* the N-unknown parameters $\underline{x}^{<est>}$. In general, $\underline{\underline{h}}$ is noninvertible and a pseudoinverse $\underline{\underline{h}}^{-g}$ must be used instead:

$$\underline{x}^{\langle est \rangle} = \underline{\underline{h}}^{-g} \cdot \underline{y}^{\langle meas \rangle} \qquad \textit{inverse problem} \tag{9.2}$$

The pseudoinverse $\underline{\underline{h}}^{-g}$ is not the "normal" inverse of the matrix $\underline{\underline{h}}$, and the products $\underline{\underline{h}} \cdot \underline{\underline{h}}^{-g}$ and $\underline{\underline{h}}^{-g} \cdot \underline{\underline{h}}$ are not necessarily equal to the identity matrix $\underline{\underline{I}}$. Let us explore the implications of this observation. The values of $\underline{y}^{<just>}$ that would be justified if the input parameters were $\underline{x}^{<est>}$ are

$$\underline{y}^{\langle just \rangle} = \underline{\underline{h}} \cdot \underline{x}^{\langle est \rangle} \tag{9.3}$$

Discrete Signals and Inverse Problems J. C. Santamarina and D. Fratta

and, replacing $\underline{x}^{<est>}$ in terms of the vector of measured values $\underline{y}^{<meas>}$ (Equation 9.1),

$$\underline{y}^{<just>} = \underline{\underline{h}} \cdot \underline{\underline{h}}^{-g} \cdot \underline{y}^{<meas>} \tag{9.4}$$

The matrix $\underline{\underline{D}} = \underline{\underline{h}} \cdot \underline{\underline{h}}^{-g} [M \times M]$ is called the *data resolution matrix*. The trace of $\underline{\underline{D}}$ is the sum of its diagonal elements and it is an indicator of the number of unknowns that can be resolved. The length of the vector of residuals

$$|\underline{e}| = (\underline{y}^{<meas>} - \underline{\underline{h}} \cdot \underline{x}^{<est>})^T \cdot (\underline{y}^{<meas>} - \underline{\underline{h}} \cdot \underline{x}^{<est>})$$

is zero when $\underline{\underline{D}} \equiv \underline{\underline{I}}$ and increases as $\underline{\underline{D}}$ deviates from the identity matrix $\underline{\underline{I}}$.

On the other hand, the measured values $\underline{y}^{<meas>}$ were gathered in a real event; thus, $\underline{y}^{<meas>} = \underline{\underline{h}} \cdot \underline{x}^{<true>}$ and Equation 9.2 becomes

$$\underline{x}^{<est>} = \underline{\underline{h}}^{-g} \cdot \underline{\underline{h}} \cdot \underline{x}^{<true>} \tag{9.5}$$

The matrix $\underline{\underline{G}} = \underline{\underline{h}}^{-g} \cdot \underline{\underline{h}}\, [N \times N]$ is called the *model resolution matrix*. Equation 9.5 indicates that the estimated i-th parameter $x_i^{<est>}$ is a linear combination of the true parameters $\underline{x}^{<true>}$, as prescribed by the elements in the i-th row of $\underline{\underline{G}}$. When $\underline{\underline{G}} \equiv \underline{\underline{I}}$, the estimated parameters $x_i^{<est>}$ are identical to the true parameters $x_i^{<true>}$.

9.2 CLASSIFICATION OF INVERSE PROBLEMS

Inverse problems can be diagnosed and classified by analyzing the available information in relation to the requested information and the characteristics data consistency.

9.2.1 Information: Rank Deficiency and Condition Number

The comparison between the number of measurements M and the number of unknowns N provides the first indication of the type of problem at hand. The problem is underdetermined if the number of unknowns N exceeds the number of equations M, that is $M > N$. The converse is not necessarily true: interrelated measurements do not contribute to the pool of available information (Section 8.5.1) and problems that appear even-determined $M = N$ or overdetermined $M > N$ may actually be underdetermined.

The rank r of a matrix is the number of linearly independent rows or columns (Section 2.2). Therefore, the rank of the transformation matrix $r[\underline{\underline{h}}]$ indicates that the number of independent measurements is $r[\underline{\underline{h}}] \leqslant \min(M, N)$. Yet, rank can be misleading. Consider the following two matrices:

$$\begin{bmatrix} 1 & 0 \\ 1 & 0 \end{bmatrix} \quad \text{and} \quad \begin{bmatrix} 1 & 0 \\ 1 & 10^{-6} \end{bmatrix}$$

The matrix on the right is rank $= 2$; yet, the second row is "almost" linearly dependent on the first row and it does not really contribute new information to the solution of the inverse problem.

Eigenvalue analysis or singular value decomposition (SVD) provide a better alternative to assess a matrix. The *condition number* κ is defined as the ratio between the eigenvalues or singular values λ with maximum and minimum absolute value:

$$\kappa = \frac{\max|\underline{\lambda}|}{\min|\underline{\lambda}|} \qquad \textit{condition number} \qquad (9.6)$$

The condition number applies only to square matrices. This is not a limitation because the computation of the pseudoinverse involves either $\underline{\underline{h}}^T \cdot \underline{\underline{h}}$ or $\underline{\underline{h}} \cdot \underline{\underline{h}}^T$ (later in this chapter). If the matrix is positive definite, all singular values are positive, and the bars for absolute value can be removed. The condition number properly captures the transition from invertible to noninvertible matrices:

	Invertible	*Ill-conditioned*	*Noninvertible*
Matrix:	$\begin{bmatrix} 1 & 0 \\ 1 & 1 \end{bmatrix}$	$\begin{bmatrix} 1 & 0 \\ 1 & 10^{-6} \end{bmatrix}$	$\begin{bmatrix} 1 & 0 \\ 1 & 0 \end{bmatrix}$
Rank:	2	2	1
Condition number:	2.6	2 000 000	∞

A matrix is noninvertible when $\kappa = \infty$. On the other hand, a matrix is *ill-conditioned* when κ is very large; in this case, numerical inaccuracies become important to the solution, and errors in the data are magnified during inversion. The magnification of numerical noise in computer algorithms with double precision takes place when $\kappa \to 10^{12}$. However, the condition number required to prevent data noise magnification can be significantly lower and it is related to the noise level in the data. (The procedure to identify the optimal condition number is outlined in Section 9.6 – for an example see Chapter 11.) The number of singular values between $\max|\underline{\lambda}|$ and the minimum acceptable singular value is a measure of the amount of available information.

Singular values are computed for the $\underline{\underline{h}}^T \cdot \underline{\underline{h}}$ square matrices for the two cases shown in Figure 9.1. Figure 9.1a presents the singular values for the regression analysis matrix $\underline{\underline{h}}$ developed to fit a fifth-order polynomial to 11 data points. There are $M = 11$ measurements but only three "meaningful" singular values if

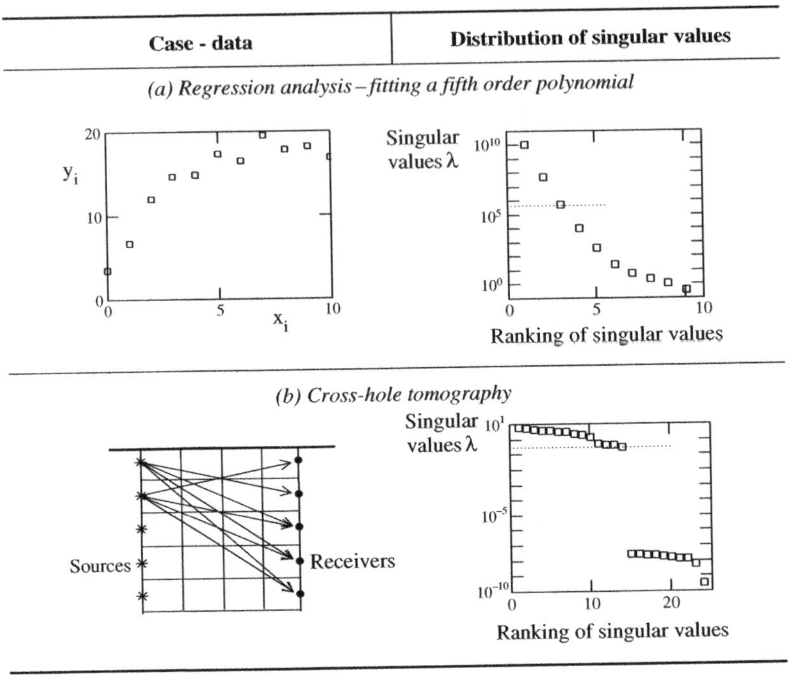

Figure 9.1 Singular values. The number p of "meaningful" singular values is not always trivial, even when a clear break is found

the condition number is limited to $\kappa = 2 \times 10^4$. Figure 9.1b shows the singular values for the matrix of travel times in cross-hole tomography. There are $M = 25$ travel time measurements but only ~ 14 meaningful singular values when the condition number is limited to $\kappa = 5 \times 10^2$.

9.2.2 Errors - Consistency

Rank and condition number permit the assessment of the transformation matrix $\underline{\underline{h}}$ even before data are acquired. Once data become available, the system of equations can be tested for consistency. The system of equations is "consistent" if there is a solution $\underline{x}$ that satisfies $\underline{y} = \underline{\underline{h}} \cdot \underline{x}$. Therefore, the rank is

$$r\left[\underline{\underline{h}}\right] = r\left[\underline{\underline{h|y}}\right] \tag{9.7}$$

where the expanded or augmented matrix $\underline{\underline{h}}|\underline{y}$ is formed by adding the vector of measurements $\underline{y}^{<meas>}$ as the N+1 column of $\underline{\underline{h}}$.

Data or model errors make the system of equations inconsistent. Therefore, there is no solution $\underline{x}$ that can satisfy all the data $\underline{y}$, and the vector of residuals is not the null vector,

$$\underline{e} = \underline{y}^{\langle meas \rangle} - \underline{\underline{h}} \cdot \underline{x}^{\langle est \rangle} \neq \underline{0} \tag{9.8}$$

9.2.3 Problem Classification

The amount of information that is available (for an acceptable condition number) relative to the number of unknowns is used to classify inverse problems into: *underdetermined, even-determined,* and *overdetermined.*

Very often the amount of available information is not the same for the different unknowns and the inverse problem is *mixed-determined.* Consider the case of atmospheric data: it is relatively easy to gather information near the earth's surface; however, data become gradually sparser at higher elevations, so that lower atmospheric layers are overdetermined whereas remote layers remain underdetermined. Mixed-determined inverse problems are frequently encountered and they tend to cause uneven data and model error magnification onto the inverted parameters.

Inverse problems using real data are *inherently inconsistent*; that is, $r[\underline{\underline{h}}] < r[\underline{\underline{h}}|\underline{y}]$. Therefore, no solution $\underline{x}$ can satisfy all equations when there are more equations than unknowns $M > N$. In this case, the inverse problem is solved by identifying a compromise solution $\underline{x}^{<est>}$ that minimizes a preselected error norm such as the L_2 norm. This leads to the family of least-squares solutions.

9.3 LEAST SQUARES SOLUTION (LSS)

The least squares solution (LSS) is the set of values $\underline{x}$ that minimizes the L_2 norm or min $(\underline{e}^T \cdot \underline{e})$, where the square root is herein omitted for simplicity (Section 8.3.2). Individual errors $e_i = y_i^{<meas>} - y_i^{<just>}$ form the vector of residuals $\underline{e} = \underline{y}^{<meas>} - \underline{\underline{h}} \cdot \underline{x}$. Therefore, the objective function Γ to be minimized becomes

$$\begin{aligned} \Gamma &= \underline{e}^T \cdot \underline{e} \\ &= \left(\underline{y}^{\langle meas \rangle} - \underline{\underline{h}} \cdot \underline{x}\right)^T \cdot \left(\underline{y}^{\langle meas \rangle} - \underline{\underline{h}} \cdot \underline{x}\right) \\ &= \underline{y}^{\langle meas \rangle^T} \cdot \underline{y}^{\langle meas \rangle} - \underline{y}^{\langle meas \rangle^T} \cdot \underline{\underline{h}} \cdot \underline{x} - \underline{x}^T \cdot \underline{\underline{h}}^T \cdot \underline{y}^{\langle meas \rangle} + \underline{x}^T \cdot \underline{\underline{h}}^T \cdot \underline{\underline{h}} \cdot \underline{x} \end{aligned} \tag{9.9}$$

The least squares solution corresponds to the minimum point of the error surface and it is found by setting the derivative of the objective function Γ with respect to $\underline{x}$ equal to zero. The derivative of Γ is computed using the rules derived in Section 2.3:

$$\frac{\partial \Gamma}{\partial \underline{x}} = \underline{0} = \underline{0} - \underline{\underline{h}}^T \cdot \underline{y}^{\langle meas \rangle} - \underline{\underline{h}}^T \cdot \underline{y}^{\langle meas \rangle} + 2 \cdot \underline{\underline{h}}^T \cdot \underline{\underline{h}} \cdot \underline{x} \tag{9.10}$$

If $\underline{\underline{h}}^T \cdot \underline{\underline{h}}$ is invertible, the solution for $\underline{x}$ returns the sought estimate $\underline{x}^{<est>}$:

$$\text{LSS} \qquad \underline{x}^{\langle est \rangle} = \left(\underline{\underline{h}}^T \cdot \underline{\underline{h}}\right)^{-1} \cdot \underline{\underline{h}}^T \cdot \underline{y}^{\langle meas \rangle} \tag{9.11}$$

This is the least squares solution.[1] The matrix $\underline{\underline{h}}^T \cdot \underline{\underline{h}}$ is square $[N \times N]$ and symmetric; it is invertible when $\underline{\underline{h}}\,[M \times N]$ has linearly independent columns so that $r\left[\underline{\underline{h}}\right] = N$.

The corresponding generalized inverse, data resolution matrix $\underline{\underline{D}} = \underline{\underline{h}} \cdot \underline{\underline{h}}^{-g}$, and model resolution matrix $\underline{\underline{G}} = \underline{\underline{h}}^{-g} \cdot \underline{\underline{h}}$ become (substitute the solution Equation 9.11 in the corresponding definitions):

$$\underline{\underline{h}}^{-g} = \left(\underline{\underline{h}}^T \cdot \underline{\underline{h}}\right)^{-1} \cdot \underline{\underline{h}}^T \tag{9.12}$$

$$\underline{\underline{D}} = \underline{\underline{h}} \cdot \left(\underline{\underline{h}}^T \cdot \underline{\underline{h}}\right)^{-1} \cdot \underline{\underline{h}}^T \tag{9.13}$$

$$\underline{\underline{G}} = \left(\underline{\underline{h}}^T \cdot \underline{\underline{h}}\right)^{-1} \cdot \left(\underline{\underline{h}}^T \cdot \underline{\underline{h}}\right) = \underline{\underline{I}} \tag{9.14}$$

---

[1] An alternative demonstration is presented to gain further insight into the solution. The goal is to determine $\underline{x}$ so that the justified values $\underline{y}^{<just>} = \underline{\underline{h}} \cdot \underline{x}$ (in the range of the transformation) lie closest to the set of measurements $\underline{y}^{<meas>}$ (which cannot be reached by the transformation). This will be the case when $\underline{y}^{<just>}$ is the "projection" of $\underline{y}^{<meas>}$ onto the range of the transformation. The vector normal to the space of y that executes the projection is $(\underline{y}^{<meas>} - \underline{y}^{<just>})$, and by definition of normality, its dot product with $\underline{y}^{<just>}$ must be equal to zero. Mathematically,

$$0 = \underline{y}^{\langle just \rangle T} \cdot \left(\underline{y}^{\langle meas \rangle} - \underline{y}^{\langle just \rangle}\right)$$

The following sequence of algebraic manipulations leads to the solution:

$$0 = \left(\underline{\underline{h}} \cdot \underline{x}\right)^T \cdot \left(\underline{y}^{\langle meas \rangle} - \underline{\underline{h}} \cdot \underline{x}\right)$$

$$0 = \underline{x}^T \cdot \underline{\underline{h}}^T \cdot \underline{y}^{\langle meas \rangle} - \underline{x}^T \cdot \underline{\underline{h}}^T \cdot \underline{\underline{h}} \cdot \underline{x}$$

$$0 = \underline{\underline{h}}^T \cdot \underline{y}^{\langle meas \rangle} - \underline{\underline{h}}^T \cdot \underline{\underline{h}} \cdot \underline{x}$$

Finally, if $\underline{\underline{h}}^T \cdot \underline{\underline{h}}$ is nonsingular, the sought estimate is $\underline{x}^{\langle est \rangle} = \left(\underline{\underline{h}}^T \cdot \underline{\underline{h}}\right)^{-1} \cdot \underline{\underline{h}}^T \cdot \underline{y}^{\langle meas \rangle}$.

As the model resolution matrix is the identity matrix, the LSS resolves $\underline{x}$, but it does not resolve the data $\underline{y}$ (Note: $\underline{\underline{D}} \equiv \underline{\underline{I}}$ when the problem is even-determined). A solved example is presented at the end of the chapter.

9.4 REGULARIZED LEAST SQUARES SOLUTION (RLSS)

The LSS applies to overdetermined problems. However, many inverse problems are mixed-determined. In this case, the inversion – if at all possible – would unevenly magnify noise in the solution, particularly on values of x_k that are least constrained due to limited information. This can be prevented by enforcing known properties on the solution $\underline{x}$.

It is possible to include available *a priori information about the solution* $\underline{x}$ during the inversion stage. This information is captured in the "regularization matrix" $\underline{\underline{R}}$ and it is added as a second criterion to be minimized in the objective function. Therefore, the objective function for the regularized least squares solution (RLSS) includes: (1) the length of the vector of residual $\underline{e}^T \cdot \underline{e}$, where $\underline{e} = \underline{y}^{<meas>} - \underline{y}^{<just>}$, and (2) the length of the regularizing criterion applied to the solution $[(\underline{\underline{R}} \cdot \underline{x})^T \cdot (\underline{\underline{R}} \cdot \underline{x})]$:

$$\begin{aligned}\Gamma &= \left(\underline{y}^{\langle meas\rangle} - \underline{\underline{h}} \cdot \underline{x}\right)^T \cdot \left(\underline{y}^{\langle meas\rangle} - \underline{\underline{h}} \cdot \underline{x}\right) + \lambda \cdot \left[\left(\underline{\underline{R}} \cdot \underline{x}\right)^T \cdot \underline{\underline{R}} \cdot \underline{x}\right] \\ &= \underline{y}^{\langle meas\rangle^T} \cdot \underline{y}^{\langle meas\rangle} - \underline{y}^{\langle meas\rangle^T} \cdot \underline{\underline{h}} \cdot \underline{x} - \underline{x}^T \cdot \underline{\underline{h}}^T \cdot \underline{y}^{\langle meas\rangle} \\ &\quad + \underline{x}^T \cdot \underline{\underline{h}}^T \cdot \underline{\underline{h}} \cdot \underline{x} + \lambda \cdot \left[\underline{x}^T \cdot \underline{\underline{R}}^T \cdot \underline{\underline{R}} \cdot \underline{x}\right] \end{aligned} \tag{9.15}$$

where λ is the nonnegative *regularization coefficient* that controls the weighted minimization of the two functionals in the objective function. (Note: assuming that R is dimensionless, the units of λ are $[\lambda] = [y^2/x^2]$.) The partial derivative of the objective function with respect to $\underline{x}$ is set equal to zero

$$\frac{\partial \Gamma}{\partial \underline{x}} = \underline{0} = \underline{0} - 2 \cdot \underline{\underline{h}}^T \cdot \underline{y}^{\langle meas\rangle} - 2 \cdot \underline{\underline{h}}^T \cdot \underline{\underline{h}} \cdot \underline{x} + 2 \cdot \lambda \cdot \underline{\underline{R}}^T \cdot \underline{\underline{R}} \cdot \underline{x} \tag{9.16}$$

The estimate of $\underline{x}$ is obtained assuming that $(\underline{\underline{h}}^T \cdot \underline{\underline{h}} + \lambda \cdot \underline{\underline{R}}^T \cdot \underline{\underline{R}})$ is invertible, and results in

$$\text{RLSS} \qquad \underline{x}^{\langle est\rangle} = \left(\underline{\underline{h}}^T \cdot \underline{\underline{h}} + \lambda \cdot \underline{\underline{R}}^T \cdot \underline{\underline{R}}\right)^{-1} \cdot \underline{\underline{h}}^T \cdot \underline{y}^{\langle meas\rangle} \tag{9.17}$$

A symmetric, positive-definite matrix is invertible (Chapter 2). Therefore, the effect of regularization is to guarantee the invertibility of $(\underline{\underline{h}}^T \cdot \underline{\underline{h}} + \lambda \cdot \underline{\underline{R}}^T \cdot \underline{\underline{R}})$ by correcting the ill-conditioning of $\underline{\underline{h}}^T \cdot \underline{\underline{h}}$ and delivering a stable solution. The corresponding generalized inverse, data resolution matrix $\underline{\underline{D}} = \underline{\underline{h}} \cdot \underline{\underline{h}}^{-g}$, and model resolution matrix $\underline{\underline{G}} = \underline{\underline{h}}^{-g} \cdot \underline{\underline{h}}$ are

$$\underline{\underline{h}}^{-g} = \left(\underline{\underline{h}}^T \cdot \underline{\underline{h}} + \lambda \cdot \underline{\underline{R}}^T \cdot \underline{\underline{R}}\right)^{-1} \cdot \underline{\underline{h}}^T \quad (9.18)$$

$$\underline{\underline{D}} = \underline{\underline{h}} \cdot \left(\underline{\underline{h}}^T \cdot \underline{\underline{h}} + \lambda \cdot \underline{\underline{R}}^T \cdot \underline{\underline{R}}\right)^{-1} \cdot \underline{\underline{h}}^T \quad (9.19)$$

$$\underline{\underline{G}} = \left(\underline{\underline{h}}^T \cdot \underline{\underline{h}} + \lambda \cdot \underline{\underline{R}}^T \cdot \underline{\underline{R}}\right)^{-1} \cdot \underline{\underline{h}}^T \cdot \underline{\underline{h}} \quad (9.20)$$

The versatile RLSS solution can provide adequate estimates even in the presence of data and model errors. The approach is also known as the Phillips–Twomey method, ridge regression, or Tikhonov–Miller regularization.

9.4.1 Special Cases

The LSS is obtained from the RLSS when the regularization coefficient is set to zero, $\lambda = 0$.

If the identity matrix is selected as the regularization matrix $\underline{\underline{R}}$, the solution is known as the *damped least squares solution* (DLSS) and Equation 9.16 becomes

$$\underline{x}^{\langle est \rangle} = \left(\underline{\underline{h}}^T \cdot \underline{\underline{h}} + \eta^2 \cdot \underline{\underline{I}}\right)^{-1} \cdot \underline{\underline{h}}^T \cdot \underline{y}^{\langle meas \rangle} \quad \textit{damped least squares solution} \quad (9.21)$$

A solved problem is presented at the end of this chapter. The effect of damping η^2 in promoting a positive-definite invertible matrix is readily seen in this solution where the main diagonal of $\underline{\underline{h}}^T \cdot \underline{\underline{h}}$ is increased by η^2. Note that (1) the value η^2 is always positive; (2) typically, a matrix is positive-definite when the elements along the main diagonal are positive and large when compared to other entries in the matrix; and (3) a positive-definite symmetric matrix is invertible – Chapter 2.

9.4.2 The Regularization Matrix

The matrix $\underline{\underline{R}}$ often results from the finite difference approximation to one of the following criteria:

- If a priori knowledge of $\underline{x}$ indicates that values should be constant, then the first derivative is minimized.

- If the local variation of $\underline{x}$ can be approximated with a straight line, then the second derivative is minimized. This is the Laplacian regularizer.
- If the values of $\underline{x}$ are expected to follow a second-degree polynomial, then the third derivative should be minimized.

This reasoning can be extended to higher-order variations. Table 9.1 presents examples of regularization matrices $\underline{\underline{R}}$ for applications in one and two dimensions.

Notice that the matrix $\underline{\underline{R}}$ is constructed with a criterion that is *minimized* in Equation 9.15. Therefore, if a priori information suggests that the solution $\underline{x}$ is "smooth", then the criterion to be minimized is "variability", which is often computed with the second derivative, as shown in Table 9.1.

The regularization matrix can also reflect physical principles that govern the phenomenon under consideration, such as heat conduction, chemical diffusion,

Table 9.1 Guidelines to construct the regularization matrix $\underline{\underline{R}}$

Expected variation of parameter x	Criterion	Kernel	A row in the regularization matrix
$x = \text{constant}$ (1D system)	$\min\left(\frac{dx}{dp}\right)$	$x_i - x_{i-1}$ Finite difference	$[0 \ldots 0\ \mathbf{1}\ {-1}\ 0 \ldots 0]$
$x = a + b \cdot p$ (1D system)	$\min\left(\frac{d^2x}{dp^2}\right)$	$x_{i+1} - 2 \cdot x_i + x_{i-1}$ Finite difference	$[0 \ldots 0\ 1\ \mathbf{-2}\ 1\ 0 \ldots 0]$
$x = a + b \cdot p + c \cdot p^2$ (1D system)	$\min\left(\frac{d^3x}{dp^3}\right)$	$x_{i+2} - 3 \cdot x_{i+1} + 3 \cdot x_i - x_{i-1}$ Finite difference	$[0 \ldots 0\ 1\ {-3}\ \mathbf{3}\ {-1}\ 0 \ldots 0]$
x linear in p and q (2D system)	$\min\left(\frac{d^2x}{dp^2} + \frac{d^2x}{dq^2}\right)$	0 1 0 1 −4 1 0 1 0	$[0 \ldots 1 \ldots 1\ \mathbf{-4}\ 1 \ldots 1 \ldots]$

Note: Entries not shown in the rows of $\underline{\underline{R}}$ are zeros. $R_{i,i}$ element shown in bold.

Systems: *1D*. Beam, layered cake model of the near surface.
2D. Cross-section of a body, topographic surfaces, digital images.

Boundaries: "Imaginary points" follow zero-gradient or constant gradient extrapolation

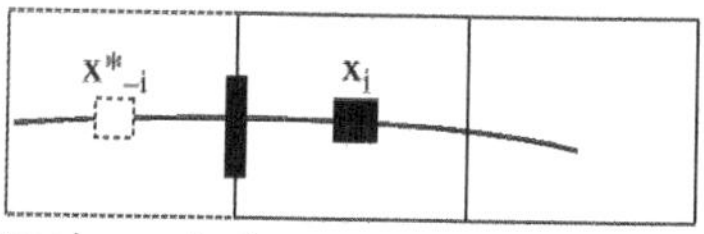

Imaginary pixel:
$x^*_{-i} = x_i$

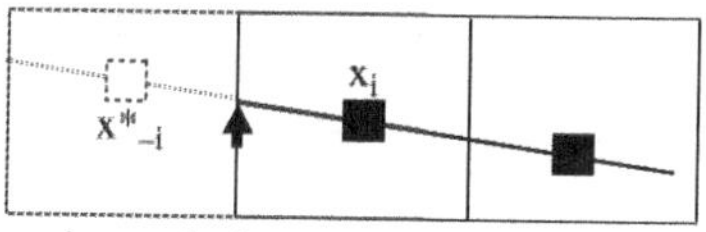

Imaginary pixel:
$x^*_{-i} = 2 \cdot x_i - x_{i+1}$

equilibrium and compatible deformations in a continuum. In this case the construction of the regularization matrix proceeds as follows:

1. Express the governing differential equation in finite differences to determine the kernel $\underline{\kappa}$.
2. The convolution of the kernel with the vector of unknowns $\underline{x}$ would result in a new vector $\underline{x}^*$ that would better satisfy the physical law. In matrix form, $\underline{x}^* = \underline{\underline{\kappa}} \cdot \underline{x}$ (Sections 4.4 and 4.5).
3. The goal of regularization is to minimize the distance $\underline{x}^* - \underline{x}$; hence, the regularization matrix is $\underline{\underline{R}} = \underline{\underline{\kappa}} - \underline{\underline{I}}$, where $\underline{\underline{I}}$ is the identity matrix.

The kernel that is used to compute rows in $\underline{\underline{R}}$ for elements x_i away from the boundary cannot be used for boundary elements. There are two possible alternatives:

- Finite difference approximation: develop asymmetric kernels using forward or backward approximations in the finite difference formulation of differential equations.
- Imaginary elements: create "imaginary" elements outside the boundary of $\underline{x}$. The value of these imaginary elements must be physically compatible with the physics of the problem. For example (see sketches in Table 9.1): (a) if there is zero-gradient or no flow across the boundary, the imaginary value x'_{-i} is symmetric and adopts the same value as the corresponding point inside the boundary, $x'_{-i} = x_i$; (b) if it is a pivoting boundary, the gradient is constant across the boundary and $x'_{-i} = 2x_{i+1} - x_i$, where x_k is the boundary value and x_i its immediate neighbor. Once imaginary elements are created, the regularization matrix is formed by running the kernel within the problem boundaries (see solved problem at the end of this Chapter).

The complete Laplacian smoothing regularization matrix for a 2D image of 6×6 pixels is shown in Figure 9.2. Symmetric imaginary pixels are assumed outside the boundary.

Regularization applies to unknown parameters $\underline{x}$ of the same kind (same units), such as the concentration of a contaminant in a 3D volume, pixel values in a 2D image, or temperature along a 1D pipeline.

9.4.3 The Regularization Coefficient λ

The optimal λ value depends on the characteristics of the problem under consideration, the quality of the data $\underline{y}$ and the adequacy of the assumed model.

-4	2	0	0	0	0	2	0	0	0	0	0	0	0	0	0	0	0	0	0	0	0	0	0	0	0	0	0	0	0	0	0	0	0	0	0
1	-4	1	0	0	0	0	2	0	0	0	0	0	0	0	0	0	0	0	0	0	0	0	0	0	0	0	0	0	0	0	0	0	0	0	0
0	1	-4	1	0	0	0	0	2	0	0	0	0	0	0	0	0	0	0	0	0	0	0	0	0	0	0	0	0	0	0	0	0	0	0	0
0	0	1	-4	1	0	0	0	0	2	0	0	0	0	0	0	0	0	0	0	0	0	0	0	0	0	0	0	0	0	0	0	0	0	0	0
0	0	0	1	-4	1	0	0	0	0	2	0	0	0	0	0	0	0	0	0	0	0	0	0	0	0	0	0	0	0	0	0	0	0	0	0
0	0	0	0	2	-4	0	0	0	0	0	2	0	0	0	0	0	0	0	0	0	0	0	0	0	0	0	0	0	0	0	0	0	0	0	0
1	0	0	0	0	0	-4	2	0	0	0	0	1	0	0	0	0	0	0	0	0	0	0	0	0	0	0	0	0	0	0	0	0	0	0	0
0	1	0	0	0	0	1	-4	1	0	0	0	0	1	0	0	0	0	0	0	0	0	0	0	0	0	0	0	0	0	0	0	0	0	0	0
0	0	1	0	0	0	0	1	-4	1	0	0	0	0	1	0	0	0	0	0	0	0	0	0	0	0	0	0	0	0	0	0	0	0	0	0
0	0	0	1	0	0	0	0	1	-4	1	0	0	0	0	1	0	0	0	0	0	0	0	0	0	0	0	0	0	0	0	0	0	0	0	0
0	0	0	0	1	0	0	0	0	1	-4	1	0	0	0	0	1	0	0	0	0	0	0	0	0	0	0	0	0	0	0	0	0	0	0	0
0	0	0	0	0	1	0	0	0	0	2	-4	0	0	0	0	0	1	0	0	0	0	0	0	0	0	0	0	0	0	0	0	0	0	0	0
0	0	0	0	0	0	1	0	0	0	0	0	-4	2	0	0	0	0	1	0	0	0	0	0	0	0	0	0	0	0	0	0	0	0	0	0
0	0	0	0	0	0	0	1	0	0	0	0	1	-4	1	0	0	0	0	1	0	0	0	0	0	0	0	0	0	0	0	0	0	0	0	0
0	0	0	0	0	0	0	0	1	0	0	0	0	1	-4	1	0	0	0	0	1	0	0	0	0	0	0	0	0	0	0	0	0	0	0	0
0	0	0	0	0	0	0	0	0	1	0	0	0	0	1	-4	1	0	0	0	0	1	0	0	0	0	0	0	0	0	0	0	0	0	0	0
0	0	0	0	0	0	0	0	0	0	1	0	0	0	0	1	-4	1	0	0	0	0	1	0	0	0	0	0	0	0	0	0	0	0	0	0
0	0	0	0	0	0	0	0	0	0	0	1	0	0	0	0	2	-4	0	0	0	0	0	1	0	0	0	0	0	0	0	0	0	0	0	0
0	0	0	0	0	0	0	0	0	0	0	0	1	0	0	0	0	0	-4	2	0	0	0	0	1	0	0	0	0	0	0	0	0	0	0	0
0	0	0	0	0	0	0	0	0	0	0	0	0	1	0	0	0	0	1	-4	1	0	0	0	0	1	0	0	0	0	0	0	0	0	0	0
0	0	0	0	0	0	0	0	0	0	0	0	0	0	1	0	0	0	0	1	-4	1	0	0	0	0	1	0	0	0	0	0	0	0	0	0
0	0	0	0	0	0	0	0	0	0	0	0	0	0	0	1	0	0	0	0	1	-4	1	0	0	0	0	1	0	0	0	0	0	0	0	0
0	0	0	0	0	0	0	0	0	0	0	0	0	0	0	0	1	0	0	0	0	1	-4	1	0	0	0	0	1	0	0	0	0	0	0	0
0	0	0	0	0	0	0	0	0	0	0	0	0	0	0	0	0	1	0	0	0	0	2	-4	0	0	0	0	0	1	0	0	0	0	0	0
0	0	0	0	0	0	0	0	0	0	0	0	0	0	0	0	0	0	1	0	0	0	0	0	-4	2	0	0	0	0	1	0	0	0	0	0
0	0	0	0	0	0	0	0	0	0	0	0	0	0	0	0	0	0	0	1	0	0	0	0	1	-4	1	0	0	0	0	1	0	0	0	0
0	0	0	0	0	0	0	0	0	0	0	0	0	0	0	0	0	0	0	0	1	0	0	0	0	1	-4	1	0	0	0	0	1	0	0	0
0	0	0	0	0	0	0	0	0	0	0	0	0	0	0	0	0	0	0	0	0	1	0	0	0	0	1	-4	1	0	0	0	0	1	0	0
0	0	0	0	0	0	0	0	0	0	0	0	0	0	0	0	0	0	0	0	0	0	1	0	0	0	0	1	-4	1	0	0	0	0	1	0
0	0	0	0	0	0	0	0	0	0	0	0	0	0	0	0	0	0	0	0	0	0	0	1	0	0	0	0	2	-4	0	0	0	0	0	1
0	0	0	0	0	0	0	0	0	0	0	0	0	0	0	0	0	0	0	0	0	0	0	0	2	0	0	0	0	0	-4	2	0	0	0	0
0	0	0	0	0	0	0	0	0	0	0	0	0	0	0	0	0	0	0	0	0	0	0	0	0	2	0	0	0	0	1	-4	1	0	0	0
0	0	0	0	0	0	0	0	0	0	0	0	0	0	0	0	0	0	0	0	0	0	0	0	0	0	2	0	0	0	0	1	-4	1	0	0
0	0	0	0	0	0	0	0	0	0	0	0	0	0	0	0	0	0	0	0	0	0	0	0	0	0	0	2	0	0	0	0	1	-4	1	0
0	0	0	0	0	0	0	0	0	0	0	0	0	0	0	0	0	0	0	0	0	0	0	0	0	0	0	0	2	0	0	0	0	1	-4	1
0	0	0	0	0	0	0	0	0	0	0	0	0	0	0	0	0	0	0	0	0	0	0	0	0	0	0	0	0	2	0	0	0	0	2	-4

Figure 9.2 Regularization matrix $\underline{\underline{R}}$ constructed to minimize the second derivative during the inversion of an image with 6×6 pixels. Each row in $\underline{\underline{R}}$ corresponds to a pixel in the image, which are numbered as shown on the right. Compare the rows in $\underline{\underline{R}}$ that correspond to pixels 29, 30, 35 and 36

1	2	3	4	5	6
7	8	9	10	11	12
13	14	15	16	17	18
19	20	21	22	23	24
25	26	27	28	29	30
31	32	33	34	35	36

Let us explore the evolution of the residual and the solution $\underline{x}^{<est>}$ for different values of λ. The following discussion relates to schematic trends shown in Figure 9.3.

Follow the Evolution of the Residual (Data Space)

Low regularization lets the solution $\underline{x}^{<est>}$ accommodate to the measured data $\underline{y}^{<meas>}$ and residuals $\underline{e} = [\underline{y}^{<meas>} - \underline{\underline{h}} \cdot \underline{x}^{<est>}]$ are small. Conversely, an increase in regularization coefficient λ constrains the solution and the residuals increase. Discard the range of λ where the solution stops justifying the data to an acceptable degree. To facilitate this decision, plot the length of the vector of residuals $\underline{e}^T \cdot \underline{e}$ versus λ (Figure 9.3b).

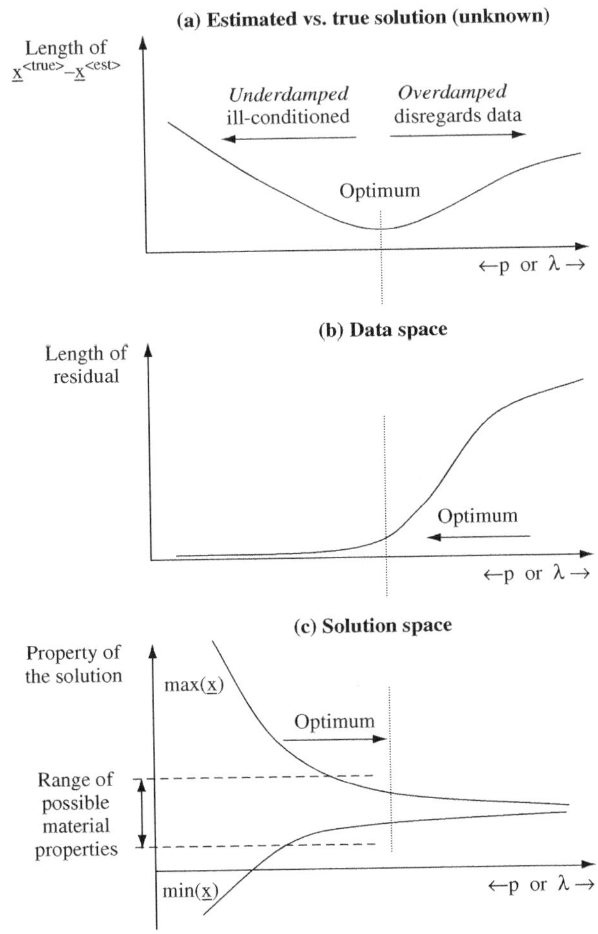

Figure 9.3 Selection of optimal inversion parameters λ (includes η^2) and p: (a) deviation of estimated solution $\underline{x}^{<est>}$ from the true $\underline{x}^{<true>}$. In real cases this plot is not known; (b) length of residuals; (c) property of the solution. The figure shows a plot of the minimum and maximum values of $\underline{x}$

Follow the Evolution of the Solution $\underline{x}^{<est>}$ (Solution Space)

The estimated solution $\underline{x}^{<est>}$ is very sensitive to data and model errors when λ is low and the problem is underregularized. However, the solution $\underline{x}^{<est>}$ is overregularized and fails to justify the data $\underline{y}^{<meas>}$ when the regularization coefficient

λ is too high; for example, an overregularized image becomes featureless when λ is high $\underline{\underline{R}}$ is constructed to minimize variability. Therefore, discard ranges of λ where the solution is physically unacceptable or uninformative. A robust approach is to plot salient characteristics of the solution versus the corresponding λ values. Consider parameters such as (Figure 9.3c):

- The characteristic to be minimized, which is captured in the regularization matrix, and summarized in the length $[(\underline{\underline{R}} \cdot \underline{x}^{<est>})^T \cdot (\underline{\underline{R}} \cdot \underline{x}^{<est>})]$.
- Extreme values $\min(\underline{x}^{<est>})$ and $\max(\underline{x}^{<est>})$. Superimpose the physically acceptable range for parameters x.
- Statistical summaries of the values in $\underline{x}$, such as mean, standard deviation, or coefficient of variation (standard deviation/mean).
- The relative magnitude of parameters (e.g. the smoothness of the spatial variation of $\underline{x}^{<est>}$ can be known a priori such as when $\underline{x}$ relates to surface elevation or light intensity).
- A prevailing trend in the solution $\underline{x}^{<est>}$ (e.g. ocean temperature decreases with depth).

Decision

This discussion highlights the *inherent trade-off between the stability of the solution (improved at high regularization) and the predictability of the data (best at low regularization)*. The selected level of regularization λ must (1) return a physically meaningful solution within the context of the engineering or science problem being studied and (2) adequately justify the data.

It is anticipated that higher levels of data or model error and uneven data coverage will require higher level of regularization λ. Details are provided in Implementation Procedure 9.1. The RLSS is demonstrated at the end of this chapter.

Implementation Procedure 9.1 Optimal regularization coefficient λ and optimal number of singular values p

1. Solve the inverse problem for different values of λ (RLSS) or p (SVD solution; section 9.6).
2. *Solution relevance*. For each solution $\underline{x}^{<est>}$, compute and plot the following parameters versus λ or p (see Figure 9.3):
   - the regularization criterion $[(\underline{\underline{R}} \cdot \underline{x}^{<est>})^T \cdot (\underline{\underline{R}} \cdot \underline{x}^{<est>})]$

- extreme values $\min(\underline{x}^{<est>})$ and $\max(\underline{x}^{<est>})$ and superimpose the physically acceptable range for parameters x
- $\text{mean}(\underline{x}^{<est>})$ and standard deviation of the solution $\underline{x}^{<est>}$
- plot the solution itself when appropriate, such as a sequence of tomograms

3. Discard the range of λ or p where the solution is physically unacceptable.
4. *Data justification*. Analyze the residuals $\underline{e} = [\underline{y}^{<meas>} - \underline{\underline{h}} \cdot \underline{x}^{<est>}]$ for each solution $\underline{x}^{<est>}$. Compute and plot the following parameters versus λ or p:
   - L_2 norm $= \underline{e}^T \cdot \underline{e}$
   - L_∞ norm $= \max(\underline{e})$
   - A 3D plot of residuals e_i vs. measurement counter i, against λ or p
5. Discard the range of λ or p where the solution stops justifying the data to an acceptable degree.
6. Plot of the trace of the data resolution matrix $\text{tr}(\underline{\underline{D}}) = \text{tr}(\underline{\underline{h}} \cdot \underline{\underline{h}}^{-g})$ versus λ or p.
7. Discard the range of λ or P that yields trace values much smaller than the number of equations M.
8. Select λ or p that returns a physically meaningful solution and adequately justifies the data.

9.5 INCORPORATING ADDITIONAL INFORMATION

Regularization permits the addition of known information about the solution $\underline{x}$. One may also have additional information about the data $\underline{y}$, the model, or an initial guess of the solution $\underline{x}_0$. The following sections explain how this information can be incorporated during inverse problem solving. In all cases, incorporating additional information should lead to a better conditioned inverse problem and more physically meaningful solutions.

9.5.1 Weighted Measurements

We often have *additional information about*:

- *the model that is assumed to construct the matrix* $\underline{\underline{h}}$. For example: a certain model is not adequate for measurements gathered in the near-field of the source or for measurements that cause nonlinear effects

- *the quality of gathered data* $\underline{y}$. For example: some measurements were gathered with a low signal-to-noise ratio, few measurements were obtained by driving the instrumentation outside its range or under difficult conditions (e.g. tight corners, difficult to reach transducer locations, or low-energy directivity angles), a subset of the measurements was recorded by a tired and inexperienced operator
- *the statistics of each measurement.* In particular, all measurements can have the same importance in the solution when the standard error is used (Section 9.8)
- *the presence of outliers in the data.* This can be identified a priori during data preprocessing or by exploring errors e_i once a preliminary inversion has been completed

This information can be incorporated by applying different weights to each measurement or equation,

$$\begin{aligned} &w_1 \left(y_1 = h_{1,1} \cdot x_1 + \ldots + h_{1,N} \cdot x_N\right) \\ &\ldots \\ &w_i \left(y_i = h_{i,1} \cdot x_1 + \ldots + h_{i,N} \cdot x_N\right) \\ &\ldots \\ &w_M \left(y_M = h_{M,1} \cdot x_1 + \ldots + h_{M,N} \cdot x_N\right) \end{aligned} \tag{9.22}$$

In matrix form

$$\underline{\underline{W}} \cdot \underline{y} = \underline{\underline{W}} \cdot \underline{\underline{h}} \cdot \underline{x} \tag{9.23}$$

where the elements in the [M × M] diagonal matrix $\underline{\underline{W}}$ are the weights assigned to each equation $W_{i,i} = w_i$. Equation 9.22 is equivalent to the system of equations $\underline{y} = \underline{\underline{h}} \cdot \underline{x}$, when the following substitutions are implemented:

$$\underline{y} \rightarrow \underline{\underline{W}} \cdot \underline{y} \quad \text{and} \quad \underline{\underline{h}} \rightarrow \underline{\underline{W}} \cdot \underline{\underline{h}}$$

Then, the LSS and RLSS pseudoinverse solutions become

W – LSS
$$\underline{x}^{\langle est \rangle} = \left(\underline{\underline{h}}^T \cdot \underline{\underline{W}}^T \cdot \underline{\underline{W}} \cdot \underline{\underline{h}}\right)^{-1} \cdot \underline{\underline{h}}^T \cdot \underline{\underline{W}}^T \cdot \underline{\underline{W}} \cdot \underline{y}^{\langle meas \rangle} \tag{9.24}$$

W – RLSS
$$\underline{x}^{\langle est \rangle} = \left(\underline{\underline{h}}^T \cdot \underline{\underline{W}}^T \cdot \underline{\underline{W}} \cdot \underline{\underline{h}} + \lambda \cdot \underline{\underline{R}}^T \cdot \underline{\underline{R}}\right)^{-1} \cdot \underline{\underline{h}}^T \cdot \underline{\underline{W}}^T \cdot \underline{\underline{W}} \cdot \underline{y}^{\langle meas \rangle} \tag{9.25}$$

To avoid any confusion, make these substitutions in the corresponding objective functions and retrace the derivations. Furthermore, because *these substitutions affect the objective function,* the meaning of the solution has to be reworded accordingly.

9.5.2 Initial Guess of the Solution

The estimate $\underline{x}^{<est>}$ can be computed starting from an initial guess of the solution $\underline{x}_0$, and redistributing the unjustified component of the measurements $(\underline{y}^{<meas>} - \underline{\underline{h}} \cdot \underline{x}_0)$ according to the generalized inverse solutions derived earlier. The new set of pseudoinverse solutions is obtained from the original set of solutions by replacing

$$\underline{y}^{\langle meas \rangle} \rightarrow \left(\underline{y}^{\langle meas \rangle} - \underline{\underline{h}} \cdot \underline{x}_0\right) \qquad \text{and} \qquad \underline{x}^{\langle est \rangle} \rightarrow \left(\underline{x}^{\langle est \rangle} - \underline{x}_0\right)$$

to obtain

$$\underline{x}_0 - \text{LSS} \qquad \underline{x}^{\langle est \rangle} = \underline{x}_0 + \left(\underline{\underline{h}}^T \cdot \underline{\underline{h}}\right)^{-1} \cdot \underline{\underline{h}}^T \cdot \left(\underline{y}^{\langle meas \rangle} - \underline{\underline{h}} \cdot \underline{x}_0\right) \tag{9.26}$$

$$\underline{x}_0 - \text{RLSS} \qquad \underline{x}^{\langle est \rangle} = \underline{x}_0 + \left(\underline{\underline{h}}^T \cdot \underline{\underline{h}} + \lambda \cdot \underline{\underline{R}}^T \cdot \underline{\underline{R}}\right)^{-1} \cdot \underline{\underline{h}}^T \cdot \left(\underline{y}^{\langle meas \rangle} - \underline{\underline{h}} \cdot \underline{x}_0\right) \tag{9.27}$$

Once again, these substitutions affect the corresponding objective functions and the solutions gain new meaning. For example, the LSS becomes the solution with minimum global distance between $\underline{x}^{<est>}$ and $\underline{x}_0$. Likewise, the criterion expressed in the regularization matrix $\underline{\underline{R}}$ now applies to $(\underline{x}^{<est>} - \underline{x}_0)$. The initial guess $\underline{x}_0$ may be the solution of a similar inversion with earlier data in a time-varying process, it may reflect previous knowledge about the solution, or it may be estimated during data preprocessing (Chapter 11).

9.5.3 Simple Model - Ockham's Criterion

Available information about model parameters may allow a reduction in the number of unknowns, in accordance with Ockham's criterion (Section 8.4),

$$\underline{y} = \underline{\underline{h}} \cdot \underline{x} = \underline{\underline{h}} \cdot \underline{\underline{O}} \cdot \underline{u} \tag{9.28}$$

The corresponding substitutions are

$$\underline{x}^{\langle est \rangle} \rightarrow \underline{u}^{\langle est \rangle} \qquad \text{and} \qquad \underline{\underline{h}} \rightarrow \underline{\underline{h}} \cdot \underline{\underline{O}}$$

The Ockham-based LSS becomes

$$\text{O} - \text{LSS} \qquad \underline{u}^{\langle est \rangle} = \left(\underline{\underline{O}}^T \cdot \underline{\underline{h}}^T \cdot \underline{\underline{h}} \cdot \underline{\underline{O}} \right)^{-1} \cdot \underline{\underline{O}}^T \cdot \underline{\underline{h}}^T \cdot \underline{y}^{\langle meas \rangle} \tag{9.29}$$

The invertibility of $(\underline{\underline{O}}^T \cdot \underline{\underline{h}}^T \cdot \underline{\underline{h}} \cdot \underline{\underline{O}})$ is not guaranteed but may be attained when a sufficiently low number of unknowns remains in $\underline{u}$. Indeed, a salient advantage in selecting simpler models is to reduce the ill-conditioning of the inverse problem.

9.5.4 Combined Solutions

The previous sections revealed methods to incorporate additional information: initial guess of the solution $\underline{x}_0$, measurements or equations with different weights $\underline{\underline{W}}$, regularization $\underline{\underline{R}}$ and Ockham's criterion $\underline{\underline{O}}$. These methods can be combined to generate new solutions. For example, a "weighted regularized least squares with initial guess" is

$$\underline{x}^{\langle est \rangle} = \underline{x}_0 + \left(\underline{\underline{h}}^T \cdot \underline{\underline{W}}^T \cdot \underline{\underline{W}} \cdot \underline{\underline{h}} + \lambda \cdot \underline{\underline{R}}^T \cdot \underline{\underline{R}} \right)^{-1} \cdot \underline{\underline{h}}^T \cdot \underline{\underline{W}}^T \cdot \underline{\underline{W}} \cdot \left(\underline{y}^{\langle meas \rangle} - \underline{\underline{h}} \cdot \underline{x}_0 \right) \tag{9.30}$$

The meaning of the solution becomes apparent when the substitutions are implemented in the objective function that is minimized.

9.6 SOLUTION BASED ON SINGULAR VALUE DECOMPOSITION

The singular value decomposition (SVD) is as a powerful method to diagnose inverse problems and to assess available information (Section 9.2). It also permits computation of the pseudoinverse of a matrix $\underline{\underline{h}}\,[M \times N]$. The solution follows immediately from $\underline{\underline{h}} = \underline{\underline{U}} \cdot \underline{\underline{\Lambda}} \cdot \underline{\underline{V}}^T$ and the orthogonality of the matrices (Section 2.2.4):

$$\underline{\underline{h}}^{-g} = \underline{\underline{V}} \cdot \underline{\underline{\Lambda}}^{-1} \cdot \underline{\underline{U}}^T \tag{9.31}$$

where the entries in the diagonal matrix $\underline{\underline{\Lambda}}$ $[M \times N]$ are the singular values λ_i of $\underline{\underline{h}} \cdot \underline{\underline{h}}^T$ or $\underline{\underline{h}}^T \cdot \underline{\underline{h}}$ in descending order, the columns in the orthogonal matrix

$\underline{\underline{U}}\,[M \times M]$ are formed by the eigenvectors $\underline{u}$ of $\underline{\underline{h}} \cdot \underline{\underline{h}}^T$ (ordered according to the eigenvalues λ in $\underline{\underline{\Lambda}}$), and the columns in matrix $\underline{\underline{V}}\,[N \times N]$ are formed by the eigenvectors $\underline{v}$ of $\underline{\underline{h}}^T \cdot \underline{\underline{h}}$ (in the same order as the eigenvalues λ in $\underline{\underline{\Lambda}}$).

In explicit form, the solution $\underline{x}^{<est>} = \underline{\underline{h}}^{-g} \cdot \underline{y}^{<meas>}$ becomes

$$\underline{x}^{<est>} = \underline{\underline{h}}^{-g} \cdot \underline{y}^{<meas>} = \sum_{i=1}^{p} \frac{\underline{v}_i \cdot \underline{u}_i^T \cdot \underline{y}^{<meas>}}{\lambda_i} \quad \text{(order p)} \tag{9.32}$$

This equation indicates that small singular values λ_i in ill-conditioned problems will magnify model errors in $\underline{\underline{h}}$ (retained in u_i and v_i) and measurement errors in $\underline{y}$. Error magnification is controlled by restricting the summation bound "p" to take into consideration the largest singular values. Then, the generalized inverse of $\underline{\underline{h}}$ obtained by keeping the first p singular values and corresponding singular vectors is

$$\text{SVSS} \qquad \underset{N \times M}{\underline{\underline{h}}^{-g}} = \underset{N \times p}{\underline{\underline{V}}^{<p>}} \cdot \underset{p \times p}{\left(\underline{\underline{\Lambda}}^{<p>}\right)^{-1}} \cdot \underset{p \times M}{\left(\underline{\underline{U}}^{<p>}\right)^{T}} \quad \text{order p} \tag{9.33}$$

The data resolution matrix $\underline{\underline{D}} = \underline{\underline{h}} \cdot \underline{\underline{h}}^{-g}$ and model resolution matrix $\underline{\underline{G}} = \underline{\underline{h}}^{-g} \cdot \underline{\underline{h}}$ can be computed for different values p to further characterize the nature of the inverse problem, and to optimize its design. A numerical example is presented at the end of this Chapter.

9.6.1 *Selecting the Optimal Number of Meaningful Singular Values p*

How many singular values should be used in the solution? The selection of an optimal value of p starts by sorting singular values to identify jumps; in the absence of jumps, select a condition number that ensures numerical stability. This is the preselected order p_0. Then follow a similar methodology to the identification of the optimal level of regularization (Section 9.4.3 – Figure 9.3):

1. Compute the solution $\underline{x}^{<est>}$ for the preselected order p_0.
2. Assess the physical meaningfulness of the solution.
3. Assess its ability to justify the data.

4. Repeat for larger and smaller values of p around p_0.
5. Select the value of p that provides the best compromise between the two criteria.

Higher data and model errors lead to lower optimal p-values. Details are provided in Implementation Procedure 9.1.

9.6.2 SVD and Other Inverse Solutions

If the measured data $\underline{y}^{<meas>}$ can be expressed as a linear combination of the $u_1 \ldots u_r$ vectors, then $\underline{y}^{<meas>}$ is in the range of the transformation, and the solution to the inverse problem is a vector in the space of $\underline{x}$ that includes the null space. Therefore, there are infinite possible solutions and some criterion will be needed to select one. However, if the measured data $\underline{y}^{<meas>}$ *cannot* be expressed as a linear combination of the $u_1 \ldots u_r$ vectors, then $\underline{y}^{<meas>}$ is *not* in the range of the transformation and there is no solution $\underline{x}$. Yet one may still identify the solution that satisfies some criterion, such as the least squares.

When the rank of $\underline{\underline{h}}$ is r, and the summation bound "p" in Equation 9.32 is equal to r, the computed estimate $\underline{x}^{<est>}$ corresponds to the LSS:

$$\underline{\underline{h}}^{-g} = \underbrace{\left(\underline{\underline{h}}^T \cdot \underline{\underline{h}}\right)^{-1} \cdot \underline{\underline{h}}^T}_{\text{LSS}} = \underbrace{\underline{\underline{V}}^{\langle r \rangle} \cdot \left(\underline{\underline{\Lambda}}^{\langle r \rangle}\right)^{-1} \cdot \left(\underline{\underline{U}}^{\langle r \rangle}\right)^T}_{\text{SVD}} \tag{9.34}$$

The RLSS and DLSS convert the matrix $\underline{\underline{h}}^T \cdot \underline{\underline{h}}\,[N \times N]$ into a positive-definite invertible matrix. When these inversion methods are analyzed using SVD, it is readily seen that regularization and damping raise the magnitude of singular values, control the ill-conditioning of the inverse problem, and "damp" the magnification of errors from the measurements $\underline{y}$ and the model $\underline{\underline{h}}$ onto the solution $\underline{x}$ (Equation 9.32).

9.7 NONLINEARITY

A nonlinear problem can be linearized around an initial estimate, as shown in Section 8.2. The iterative Newton-type algorithm includes the appropriate updating of the transformation matrix. In the q + 1 iteration, the estimate $(\underline{x}^{<est>})_{q+1}$ is computed from the q-th estimate $(\underline{x}^{<est>})_q$ as

$$\left(\underline{x}^{\langle est \rangle}\right)_{q+1} = \left(\underline{x}^{\langle est \rangle}\right)_q + \left(\underline{\underline{h}}^{-g}\right)_q \cdot \left[\underline{y}^{\langle meas \rangle} - \left(\underline{\underline{h}}\right)_q \cdot \left(\underline{x}^{\langle est \rangle}\right)_q\right] \tag{9.35}$$

where $(\underline{\underline{h}})_q$ and $(\underline{\underline{h}}^{-g})_q$ are computed taking into consideration the q-th estimate $(\underline{x}^{<est>})_q$. Convergence difficulties and the likelihood of nonuniqueness are exacerbated in nonlinear problems, and different solutions may be reached when starting from different initial estimates. Other approaches are discussed in Chapter 10.

9.8 STATISTICAL CONCEPTS – ERROR PROPAGATION

Measurements $\underline{y}^{<meas>}$, any initial guess $\underline{x}_0$ and the assumed model are uncertain. Therefore, the inverse problem can be stated in probabilistic terms. The solutions that are obtained following probabilistic approaches are mathematically similar to those obtained earlier.

9.8.1 Least Squares Solution with Standard Errors

Measurements have the same importance in the L_2 norm when the standard error e_i is used (Section 8.3):

$$e_i = \frac{y_i^{<meas>} - y_i^{<pred>}}{\sigma_i} \qquad \textit{standard error} \tag{9.36}$$

where σ_i is the standard deviation for the i-th measurement. The vector of residuals becomes $\underline{e} = \underline{\underline{\Omega}} \cdot (\underline{y}^{<meas>} - \underline{\underline{h}} \cdot \underline{x})$ and the diagonal matrix $\underline{\underline{\Omega}}\,[M \times M]$ is formed with the inverse values of the standard deviations $\Omega_{i,i} = 1/\sigma_i$. Then, the objective function Γ for the LSS is

$$\begin{aligned}\Gamma &= \underline{e}^T \cdot \underline{e} \\ &= \left(\underline{y}^{\langle meas\rangle} - \underline{\underline{h}} \cdot \underline{x}\right)^T \cdot \underline{\underline{\Omega}}^T \cdot \underline{\underline{\Omega}} \cdot \left(\underline{y}^{\langle meas\rangle} - \underline{\underline{h}} \cdot \underline{x}\right)\end{aligned} \tag{9.37}$$

Finally, setting the derivative of the objective function Γ with respect to $\underline{x}$ equal to zero returns the sought estimate $\underline{x}^{<est>}$:

$$\underline{x}^{\langle est\rangle} = \left(\underline{\underline{h}}^T \cdot \underline{\underline{\Omega}}^T \cdot \underline{\underline{\Omega}} \cdot \underline{\underline{h}}\right)^{-1} \cdot \underline{\underline{h}}^T \cdot \underline{\underline{\Omega}}^T \cdot \underline{\underline{\Omega}} \cdot \underline{y}^{\langle meas\rangle} \tag{9.38}$$

The matrix $\underline{\underline{\Omega}}$ is diagonal; therefore, $\underline{\underline{\Omega}}^T \cdot \underline{\underline{\Omega}}$ is also a diagonal matrix with entries $1/\sigma_i^2$. When all measurements exhibit the same standard deviation, this solution becomes the previously derived LSS.

Equation 9.38 is the same as Equation 9.24 for the weighted LSS solution, where $\underline{\underline{\Omega}} = \underline{\underline{W}}$; by extension, the other solutions obtained with weighted measurements can be used to take into consideration the standard deviation of measurements.

These results can be generalized to correlated measurements: the matrix $\underline{\underline{\Omega}}^T \cdot \underline{\underline{\Omega}}$ (or $\underline{\underline{W}}^T \cdot \underline{\underline{W}}$ in Equations 9.24 and 9.25) becomes the inverse of the covariance matrix where the main diagonal elements represent the width of the distribution and off-diagonal elements capture the pairwise correlation between measurements. Then the expression for the weighted LSS is mathematically analogous to the "maximum likelihood solution". Finally, maximum entropy methods result in mathematical expressions comparable to generalized regularization procedures.

9.8.2 Gaussian Statistics – Outliers

In a broad least squares sense, the statistics of the measurements $\underline{y}^{<meas>}$, the transformation (entries in $\underline{\underline{h}}$) and an initial guess $\underline{x}_0$ are presumed Gaussian. The least squares criterion is a poor choice if Gaussian statistics are seriously violated, for example, when there are few large errors in the measurements. In such a case:

- Improve the data at the lowest possible level, starting with a proper experimental design (Chapters 4 and 5).
- Identify and remove outliers during data preprocessing prior to inversion (see example in Chapter 11).
- Guide the evolution of the inversion with the more robust L_1 norm rather than the L_2 norm (Chapter 10).
- Redefine the objective function Γ to consider proper statistics.
- Identify and downplay outliers during inversion.

Let us explore the last two options. Many inverse problems relate to nonnegative quantities, yet the tail of the normal Gaussian distribution extends into negative values. Examples include mass, precipitation, traffic, population, conduction, diffusion, strength, and stiffness. Often, such parameters are better represented by the log-normal distribution and the error function is computed in terms of log(y) rather than in terms of y. (Note: y is log-normal when log(y) is Gaussian distributed – Section 8.3.) If regularization is used, the matrix $\underline{\underline{R}}$ must be designed taking into consideration values $z_k = \log(x_k)$ rather than x_k. Once again, the meaning of the computed solution becomes apparent when the redefined objective function is carefully expressed in words.

Finally, outliers can be identified during inversion to reduce their impact on the final solution. The inversion solution is as follows:

1. Invert the data and obtain a first estimate of the solution $\underline{x}^{<est-1>}$.
2. Compute the error between measured and justified values:

$$\underline{e} = \underline{y}^{<meas>} - \underline{y}^{<just-1>} = \underline{y}^{<meas>} - \underline{\underline{h}} \cdot \underline{x}^{<est-1>}$$

3. Form a histogram of e_i values and explore whether it satisfies Gaussian statistics.
4. Identify outliers $y_i^{<meas>}$ that exhibit large deviations from the mean.
5. Underweight and even remove those measurements and obtain a new estimate $\underline{x}^{<est-2>}$ using weighted solutions such as W-LSS or W-RLSS (Section 9.5.1).

This approach must be applied with caution to avoid biasing the inversion; place emphasis on large deviations, typically two or more standard deviations from the mean.

9.8.3 Accidental Errors

Consider a function $t = f(s_1, \ldots, s_N)$, where the variables s_i follow Gaussian statistics. The mean of t is the function f of the means,

$$\mu_t = f(\mu_1, \mu_2, \ldots, \mu_N) \tag{9.39}$$

The standard deviation σ_t is estimated using a first-order Taylor expansion of the function f about the mean values μ_i, assuming that s-values are independent (the covariance coefficients are zero),

$$\sigma_t^2 = \sum_i \left(\left. \frac{\partial f}{\partial s_i} \right|_{\underline{s}=\underline{\mu}} \right)^2 \sigma_{s_i}^2 \tag{9.40}$$

For the linear inverse problem in matrix form, $\underline{x} = \underline{\underline{h}}^{-g} \cdot \underline{y}$, these equations become (Note: the values of the estimated parameters $\underline{x}^{<est>}$ are correlated through the transformation $\underline{\underline{h}}^{-g}$ as indicated by the model resolution matrix in Equation 9.5, however measurements $\underline{y}^{<meas>}$ are assumed uncorrelated),

$$\underline{\mu}_x = \underline{\underline{h}}^{-g} \cdot \underline{\mu}_y \tag{9.41}$$

$$\underline{\sigma}_x^2 = \underline{\underline{h2}}^{-g} \cdot \underline{\sigma}_y^2 \tag{9.42}$$

Each entry in $\underline{\underline{h2}}^{-g}$ is the square of the corresponding entry in $\underline{\underline{h}}^{-g}$ for the selected inversion solution, such as LSS, RLSS or their modifications. If the standard deviation of the measurements is constant for all measurements and equal to "c", Equation 9.42 becomes

$$\underline{\sigma}_x^2 = c^2 \cdot \underline{\underline{h2}}^{-g} \cdot \underline{1} \qquad (9.43)$$

where all entries in the vector $\underline{1}$ are equal to 1.0. In other words, the standard deviation of the k-th unknown x_k is equal to c-times the square root of the sum of all elements in the k-th row of $\underline{\underline{h2}}^{-g}$. Equations 9.42 and 9.43 can be applied to linearized inverse problems where the matrix $\underline{\underline{h}}$ corresponds to the current q-th iteration.

9.8.4 Systematic and Proportional Errors

Systematic errors, such as a constant trigger delay in recording devices, do not cancel out by signal stacking or within LSS. The only alternative is to detect them and to correct the data. A constant systematic error ε in all measurements alters the estimated value of the unknowns by a quantity $\underline{\Delta x}$:

$$\underline{x}^{\langle est \rangle} + \underline{\Delta x} = \underline{\underline{h}}^{-g} \cdot \left(\underline{y} + \underline{\varepsilon}\right) \quad \Rightarrow \quad \underline{\Delta x} = \underline{\underline{h}}^{-g} \cdot \underline{\varepsilon} = \varepsilon \cdot \underline{\underline{h}}^{-g} \cdot \underline{1} \qquad (9.44)$$

Therefore, the error in the k-th unknown Δx_k is ε times the sum of all elements in the k-th row of the generalized inverse $\underline{\underline{h}}^{-g}$.

Proportional errors occur when a measured value is equal to a constant α times the true value, $y_i^{<meas>} = \alpha_i \cdot y_i^{<true>}$. Often, a proportional error reflects improper transducer calibration and it may be present in conjunction with a systematic offset. In the general case when all α_i-values are different (for example, when each transducer has its own calibration), the estimate of the solution becomes

$$\underline{x}^{\langle est \rangle} = \underline{\underline{h}}^{-g} \cdot \underline{\underline{diag(\alpha)}} \cdot \underline{y}^{\langle true \rangle} \qquad (9.45)$$

where $\underline{\underline{diag(\alpha)}}$ is the diagonal matrix formed by α_i values in the main diagonal. If all α-values are the same, $\alpha_1 = \ldots = \alpha_n = \alpha$, then $\underline{x}^{<est>} = \alpha \cdot \underline{\underline{h}}^{-g} \cdot \underline{y}$, and the solution that is computed is α times the solution estimated without the proportional error in the data.

9.8.5 Error Propagation – Regularization and SVD Solutions

Poorly conditioned inversions are characterized by high values in $\underline{\underline{h}}^{-g}$. Therefore, Equations 9.42, 9.44 and 9.45 predict that errors will be preferentially magnified in some x-values more than others. Magnification is related to the amount of information that is available to determine each x-value: poorly constrained x-values will magnify data noise the most.

Regularization and SVD solutions (Equations 9.17 and 9.33) reduce the effects of ill-conditioning, limit the high values in $\underline{\underline{h}}^{-g}$ and control error propagation as inherently predicted when these solutions are replaced in Equations 9.42, 9.44 and 9.45. These observations are further explored in Chapter 11 in the context of tomographic imaging.

A strong correlation between $\underline{x}^{<est>}$ and the vectors $\underline{\underline{h}}^{-g} \cdot \underline{1}$ and $\underline{\underline{h2}}^{-g} \cdot \underline{1}$ that contain the row-sums in matrices $\underline{\underline{h}}^{-g}$ and $\underline{\underline{h2}}^{-g}$ should be carefully scrutinized: make sure that the solution is not determined by the combined effects of high data error and uneven distribution of information.

9.9 EXPERIMENTAL DESIGN FOR INVERSE PROBLEMS

The transformation matrix $\underline{\underline{h}}$ *is a function of the test design and data collection strategy only*. It does not depend on the data $\underline{y}$ but on sensor location and the spatial/temporal distribution of measurements. Therefore, the following information is known and can be analyzed as soon as the experiment is designed, and before expensive and time-consuming data gathering:

- transformation matrix $\underline{\underline{h}}$
- the number of meaningful singular values that keep ill-conditioning under control
- generalized inverse $\underline{\underline{h}}^{-g}$ (with some level of regularization or formed with the meaningful singular values)
- data and model resolution matrices $\underline{\underline{D}}$ and $\underline{\underline{G}}$
- the spatial distribution of information – a simple estimate is related to the column-sum of the transformation matrix $\underline{1}^T \cdot \underline{\underline{h}}$
- the vectors $\underline{\underline{h}}^{-g} \cdot \underline{1}$ and $\underline{\underline{h2}}^{-g} \cdot \underline{1}$ that contain the row-sums in matrices $\underline{\underline{h}}^{-g}$ and $\underline{\underline{h2}}^{-g}$

Conversely, this information can be used to improve the design of experiments. Preliminary guidelines are summarized in Implementation Procedure 9.2. Details are presented in Section 11.2.

Implementation Procedure 9.2 Preliminary guidelines for the design of experiments leading to inverse problems

1. *Design the test to attain maximum information and stable solutions.* Conceive different realizable experimental configurations. For each configuration:

   - Identify/compute and tabulate the number of measurements M, the number of unknowns N, and the trace of the $\underline{\underline{h}}^T \cdot \underline{\underline{h}}$ matrix.
   - Compute the singular values of the transformation $\underline{\underline{h}}$. Plot sorted singular values and identify the number p of singular values that satisfy an acceptable condition number.
   - Monitor the spatial distribution of information by computing the column-sum of the transformation matrix $\underline{1}^T \cdot \underline{\underline{h}}$.
   - Compute the row-sums $\underline{\underline{h}}^{-g} \cdot \underline{1}$ and $\underline{\underline{h2}}^{-g} \cdot \underline{1}$ to explore the propagation of systematic and accidental errors onto each parameter in the solution $\underline{x}^{<est>}$. For this step, compute the generalized inverse for various realistic levels of regularization.
   - Compute the data resolution matrix $\underline{\underline{D}} = \underline{\underline{h}} \cdot \underline{\underline{h}}^{-g}$ and the model resolution matrix $\underline{\underline{G}} = \underline{\underline{h}}^{-g} \cdot \underline{\underline{h}}$. Analyze their resemblance with the identity matrix and calculate their trace.
   - Prevent spatial and temporal aliasing in measurements.

2. For a similar number of measurements M, favor test configurations that provide high amount of information, evenly distributed and with controlled error propagation.

3. Utilize the insight gained from the initial test configurations to generate new ones, as needed.

4. *Design the test to gather high-quality data.* Carefully select transducers and peripheral electronics. Create testing conditions that minimize noise. Implement signal processing algorithms that facilitate measuring the data with minimum error. Remove the effects of transducers and peripheral electronics from the measurements.

9.10 METHODOLOGY FOR THE SOLUTION OF INVERSE PROBLEMS

The elegant close-form solutions obtained using the L_2 norm facilitate the analysis of inverse problems, provide diagnostic tools to identify difficulties and limitations, and permit incorporating information in the form of an initial guess, regularization, relative weights, and model characteristics. These solutions apply to linear problems and are extended to linearized nonlinear problems within the context of iterative algorithms.

We must pay special attention to the choices that are made and track noise magnification during the inversion so that the solution is controlled neither by data errors and model errors, nor by our own preconceptions. Guidelines for the solution of inverse problems are summarized in Implementation Procedure 9.3. A comprehensive methodology is proposed in Chapter 11.

Implementation Procedure 9.3 Preliminary guidelines for the solution of inverse problems in matrix form

1. Properly design the experiment (Implementation Procedure 9.2).
2. Gather high-quality data. While conducting the experiment, identify measurements that present unique difficulty and repeat doubtful measurements.
3. Accumulate additional information that may be later incorporated during inversion.
4. Select a model to relate the unknowns $\underline{x}$ to the measurements $\underline{y}^{<meas>}$ that adequately captures all essential aspects of the system or process.
5. Favor simplicity – limit the number of unknowns in the representation of the problem.
6. Preanalyze the data $\underline{y}^{<meas>}$ to identify trends and outliers. This step may help define an initial guess of the solution $\underline{x}_0$.
7. Implement more than one pseudoinverse.
8. Vary inversion parameters such as the regularization coefficient λ or the number of singular values p while monitoring changes in the solution and in the residuals (Implementation Procedure 9.1).

9. Gradually incorporate additional information, such as an initial guess $\underline{x}_0$, information about the solution (regularization $\underline{\underline{R}}$) and information about the data ($\underline{\underline{W}}$).

10. Compute the residuals $\underline{e} = \underline{y}^{<meas>} - \underline{\underline{h}} \cdot \underline{x}^{<est>}$. Look for trends in the mean, median, extreme values, and histogram.

11. Underweight or remove equations that are clear outliers and rerun the inversion. This can be a dangerous step!

12. Compute the column-sum $\underline{1}^T \cdot \underline{\underline{h}}$, the row-sums $\underline{\underline{h}}^{-g} \cdot \underline{1}$ and $\underline{\underline{h2}}^{-g} \cdot \underline{1}$. Plot these vectors versus $\underline{x}^{<est>}$. Carefully scrutinize any apparent correlation.

13. The final solution $\underline{x}^{<est>}$ must justify the data and be physically meaningful.

Note: Chapter 11 presents a more comprehensive approach and examples.

The generalized inverse expressions derived in this chapter include various matrix operations such as addition, multiplication, transpose, inverse, eigenvectors and eigenvalues. Efficient computer implementations are developed to reduce data storage and processing time. These algorithms recognize the inherent characteristics of the matrices involved, which can be positive-definite, sparse, diagonal, symmetric, Toepliz, and so forth.

9.11 SUMMARY

- The goal of inverse problem solving is to identify the parameters of a physically meaningful solution $\underline{x}^{<est>}$ that can adequately justify the data $\underline{y}$ given an acceptable model that is captured in $\underline{\underline{h}}$.
- A problem is "well-posed" when a unique and stable solution exists. This is a rare situation in real inverse problems.
- Elegant expressions can be obtained for the solution of discrete inverse problems expressed in matrix form. The L_2 error norm plays a preponderant role in these derivations. The least squares criterion is a poor choice if Gaussian statistics are seriously violated, for example, when there are few large errors in the measurements.
- Additional information available to the analyst may be considered. Information about the solution $\underline{x}$ is included through the regularization matrix $\underline{\underline{R}}$, Ockham's matrix $\underline{\underline{O}}$, or as an initial guess $\underline{x}_0$. Information about the measurements or the model is incorporated through weights $\underline{\underline{W}}$.

- SVD is a powerful tool to diagnose the transformation matrix $\underline{\underline{h}}$. The relative size of singular values gives an indication of rank deficiency in the transformation, shows how close the system of equations is to a system of lower rank, and provides a reliable indicator of ill-conditioning. The pseudoinverse computed by SVD explicitly shows the amplification of data and model error caused by small singular values.
- Well-posed problems satisfy the requirements of existence, uniqueness, and stability. Sufficient information is required to secure the first two requirements. Stability can be controlled with proper regularization. Regularization makes the matrix $\underline{\underline{h}}^T \cdot \underline{\underline{h}}$ invertible and increases the size of small singular values.
- Ill-conditioning is determined by experimental design, rather than by the accuracy of data. Consider various viable distributions of transducers and measurements and evaluate them to select the optimal one.
- Data errors are magnified during inversion. Therefore, experimental design must also attempt to minimize measurement errors through careful selection of transducers and peripheral electronics, calibration of the measurement system, and noise control. Outliers can have a major impact on the quality of the solution and should be removed during data preprocessing or downweighted as part of an iterative inversion strategy.
- Model errors are amplified as well. Focus on the physics of the problem. Select the simplest model that can properly justify the data.
- *Successful inverse problem solving is strongly dependent on the analyst.* The analyst designs the experiment, chooses the physical model, selects the inversion strategy, recognizes and incorporates additional information, and identifies optimal inversion parameters such as the degree of regularization or the number of singular values.
- *Remain skeptical.* Make sure that the solution is not a consequence of your preconceptions and choices, but that it justifies the data and truly reflects the nature of the phenomenon or system under study.

FURTHER READING

Aster, R., Borchers, B., and Thurber, R. (2004). Parameter Estimation and Inverse Problems. Academic Press. 320 pages.

Bui, H. D. (1994). Inverse Problems in the Mechanics of Materials: An Introduction. CRC Press, Boca Raton, Fla. 204 pages.

Curtis, A. (2004). Theory of Model-Based Geophysical Survey and Experimental Design: Part 1-Linear Problems. The Leading Edge. October. pp.997–1004.

Demmel, J. W. (1997). Applied Numerical Linear Algebra. Society for Industrial and Applied Mathematics, Philadelphia. 419 pages.

Golub, G. H., Hansen, P. C., and O'Leary, D. P. (1999). Tikhonov Regularization and Total Least Squares. SIAM J. Matrix Anal. Appl. Vol. 21, No. 1, pp. 185–194.

Lawson, C. L. and Hanson, R. J. (1974). Solving Least Squares Problems. Prentice-Hall, Englewood Cliffs, NJ.

Menke, W. (1989). Geophysical Data Analysis: Discrete Inverse Theory. International Geophysics Series. Academic Press, San Diego. 289 pages.

Parker, R. L. (1994), Geophysical Inverse Theory, Princeton University Press, Princeton. 386 pages.

Penrose, R. A. (1955). "A Generalized Inverse for Matrices". Proc. Cambridge Phil. Soc. Vol 51, pp. 406–413.

Tarantola, A. (1987). Inverse Problems Theory. Elsevier, Amsterdam. 613 pages.

SOLVED PROBLEMS

P9.1 *Underdetermined problems: the minimum length solution (MLS).* There is an infinite number of solutions that result in null prediction error $\underline{e} = \underline{0}$ when the problem is underdetermined and consistent (number of equations $M < N$ number of unknowns; and $r[\underline{\underline{h}}] = r\underline{\underline{h}}|\underline{y}]$). Identify the MLS.

Solution: The objective function Γ that is used to identify the *minimum length solution* (MLS) minimizes the Pythagorean length of the solution $\underline{x}^T \cdot \underline{x} = x_1^2 + \cdots + x_N{}^2$ subject to the constraint of error minimization (see constrained minimization using Lagrange multipliers in Section 2.3): $\Gamma(\underline{x}) = \underline{x}^T \cdot \underline{x} + \underline{\lambda}^T \cdot \left(\underline{y}^{\langle meas \rangle} - \underline{\underline{h}} \cdot \underline{x}\right)$.

There are N-unknown values in $\underline{x}$ and M-unknown Lagrange multipliers in $\underline{\lambda}$. The resulting system of N + M simultaneous equations is

| N-equations $\underline{0} = 2 \cdot \underline{x} - \underline{\underline{h}}^T \cdot \underline{\lambda}$ (partial derivatives of Γ with respect to $\underline{x}$)
| M-constraints $\underline{0} = \underline{y}^{<meas>} - \underline{\underline{h}} \cdot \underline{x}$

Replacing $\underline{x}$ from the first set of equations into the second set, and assuming that $\underline{\underline{h}} \cdot \underline{\underline{h}}^T$ is invertible, the vector of Lagrange multipliers is $\underline{\lambda} = 2 \cdot (\underline{\underline{h}} \cdot \underline{\underline{h}}^T)^{-1} \cdot \underline{y}^{<meas>}$. Finally, replacing $\underline{\lambda}$ in the first set of equations, the MLS estimate $\underline{x}^{<est>}$ is

$$\boxed{\text{MLS} \quad \underline{x}^{\langle est \rangle} = \underline{\underline{h}}^T \cdot \left(\underline{\underline{h}} \cdot \underline{\underline{h}}^T\right) \cdot \underline{y}^{\langle meas \rangle}}$$

Is the MLS estimate physically meaningful to your application?

P9.2 *Least squares solution.* Given the following stress-strain data, identify the model parameters for a linear elastic behavior $\sigma_i = \sigma_0 + E \cdot \varepsilon_i$ where E is Young's modulus and the sitting error σ_0 is due to the early localized deformation at end platens.

Strain: $\underline{\underline{\varepsilon}}^T = [0\,1\,2\,3\,4\,5\,6\,7] \cdot 10^{-3}$

Stress: $\underline{\underline{\sigma}}^T = [5.19 \quad 6.61 \quad 8.86 \quad 11.8 \quad 13.6 \quad 15.5 \quad 17.6 \quad 19.7] \cdot 10^6 \cdot \text{kPa}$

Solution: The LSS transformation matrix is (the transpose is shown to facilitate the display):

$$\underline{\underline{h}}^T = \begin{bmatrix} 1 & 1 & 1 & 1 & 1 & 1 & 1 & 1 \\ 0 & 0.001 & 0.002 & 0.003 & 0.004 & 0.005 & 0.006 & 0.007 \end{bmatrix}$$

The LSS generalized inverse is $\underline{\underline{h}}^{-g} = (\underline{\underline{h}}^T \cdot \underline{\underline{h}})^{-1} \cdot \underline{\underline{h}}^T$ and estimated model parameters are $\underline{x}^{<est>} = \underline{\underline{h}}^{-g} \cdot \underline{y}^{<meas>}$.

Results: $\sigma_{0est} = 4.93 \cdot 10^8 \cdot \text{Pa}$ and $E_{est} = 2.11 \cdot 10^{11} \cdot \text{Pa}$

Residuals $\underline{e} = \underline{y}^{<meas>} - \underline{\underline{h}} \cdot \underline{x}^{<est>}$. Norm of residuals $\underline{e}^T \cdot \underline{e} = 6.932 \cdot 10^{15} \cdot \text{Pa}^2$

Model resolution matrix is $\underline{\underline{G}} = (\underline{\underline{h}}^T \cdot \underline{\underline{h}})^{-1} \cdot \underline{\underline{h}}^T \cdot \underline{\underline{h}} = \begin{bmatrix} 1 & 0 \\ 0 & 1 \end{bmatrix}$

Therefore, the LSS resolves the unknowns $\underline{x}$.

P9.3 *Singular value decomposition.* Compute the SVD of

$$\underline{\underline{h}} = \begin{bmatrix} 1 & 4 \\ 2 & 5 \\ 3 & 6 \end{bmatrix} \qquad (M = 3, N = 2)$$

Solution: the eigenvalues $\leftrightarrow$ eigenvectors of matrices $h \cdot h^T$ and $h^T \cdot h$ are

$$\underline{\underline{h}} \cdot \underline{\underline{h}}^T = \begin{bmatrix} 17 & 22 & 27 \\ 22 & 29 & 36 \\ 27 & 36 & 45 \end{bmatrix}$$

$$90.403 \leftrightarrow \begin{bmatrix} 0.429 \\ 0.566 \\ 0.704 \end{bmatrix} \quad 0.597 \leftrightarrow \begin{bmatrix} 0.806 \\ 0.112 \\ -0.581 \end{bmatrix} \quad \sim 0 \leftrightarrow \begin{bmatrix} -0.408 \\ 0.816 \\ -0.408 \end{bmatrix}$$

$$\underline{\underline{h}}^T \cdot \underline{\underline{h}} = \begin{bmatrix} 14 & 32 \\ 32 & 77 \end{bmatrix} \quad 90.403 \leftrightarrow \begin{bmatrix} 0.386 \\ 0.922 \end{bmatrix} \quad 0.597 \leftrightarrow \begin{bmatrix} -0.922 \\ 0.386 \end{bmatrix}$$

Matrices $\underline{\underline{\Lambda}}$, $\underline{\underline{V}}$, and $\underline{\underline{U}}$ are

$$\underline{\underline{\Lambda}} = \begin{bmatrix} 9.508 & 0 \\ 0 & 0.773 \\ 0 & 0 \end{bmatrix} \quad \underline{\underline{U}} = \begin{bmatrix} 0.429 & 0.806 & -0.408 \\ 0.566 & 0.112 & 0.816 \\ 0.704 & -0.581 & -0.408 \end{bmatrix} \quad \underline{\underline{V}} = \begin{bmatrix} 0.386 & -0.922 \\ 0.922 & 0.386 \end{bmatrix}$$

(size M × N) (size M × M) (size N × N)

Matrices $\underline{\underline{U}}$ and $\underline{\underline{V}}$ are orthogonal, i.e. the inverse equals the transpose:

$$\underline{\underline{U}}^T \cdot \underline{\underline{U}} = \begin{bmatrix} 1 & 0 & 0 \\ 0 & 1 & 0 \\ 0 & 0 & 1 \end{bmatrix} \quad \text{and} \quad \underline{\underline{V}}^T \cdot \underline{\underline{V}} = \begin{bmatrix} 1 & 0 \\ 0 & 1 \end{bmatrix}$$

Verification of SVD: $\underline{\underline{U}} \cdot \underline{\underline{\Lambda}} \cdot \underline{\underline{V}}^T = \begin{bmatrix} 1 & 4 \\ 2 & 5 \\ 3 & 6 \end{bmatrix}$ which is equal to $\underline{\underline{h}}$

Generalized inverse for $p = 2$:

$$\underline{\underline{h}}^{-g} = \underline{\underline{V}}^{<p>} \cdot \left(\underline{\underline{\Lambda}}^{<p>}\right)^{-1} \cdot \left(\underline{\underline{U}}^{<p>}\right)^T = \begin{bmatrix} -0.944 & -0.111 & 0.722 \\ 0.444 & 0.111 & -0.222 \end{bmatrix}$$

and $\underline{\underline{h}}^{-g} \cdot \underline{\underline{h}} = \begin{bmatrix} 1 & 0 \\ 0 & 1 \end{bmatrix}$

P9.4 *Deconvolution as a least squares inverse problem.* Given the measured output signal $\underline{y}^{<meas>}$ and the impulse response $\underline{h}$, determine the input signal $\underline{x}$.

Output signal: $\underline{y} = [\, 0 \;\; 0 \;\; 2 \;\; -1 \;\; 0.5 \;\; -0.5 \;\; 0 \;\; 0 \,]^T$

Impulse response: $\underline{h} = [\, 0 \;\; -1 \;\; 0.5 \;\; -0.25 \;\; 0 \;\; 0 \;\; 0 \;\; 0 \,]^T$

Solution: the matrix $\underline{\underline{h}}$ is assembled with time-shifted copies of the vector $\underline{h}$ (Section 4.5):

$$\underline{\underline{h}} = \begin{bmatrix} 0 & 0 & 0 & 0 & 0 & 0 & 0 & 0 \\ -1.0 & 0 & 0 & 0 & 0 & 0 & 0 & 0 \\ 0.50 & -1.0 & 0 & 0 & 0 & 0 & 0 & 0 \\ -0.25 & 0.50 & -1.0 & 0 & 0 & 0 & 0 & 0 \\ 0 & -0.25 & 0.50 & -1.0 & 0 & 0 & 0 & 0 \\ 0 & 0 & -0.25 & 0.50 & -1.0 & 0 & 0 & 0 \\ 0 & 0 & 0 & -0.25 & 0.50 & -1.0 & 0 & 0 \\ 0 & 0 & 0 & 0 & -0.25 & 0.50 & -1.0 & 0 \end{bmatrix}$$

Clearly, the matrix h is noninvertible (zeros on main diagonal). A damped least squares approach is adopted, that is, the RLSS

$$\underline{x}^{\langle est \rangle} = \left(\underline{\underline{h}}^T \cdot \underline{\underline{h}} + \lambda \cdot \underline{\underline{R}}^T \cdot \underline{\underline{R}}\right)^{-1} \cdot \underline{\underline{h}}^T \cdot \underline{y}^{\langle meas \rangle}$$

where $\underline{\underline{R}} \equiv \underline{\underline{I}}$. The optimal value of λ is found following the methodology outlined in Implementation Procedure 9.2. First, solution $\underline{x}^{<est>}$ is computed for a range of λ-values; then, results are analyzed as follows:

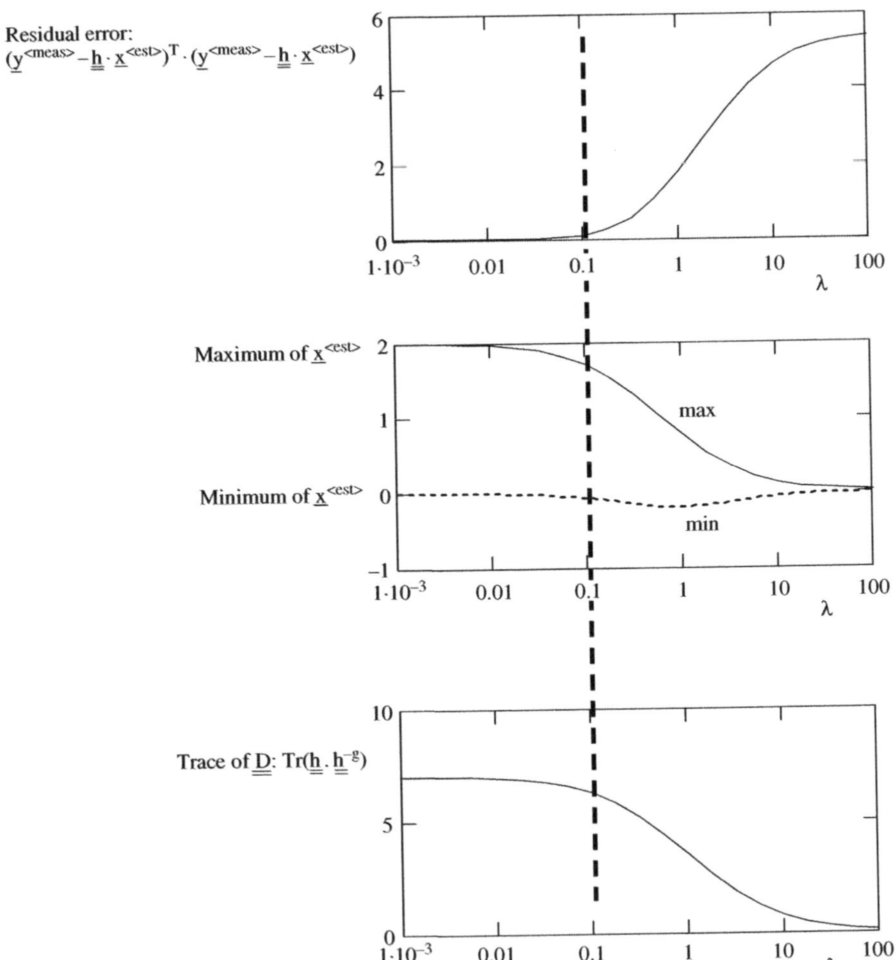

Estimated $\underline{x}^{<est>}$ and true values $\underline{x}^{<true>}$ are compared next (Note: $\underline{x}^{<true>}$ is not known in real problems):

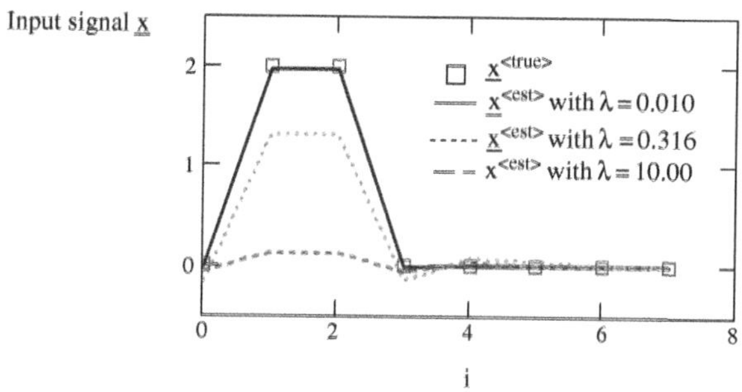

P9.5 *Regularization matrix.* Consider the beam below. The support on the left allows vertical displacement but it does not allow rotation, while the support on the right allows rotation but it prevents vertical displacement. Construct the regularization matrix $\underline{\underline{R}}$ that will be used to invert for the load distribution on the beam knowing the vertical displacements measured at points #1 through #11.

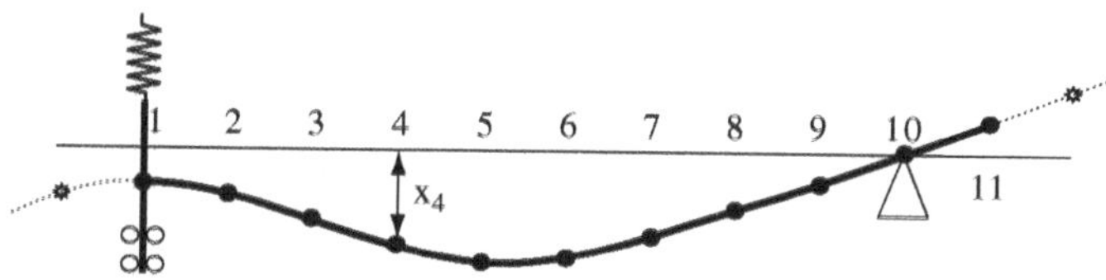

Solution: the deformed shape is assumed smooth (a priori information), and the adopted regularization criterion is the minimization of the second derivative. The corresponding regularization kernel is (1 −2 1). The kernel is applied at end points #1 and #11 by assuming imaginary points that are compatible with boundary conditions. The imaginary point on the left is computed assuming zero rotation or symmetric boundary, $x_L = x_2$. The imaginary point on the right is computed assuming constant gradient $x_R - x_{11} = x_{11} - x_{10}$, therefore $x_R = 2x_{11} - x_{10}$. Each row in the regularization matrix corresponds to the kernel applied at different points along the beam, starting at the left boundary point #1, and ending at the right boundary point #11. The resulting regularization matrix $\underline{\underline{R}}$ is shown below.

Why is the last row zero? When the kernel is centered at node #11,

$$x_{10} - 2x_{11} + x_R = x_{10} - 2x_{11} + (2x_{11} - x_{10}) = 0$$

Indeed, a constant slope has zero second derivative! Therefore constant gradient extrapolation combines with the Laplacian kernel to return a zero-row for the x_{11} entry. In this case, the last row should be computed with the forward second derivative, without assuming imaginary points, to avoid zero rows in $\underline{\underline{R}}$.

−2	2	0	0	0	0	0	0	0	0	0	*at point #1*
−1	−2	−1	0	0	0	0	0	0	0	0	*at point #2*
0	−1	−2	−1	0	0	0	0	0	0	0	*at point #3*
0	0	−1	−2	−1	0	0	0	0	0	0	*at point #4*
0	0	0	−1	−2	−1	0	0	0	0	0	*at point #5*
0	0	0	0	−1	−2	−1	0	0	0	0	*at point #6*
0	0	0	0	0	−1	−2	−1	0	0	0	*at point #7*
0	0	0	0	0	0	−1	−2	−1	0	0	*at point #8*
0	0	0	0	0	0	0	0	−2	−1	0	*at point #9*
0	0	0	0	0	0	0	0	−1	−2	−1	*at point #10*
0	0	0	0	0	0	0	0	0	0	0	*at point #11*

ADDITIONAL PROBLEMS

P9.6 *Conditions for the pseudoinverse.* Verify whether the RLSS and SVD solutions fulfill the Moore–Penrose's conditions:

$$\underline{\underline{h}} \cdot \underline{\underline{h}}^{-g} \cdot \underline{\underline{h}} = \underline{\underline{h}}$$

$$\underline{\underline{h}}^{-g} \cdot \underline{\underline{h}} \cdot \underline{\underline{h}}^{-g} = \underline{\underline{h}}^{-g}$$

$$\left(\underline{\underline{h}} \cdot \underline{\underline{h}}^{-g}\right)^{T} = \underline{\underline{h}} \cdot \underline{\underline{h}}^{-g}$$

$$\left(\underline{\underline{h}}^{-g} \cdot \underline{\underline{h}}\right)^{T} = \underline{\underline{h}}^{-g} \cdot \underline{\underline{h}}$$

P9.7 *Pseudoinverse: hand calculation.* Given the matrix $\underline{\underline{h}}$, compute: (a) the rank of $\underline{\underline{h}}$; (b) the singular values of $\underline{\underline{h}}^T \cdot \underline{\underline{h}}$; (c) the least square pseudoinverse; (d) the corresponding data and model resolution matrices; (e) the singular values of $(\underline{\underline{h}}^T \cdot \underline{\underline{h}} + 0.1 \cdot \underline{\underline{I}})$; and (f) conclude on the effects of damping in the DLSS. Observations apply to the RLSS as well.

$$\underline{\underline{h}} = \begin{bmatrix} 1 & 0 & 1 \\ 0 & 1 & 1 \\ 1 & 0 & 4 \\ 1 & 2 & 0 \end{bmatrix} \quad \text{for M} = 4 \text{ measurements and N} = 3 \text{ unknowns}$$

P9.8 *Minimum length solution.* Extend the MLS solution (Problem 9.1) to incorporate an initial guess $\underline{x}_0$. Make the corresponding substitutions in the objective function and clearly state the meaning of the solution.

P9.9 *Error propagation.* Demonstrate Equation 9.42 that predicts the propagation of uncertainty in the measurements onto the model parameters (Hint: work equation by equation.) Explore the matrix $\underline{\underline{h2}}^{-g}$ for a problem of your interest.

P9.10 *Application: ARMA model.* Consider the ARMA solution for an oscillator developed in Section 7.8 (a) Forward simulation: Assume a sinusoidal input (frequency $f = 5\,\text{Hz}$) and compute the response for a single DoF with spring constant k increasing linearly with time t: $k(t) = 1[\text{kN} \cdot \text{m}^{-1} \cdot \text{s}^{-1}] \cdot t[\text{s}]$. (b) Use the LSS to invert for the time-varying system response with a two-term AR and two-term MA model. (c) Repeat the solution for a five-term AR and five-term MA model. (d) Conclude.

P9.11 *Wiener filters.* Wiener filters can be used to compute deconvolution in a least squares sense when the matrix $\underline{\underline{h}}$ is not square. Simulate a signal $\underline{x}$ and a shorter array for the impulse response $\underline{h}$. Compute the convolution $\underline{y} = \underline{x} * \underline{h}$. Then, compute the deconvolution of $\underline{h}$ and $\underline{y}$ as a least squares inversion (see Solved Problem 9.4). Add noise to $\underline{y}$ and repeat. Extend the methodology to system identification. Analyze advantages and disadvantages with respect to the determination of the frequency response in the frequency domain outlined in Implementation Procedure 6.6.

P9.12 *Application: beam on elastic foundation.* Consider the deflection of an infinite beam on a bed of linear springs all with the same spring constant. Design an experiment to gather data for the following two inverse problems: (1) apply a known load and infer the beam stiffness and the spring constant; and (2) measure the deformed shape of the beam and infer the position and magnitude of the applied load. Simulate a data set using the forward model $\underline{y} = \underline{\underline{h}} \cdot \underline{x}$. Add random and systematic noise to the simulated data set. (Note: The relevant equations to construct the transformation matrix $\underline{\underline{h}}$ can be found in mechanics books.)

P9.13 *Application of your interest: RLSS and SVD solution.* Describe the measured data $\underline{y}^{<meas>}$ and the unknown parameters $\underline{x}$. Then develop an appropriate methodology to identify the optimal number of singular values p and the optimal value for regularization coefficient λ. Take into consideration data justification and the physical meaning of the solution. Provide specific measures and decision criteria. Obtain a data set, test the methodology and explore the effect of data noise on the optimal values of p and λ.

10
Other Inversion Methods

Elegant solutions for linear inverse problems are presented in Chapter 9. Their salient characteristics are summarized next and compared against alternative requirements for inverse problem solving.

Methods in Chapter 9:	*Alternative requirements:*
Operations are implemented in the space of the solution and the data	Study solutions in the Fourier space Consider parametric representations
Solutions are based on L_2 norm	Implement the L_1 norm (noisy data), or the L_∞ norm (uneven information density)
Solutions presume Gaussian statistics	Accommodate any statistics
Methods apply to linear problems or problems that are linearized	Explore algorithms that can be applied to both linear or nonlinear problems
They involve memory-intensive matrix-based data structures	Can be implemented in effective matrix-free algorithms. Minimize storage requirements
Solutions exhibit convergence difficulties in linearized problems	Capable of finding the optimal solution in nonconvex error surfaces
They are able to incorporate additional information	Retain flexibility to incorporate additional information

Discrete Signals and Inverse Problems J. C. Santamarina and D. Fratta

The methods explored in this chapter include transformed problem postulation, matrix-free iterative solutions of the system of equations, fully flexible inversion by successive forward simulations, and heuristic methods from the field of artificial intelligence. These algorithms bring advantages and associated costs; trade-offs are identified in each case.

10.1 TRANSFORMED PROBLEM REPRESENTATION

Three methods are explored in this section: the parametric representation of a presumed solution, the low-pass frequency representation of the solution space, and the Fourier transform of the solution and data spaces.

10.1.1 Parametric Representation

Imagine trying to find a small ringing buzzer in a dark room 10 m × 10 m × 4 m (Figure 10.1). You walk along the walls and gather $M = 100$ measurements of sound amplitude. Then, you attempt to invert the data to infer the location of the buzzer. Inversion using methods in Chapter 9 starts by dividing the volume of the room into small voxels (0.1m × 0.1m × 0.1m) and invert boundary measurements y_i along the walls to determine the sound level x_k generated at each voxel. Clearly,

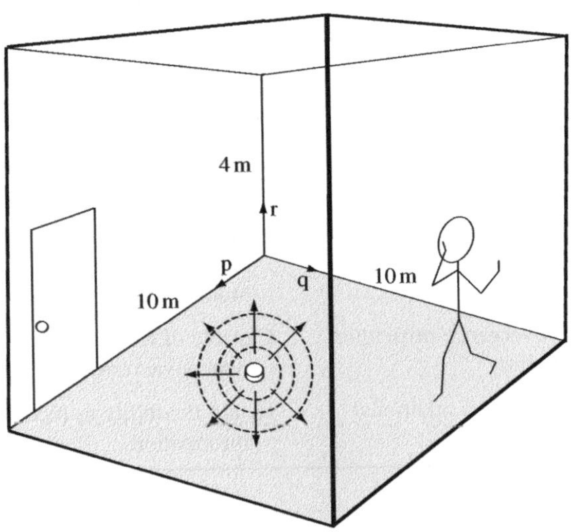

Figure 10.1 Finding a buzzer in a dark room

the inverted $\underline{x}^{<est>}$ should justify the measurements $\underline{y}$ given a reasonable physical model in $\underline{\underline{h}}$, such as the 3D geometric attenuation of sound. However, there are only M = 100 measurements and N = 400 000 unknowns (room volume/voxel volume). While a larger voxel size would reduce the number of unknowns, it would also cause a decrease in resolution. On the other hand, regularization would reduce ill-conditioning, but it would smear the source. Certainly, the voxel-based representation of the solution space is inappropriate in this case.

Instead, the problem can be cast as the identification of the spatial location p_{buz}, q_{buz}, and r_{buz} of a single source. There are only three unknowns in this representation. (Note: the buzzer intensity is also unknown but the problem can be cast in terms of normalized amplitudes.) The solution would proceed as follows: guess the buzzer position $(p_{buz}, q_{buz}, r_{buz})^{<0>}$, predict the sound amplitudes along the wall $\underline{y}^{<pred>}$, compare with measured amplitudes $\underline{y}^{<meas>}$ and modify the estimate of the buzzer position until measured and predicted amplitudes minimize some error norm (any norm could be selected).

Trade-offs

A priori information about the system is readily used; for example, a single point source is assumed above. This is a strong restriction on the solution but it permits high resolution to be attained in the inverted results. The solution cannot be computed as a one-time matrix inversion problem, but it requires successive forward simulations (disadvantages with this approach are addressed in Section 10.3). In some cases, the number of unknowns can be reduced even further if the problem is expressed in terms of dimensionless π ratios (Buckingham's π theorem, Section 8.4). Examples are presented in Chapter 11 in the context of tomographic imaging.

10.1.2 Flexible Narrow-band Representation

The parametric representation suggested above is too restrictive when there is limited information about the solution or if the physical phenomenon is not well defined, as in the case of a diffused sound source or the evolution of a chemical reaction in a body.

Flexible representations of the solution space can be attained with a limited number of unknowns within the spirit of Ockham's criterion. Consider the tomographic problem or other similar boundary value problems of the form $\underline{y} = \underline{\underline{h}} \cdot \underline{x}$. The unknown field of slowness can be approximated with a Fourier series with a limited number of terms "c" (procedure in Sections 8.4.2 and 9.5.3). Then, the pixel values $\underline{x}$ [N × 1] are expressed as

$$\underline{x} = \underline{\underline{S}} \cdot \underline{X} \tag{10.1}$$

where the vector $\underline{X}$ [c × 1] contains the c unknown coefficients in the series, and each entry in the matrix $\underline{\underline{S}}$ [N × c] depends on the known coordinates p and q of each pixel. The original inverse problem $\underline{y} = \underline{\underline{h}} \cdot \underline{x}$ becomes $\underline{y} = \underline{\underline{h}} \cdot \underline{\underline{S}} \cdot \underline{X}$. The least squares solution of this problem is (Section 9.3)

$$\underline{X}^{<est>} = (\underline{\underline{S}}^T \cdot \underline{\underline{h}}^T \cdot \underline{\underline{h}} \cdot \underline{\underline{S}})^{-1} \cdot \underline{\underline{S}}^T \cdot \underline{\underline{h}}^T \cdot \underline{y} \tag{10.2}$$

and from Equation 10.1,

$$\underline{x}^{<est>} = \underline{\underline{S}} \cdot \underline{X}^{<est>} = \underline{\underline{S}} \cdot (\underline{\underline{S}}^T \cdot \underline{\underline{h}}^T \cdot \underline{\underline{h}} \cdot \underline{\underline{S}})^{-1} \cdot \underline{\underline{S}}^T \cdot \underline{\underline{h}}^T \cdot \underline{y} \tag{10.3}$$

This transformation has advantages when the number of unknowns is reduced from the original problem, $c < N$. The invertibility of the matrix $\underline{\underline{S}}^T \cdot \underline{\underline{h}}^T \cdot \underline{\underline{h}} \cdot \underline{\underline{S}}$ [c × c] should not be an issue because the order of the Fourier series "c" is selected to avoid ill-conditioning.

10.1.3 Solution in the Frequency Domain

The flexible narrow-band representation introduced above is a preamble to the solution of inverse problems in the frequency domain discussed here. The procedure consists of transforming the inverse problem to the frequency domain, assembling the solution in the frequency domain, and computing its inverse Fourier transform to obtain the solution in the original space. Although this may sound impractical at first, the advantages of implementing convolution in the frequency domain suggest otherwise (Section 6.3). This formulation applies to problems in which boundary measurements $\underline{y}$ are line integrals of the unknown field parameter x(p, q) that varies in the p–q space

$$y = \int_{\substack{\text{across}\\\text{space}}} x(p, q)\, dr \tag{10.4}$$

Tomography is a clear example: the field parameter is either slowness or attenuation, and boundary measurements are travel time or amplitude.

Let us assume a straight ray propagation model and define a *parallel projection* as the line integral of the parameter x(p, q) along parallel rays, as shown in Figure 10.2. To facilitate the visualization of the problem, assume a body with some opacity in the p–q space x(p, q), a long fluorescent tube on one side at an angle α, and a screen on the other side also at an angle α. The shadow on the screen is the parallel projection of the body in the α-orientation.

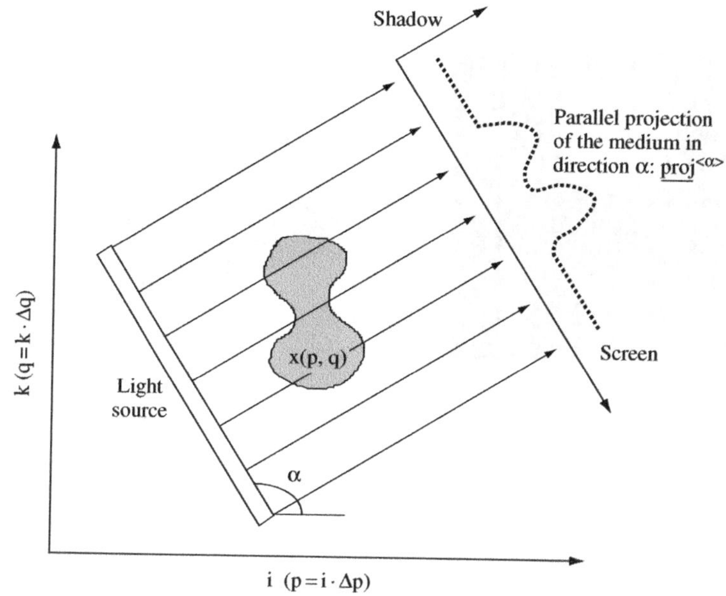

Figure 10.2 Parallel projection. The parallel projection or "shadow" obtained from illuminating a medium with an inclusion x(p, q) is the line integral of the host medium and the inclusion, from the source on one boundary to the screen on the other boundary

The Fourier Slice Theorem

The 1D Fourier transform of a parallel projection at an angle α is equal to a concentric slice taken at an angle α of the 2D Fourier transform of the p-q space.

Figure 10.3 presents a graphical confirmation of the Fourier slice theorem. The image involves a square medium discretized in 32×32 pixels and an inclusion. The 2D transform of the image is shown at the top right-hand side. The three frames in the first row present the horizontal projection, the 1D Fourier transform of this projection, and the slice taken from the 2D transform of the image in a direction parallel to the projection. The second row presents the same sequence of frames for the vertical projection. The equality between the 1D transform of the projections and the corresponding slices of the 2D transform of the image confirms the theorem.

Let us now proceed with an analytical demonstration. The 2D Fourier transform of the medium is (Equation 5.37 – Section 5.8)

$$X_{u,v} = \sum_{i=0}^{M-1} \left[\sum_{k=0}^{N-1} x_{i,k} \cdot e^{-j\left(v \cdot \frac{2\pi}{N} \cdot k\right)} \right] \cdot e^{-j\left(u \cdot \frac{2\pi}{M} \cdot i\right)} \tag{10.5}$$

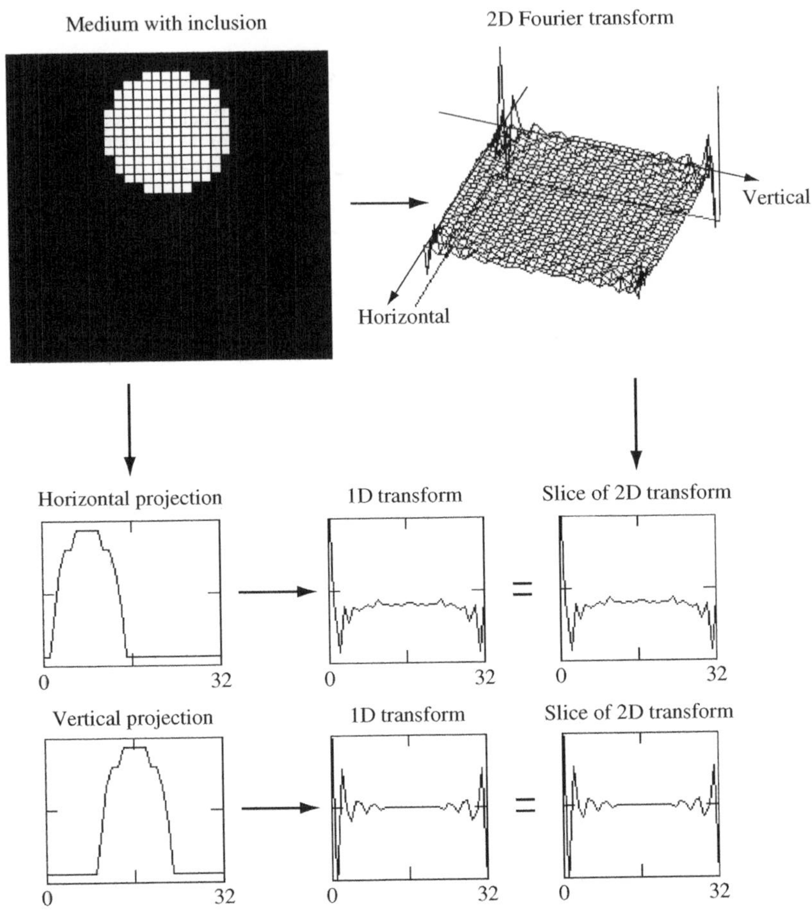

Figure 10.3 Graphical verification of the Fourier slice theorem. Observe the identity between the 1D Fourier transform of the projections and the corresponding slices of the 2D Fourier transform of the medium with the inclusion

For simplicity and without loss of generality, consider the parallel projection along the q-axis (Figure 10.4). According to the Fourier slice theorem, this should correspond to the slice of the 2D Fourier transform of the object for $v = 0$,

$$X_{u,0} = \sum_{i=0}^{M-1} \left[\sum_{k=0}^{N-1} x_{i,k} \right] \cdot e^{-j\left(u \cdot \frac{2\pi}{M} \cdot i\right)} \tag{10.6}$$

However, the term in brackets is the summation of $x_{i,k}$ on k, which is the parallel projection of $\underline{\underline{x}}$ along the q-axis. Therefore, this proves that the slice of the 2D

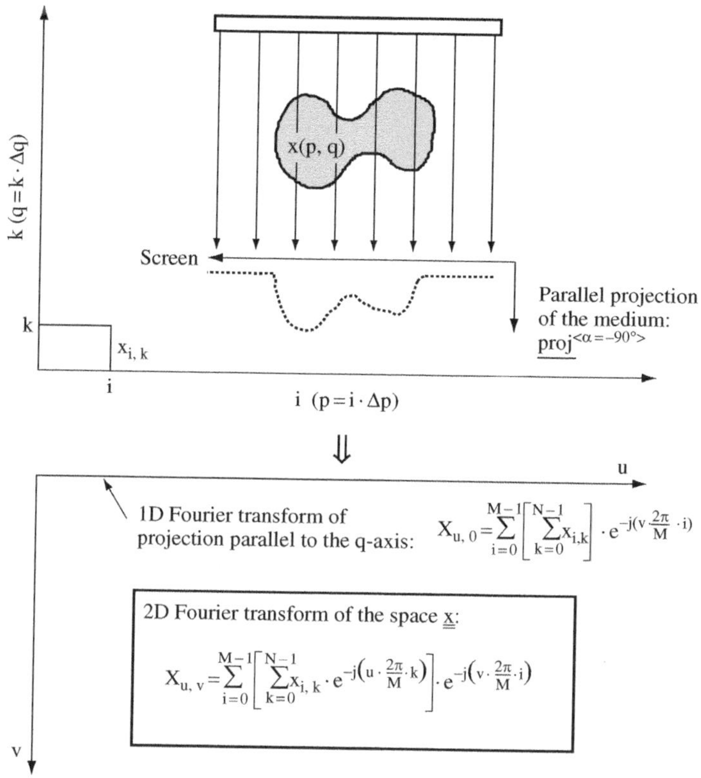

$$X_{u,0} = \sum_{i=0}^{M-1} \left[\sum_{k=0}^{N-1} x_{i,k} \right] \cdot e^{-j\left(v \cdot \frac{2\pi}{M} \cdot i\right)}$$

$$X_{u,v} = \sum_{i=0}^{M-1} \left[\sum_{k=0}^{N-1} x_{i,k} \cdot e^{-j\left(u \cdot \frac{2\pi}{M} \cdot k\right)} \right] \cdot e^{-j\left(v \cdot \frac{2\pi}{M} \cdot i\right)}$$

Figure 10.4 Analytical demonstration of the Fourier slice theorem. The DFT of the parallel projection of the medium in the q-direction yields the 2D transform of $\underline{\underline{x}}$ for $v = 0$

discrete Fourier transform of the object $\underline{\underline{x}}$ is the 1D Fourier transform of the corresponding parallel projection.

Inversion

One approach to invert the boundary measurements follows directly from the Fourier slice theorem: (1) Compute the 1D discrete Fourier transform of measured projections, (2) assemble the 2D discrete Fourier transform of the medium $\underline{\underline{X}}$ as prescribed in the Fourier slice theorem ($\kappa_p - \kappa_q$ space), and (3) compute the 2D inverse discrete Fourier transform of $\underline{\underline{X}}$ to determine $\underline{\underline{x}}$ which is the sought variation of the parameter $x_{i,k}$ in the p–q space (Implementation Procedure 10.1).

Implementation Procedure 10.1 Inversion in transformed space – Fourier slice theorem

Basic procedure

1. Determine parallel projections at different angles α: $\underline{proj}^{<\alpha>}$. This is an array of parallel measurements in which sources and receivers are aligned at an angle α with respect to a reference axis. Each measurement is a line integral of the spatial parameter x(p, q).
2. Compute the 1D-DFT of each projection: $\underline{PROJ}^{<\alpha>} = DFT(\underline{proj}^{<\alpha>})$.
3. Assemble the projections $\underline{PROJ}^{<\alpha>}$ in the Fourier space, according to their orientations α, along a radial line from the origin ($u = 0, v = 0$). Interpolate the values from the radial line to the Cartesian grid (u, v). This is the 2D Fourier transform of the image $\underline{\underline{X}}$.
4. Compute $\underline{\underline{x}} = IDFT(\underline{\underline{X}})$. This is the discrete version $\underline{\underline{x}}$ in the original domain.

Filtered back-projection

1. Determine projections, $\underline{proj}^{<\alpha>}$. Compute the 1D-DFT of each projection, $\underline{PROJ}^{<\alpha>} = DFT(\underline{proj}^{<\alpha>})$.
2. Multiply each $\underline{PROJ}^{<\alpha>}$ by a linearly increasing high-pass filter. The value of the filter at frequency κ is equivalent to the width of the wedge between projections in the Fourier space. For example, if there are "g" equally spaced projections, the filter at wavenumber κ has a value $2\pi \cdot \kappa/g$. This is the filtered transformed projection $\underline{FPROJ}^{<\alpha>}$.
3. Compute the inverse 1-D Fourier transform of $\underline{FPROJ}^{<\alpha>}$ to obtain the filtered projection $\underline{fproj}^{<\alpha>}$ in the space of the image.
4. Smear the inverted filtered projections $\underline{fproj}^{<\alpha>}$ onto the p–q space, along the ray paths, interpolating among cells in the p–q grid.
5. Add the contribution of all filtered back-projections on each pixel in the p–q space to obtain the sought solution $\underline{\underline{x}}$.

The following observations permit the development of an even more effective algorithm:

- The matrix $\underline{\underline{X}}$ in the frequency domain is assembled by gradually adding 1D transformed projections. Given the linearity property of the Fourier transform, the field $\underline{\underline{x}}$ in the original domain can be constructed as a summation of inverted transformed projections.
- The assembled 1D transformed projections fan out in $\kappa_p - \kappa_q$ space, and they are independent in the $\kappa_p - \kappa_q$ space except from shared static DC component at $(u = 0, v = 0)$.
- As the wavenumber increases, the separation between records increases. Therefore, there is an uneven coverage of the frequency domain (high coverage close to the static component but decreasing away from it). This can be corrected by multiplying the transformed projection by a function that increases linearly with the wavenumber κ. This linear high-pass filter cancels the shared static component at the origin; hence, filtered transformed projections become independent of each other.

The *filtered back projection* algorithm outlined in Implementation Procedure 10.1 improves the original procedure introduced earlier following these observations.

Trade-offs

The filtered back projection algorithm starts forming the inverted solution $\underline{\underline{x}}^{<est>}$ as soon as the first projection is obtained, and it only requires 1D Fourier transforms. Therefore, inversions are computed very fast and with significant savings in memory requirements. A disadvantage in this algorithm is the need to interpolate diagonal entries along the projection directions onto the 2D Cartesian space of the solution.

10.2 ITERATIVE SOLUTION OF SYSTEM OF EQUATIONS

Matrix inversion can be avoided by solving the system of equations $\underline{y}^{<meas>} = \underline{\underline{h}} \cdot \underline{x}^{<est>}$ using iterative algorithms (also known as the Kaczmarz solution of simultaneous equations). Iterative algorithms gradually correct the estimate of $\underline{x}$ in order to reduce the discrepancy between the measured data $\underline{y}^{<meas>}$ and the predictions $\underline{y}^{<pred>}$ made with the current estimate $\underline{x}^{<s>}$ after the s-th iteration.

10.2.1 Algebraic Reconstruction Technique (ART)

ART updates the vector of unknown parameters $\underline{x}$ by redistributing the residual every time a new measurement is analyzed. Updating starts with the first measurement $i = 1$, proceeds until the last M-th measurement is taken into consideration, and continues starting with first measurement again. The algorithm is stopped when residuals reach a predefined convergence criterion. Each updating of the solution $\underline{x}$ is considered one iteration.

The residual for the i-th measurement after the s-th iteration $(e_i)^{<s>}$ is computed with the s-th estimate $\underline{x}^{<s>}$:

$$\begin{aligned}(e_i)^{<s>} &= y_i^{<meas>} - \left(y_i^{<pred>}\right)^{<s>} && \textit{for the i-th measurement} \\ &= y_i^{<meas>} - \sum_k h_{i,k} \cdot x_k^{<s>} && \textit{after the s-th iteration} \end{aligned} \qquad (10.7)$$

This residual is distributed among values $\underline{x}^{<s>}$ according to their participation in the i-th measurement. The new $s+1$ estimate of the solution $\underline{x}$ is computed as

$$x_k^{<s+1>} = x_k^{<s>} + (e_i)^{<s>} \cdot \frac{h_{i,k}}{\sum_k \left(h_{i,k}\right)^2} \qquad \begin{aligned}&\textit{based on the i-th measurement}\\ &\textit{after the s-th iteration}\end{aligned} \qquad (10.8)$$

Note that when the i-th measurement is being considered, the k-th unknown x_k is updated only if it is affected by the i-th measurement; in other words, if $h_{i,k} \neq 0$.

Implementation Procedure 10.2 presents the step-by-step algorithm for ART. The solution of a simple system of equations using ART is presented at the end of this chapter together with insightful error evolution charts.

Implementation Procedure 10.2 Iterative solution of equations – ART

Algorithm

1. Start with an initial guess of the solution $\underline{x}^{<0>}$.
2. For the first iteration $s = 1$, consider the first measurement $i = 1$. Update the values of the solution $\underline{x}$ as prescribed below.
3. Repeat for other measurements. If the last measurement has been considered $i = M$, start from the first measurement $i = 1$ again. The counter for iterations s continues increasing.

4. Monitor the residual $\underline{e}$ and the solution $\underline{x}$. Stop iterations when the residual reaches a predefined level or when the solution fluctuates about fixed values.

Updating procedure

During the s-th iteration, the solution $\underline{x}$ is updated to justify the i-th measurement as follows:

- Predict the value of the i-th measurement $y_i^{\langle pred \rangle}$ given the current estimate of the solution $\underline{x}^{\langle s \rangle}$

$$\left(y_i^{\langle pred \rangle}\right)^{\langle s \rangle} = \sum_k h_{i,k} \cdot \left(x_k^{\langle est \rangle}\right)^{\langle s \rangle}$$

- Compute the error e_i between the i-th measurement $y_i^{\langle meas \rangle}$ and the predicted value $y_i^{\langle pred \rangle}$

$$(e_i)^{\langle s \rangle} = y_i^{\langle meas \rangle} - \left(y_i^{\langle pred \rangle}\right)^{\langle s \rangle}$$

- Distribute this error so that the new estimate of $\underline{x}$ for the s+1 iteration becomes

$$(x_k)^{\langle s+1 \rangle} = \left(x_k^{\langle est \rangle}\right)^{\langle s \rangle} + (e_i)^{\langle s \rangle} \cdot \frac{h_{i,k}}{\sum_k \left(h_{i,k}\right)^2} \qquad \textit{(update all N unknowns)}$$

Note: the k-th unknown remains the same if it is not involved in the i-th measurement so that $h_{i,k} = 0$.

10.2.2 Simulataneous Iterative Reconstruction Technique (SIRT)

SIRT also distributes the residual onto the solution $\underline{x}$, but $\underline{x}$ is not updated until all measurements have been considered. Then, each value x_k is updated "simultaneously" considering all corrections computed for all M measurements or equations. Each simultaneous updating of the solution $\underline{x}$ is considered one iteration.

In particular, the correction of the k-th unknown x_k owing to the i-th equation can be made proportional to the value of $h_{i,k}$ relative to the sum of all coefficients in the k-th column of $\underline{\underline{h}}$: $h_{i,k}/\sum_k h_{i,k}$. Within this weighting scheme, the equation to update the solution in SIRT becomes

$$x_k^{<s+1>} = x_k^{<s>} + \sum_i (e_i)^{<s>} \cdot \frac{h_{i,k}}{\sum_k (h_{i,k})^2} \frac{h_{i,k}}{\sum_i h_{i,k}} \quad \textit{based on all M measurements after the s-th iteration} \qquad (10.9)$$

Because matrix multiplication involves summation, this equation can be rewritten in matrix form:

$$\begin{aligned} \underline{x}^{<s+1>} &= \underline{x}^{<s>} + \underline{\underline{\Psi}} \cdot \underline{\underline{\Theta}}^T \cdot \underline{\underline{\Lambda}} \cdot \underline{e}^{<s>} \\ &= \underline{x}^{<s>} + \underline{\underline{\Pi}} \cdot \underline{e}^{<s>} \end{aligned} \qquad (10.10)$$

where

$\underline{x}^{<s>}$ and $\underline{x}^{<s+1>}$ — *solution vectors* $[N \times 1]$ *after* s *and* s + 1 *iterations*

$\Psi_{k,k} = \left[\sum_i h_{i,k}\right]^{-1}$ — *diagonal matrix* $[N \times N]$

$\Theta_{i,k} = (h_{i,k})^2$ — *matrix of size* $[M \times N]$

$\Lambda_{i,i} = \left[\sum_k (h_{i,k})^2\right]^{-1}$ — *diagonal matrix* $[M \times M]$

$\underline{e}^{<s>}$ — *vector of residuals* $[M \times 1]$ *after the* s*-th iteration*

M is the number of equations and N is the number of unknown parameters. Equation 10.10 is written in terms of three matrices $\underline{\underline{\Psi}}$, $\underline{\underline{\Theta}}$ and $\underline{\underline{\Lambda}}$ to highlight the correspondence with Equation 10.9. However, the product $\underline{\underline{\Pi}} = \underline{\underline{\Psi}} \cdot \underline{\underline{\Theta}}^T \cdot \underline{\underline{\Lambda}}$ is a matrix that depends on the entries $h_{i,k}$ and is computed once. While Equation 10.10 is based on matrix operations, it does not involve computing the inverse of a matrix. The algorithm proceeds as follows: (1) compute the residuals $\underline{e}^{<s>}$ for a given estimate $\underline{x}^{<s>}$, (2) update the estimate as $\underline{x}^{<s+1>} = \underline{x}^{<s>} + \underline{\underline{\Pi}} \cdot \underline{e}^{<s>}$, and (3) repeat. A solved example is presented at the end of this chapter.

10.2.3 Multiplicative Algebraic Reconstruction Technique (MART)

Iterative algorithms can also be developed to reduce deviations from 1.0 when the measured value $y_i^{<meas>}$ is divided by the predicted value $y_i^{<pred>}$

$$\frac{y_i^{<meas>}}{\left(y_i^{<pred>}\right)^{<s>}} = \frac{y_i^{<meas>}}{\sum_k h_{i,k} \cdot x_k^{<s>}} \quad \textit{for the i-th measurement after the s-th iteration} \qquad (10.11)$$

The MART algorithm updates the estimate of $\underline{x}$ to satisfy $y_i^{<meas>}/y_i^{<pred>} = 1.0$:

$$x_k^{<s+1>} = \left(\frac{y_i^{<meas>}}{\sum_k h_{i,k} \cdot x_k^{<s>}}\right) \cdot x_k^{<s>} \quad \text{update } x_k \text{ only if } h_{i,k} \neq 0 \qquad (10.12)$$

Note that the x_k should not be updated if $h_{i,k} = 0$ when the i-th measurement is being considered. Furthermore, factorial updating requires that the initial guess of the solution to be nonzero, $x_i^{<s=0>} \neq 0$ for all i. A solved problem at the end of this chapter shows the first few iterations and the evolution of the solution and the residual.

10.2.4 Convergence in Iterative Methods

Iterations proceed until an acceptable residual is obtained or a convergence criterion is fulfilled. The simultaneous updating implemented in SIRT results in more stable convergence than ART and MART (see Solved Problems at the end of this chapter). Data inconsistency prevents standard iterative algorithms from converging to a unique solution: once the minimum residual is reached, the estimated parameters fluctuate as iterations progress.

If data are noisy, the amount of updating can be decreased to facilitate convergence. On the other hand, if the degree of inconsistencies is small, the rate of convergence can be "accelerated" by overcorrecting. Acceleration or deceleration is controlled by a coefficient imposed on the second term in Equations 10.8 and 10.9 for the ART and SIRT algorithms. The same effect is achieved in MART by adding an exponent to the parentheses in Equation 10.12. Acceleration strategies must be cautiously selected. In particular, the final fluctuations when the solution converges will be exacerbated if the acceleration coefficient increases with the number of iterations "s".

The rate of convergence in ART can be improved when the sequence of equations is arranged so that subsequent equations i and i + 1 are most dissimilar. This observation suggests the reordering of equations to optimize convergence.

10.2.5 Nonlinear Problems

The solution of nonlinear problems with iterative algorithms faces difficulties related to nonconvex error surfaces and multiple minima. Weakly nonlinear problems can be linearized using a first-order Taylor approximation within a sequential linearization and inversion scheme: (1) compute the coefficients $h_{i,k}^{<0>}$ assuming a linear model; (2) solve the system of equations $\underline{y}^{<meas>} = \underline{\underline{h}}^{<0>} \cdot \underline{x}$ using the selected iterative algorithm; the vector of unknowns $\underline{x}$ is the first estimate of $\underline{x}^{<0>}$; (3) determine the new coefficients $h_{i,k}^{<1>}$ using the nonlinear model and $\underline{x}^{<0>}$; and, (4) repeat from step 2 until the preestablished convergence criterion is fulfilled.

10.2.6 Incorporating Additional Information in Iterative Solutions

Information available to the analyst can be included as additional equations to the original system of equations $\underline{y} = \underline{\underline{h}} \cdot \underline{x}$:

$$\begin{bmatrix} y_1 \\ \dots \\ y_M \\ c_1 \\ \dots \end{bmatrix} = \begin{bmatrix} h_{1,1} & h_{1,2} & \dots & h_{1,N} \\ \dots & \dots & \dots & \dots \\ h_{M,1} & h_{M,2} & \dots & h_{M,N} \\ r_{1,1} & r_{1,2} & \dots & r_{1,N} \\ \dots & \dots & \dots & \dots \end{bmatrix} \cdot \begin{bmatrix} x_1 \\ x_2 \\ \dots \\ x_N \end{bmatrix} \tag{10.13}$$

where the coefficients r and c capture the additional information or constraints, such as solution smoothness. Equations can be weighted. Finally, an initial guess $\underline{x}_0$ can be incorporated by inverting the system of equations $\underline{\Delta y} = \underline{\underline{h}} \cdot \underline{\Delta x}^{<est>}$ to obtain $\underline{\Delta x}^{<est>}$, where $\underline{\Delta y} = \underline{y}^{<meas>} - \underline{\underline{h}} \cdot \underline{x}_0$. When an initial guess is used, the final estimate is $\underline{x}^{<est>} = \underline{x}_0 + \underline{\Delta x}^{<est>}$.

10.3 SOLUTION BY SUCCESSIVE FORWARD SIMULATIONS

Inverse problems can be solved by trial and error through successive forward simulations. This completely general approach is summarized as follows:

- Generate an estimate of the solution $\underline{x}^{<est>}$.
- Use forward simulation to compute $\underline{y}^{<pred>} = f(\underline{x}^{<est>})$.
- Determine the residual between $\underline{y}^{<pred>}$ and $\underline{y}^{<meas>}$ using a selected error norm.

- Generate another estimate and repeat.
- Continue until a physically meaningful estimate is found that adequately justifies the data.

The main disadvantage in inverse problem solving by successive forward simulations is the massive demand for computer resources. Therefore, the applicability of this approach is enhanced when considering the following improvements. (The algorithm is outlined in Implementation Procedure 10.3.)

Implementation Procedure 10.3 Successive forward simulations

Goal

To find a physically meaningful solution $\underline{x}$ given a set of measurements $\underline{y}^{<meas>}$ by successive forward simulations with a physical model that relates $\underline{x}$ to $\underline{y}$.

Procedure

1. Make an initial guess of $\underline{x}^{<0>} = (x_1^{<0>}, \ldots, x_k^{<0>}, \ldots, x_n^{<0>})$.
2. Compute the predicted values of $\underline{y}^{<pred>}$ using the forward simulator.
3. Compute the residuals between the predicted and the measured values of $\underline{y}$ with an error definition that weights measurements equally (Section 8.3).
4. Evaluate the norm of residuals. Select the L_1 norm to lower the sensitivity to outliers, the L_2 if Gaussian conditions are expected, or the L_∞ norm when low-noise data are available and the information density is uneven.
5. Generate a new solution and repeat from step 2. A new solution may be obtained at random (Monte Carlo) or it can evolve from the previous solution guided by some a priori information about the solution or by the local gradients in the error norm surface.
6. Repeat steps 2–5 until the norm of the residuals between the measured $\underline{y}^{<meas>}$ and predicted $\underline{y}^{<pred>}$ values is acceptable (data are justified or a minimum is reached) and a physically meaningful solution $\underline{x}$ is obtained.

Notes:

- *The method is applicable when the number of unknown parameters is small.*
- *When a minimum is reached, explore the space of the solution in its vicinity to verify that the global minimum was found.*

1. *Reduce the number of unknowns.* The technique can be efficiently combined with transformed solution spaces that reduce the number of unknowns, such as parametric characterization (Section 10.1). As the number of unknown parameters decreases and the model definition increases, the general inverse problem turns into parameter identification.
2. *Start from a suitable initial guess.* Data preprocessing helps identify an initial guess of the solution (details in Chapter 11). When the inverse problem is not one of a kind but repetitive, forward simulation permits assembling a library of "solved cases". Then, a suitable initial guess is identified by matching the measured data $\underline{y}^{<meas>}$ against the simulated data in stored cases.
3. *Use fast forward simulators.* The physical model $f(\underline{x})$ that is used in the forward simulation can be as complex as needed (for example, a nonlinear finite element simulation with intricate constitutive equations). However, it is often possible to replace complex simulators with effective models that properly capture the governing processes and parameters (for example, Green's functions).
4. *Implement a meaningful evaluation scheme.* The goodness of a given prediction $\underline{y}^{<pred>}$ can be assessed in relation to $\underline{y}^{<meas>}$ using any error definition and norm (Section 8.3). Furthermore, estimated solutions $\underline{x}^{<est>}$ can also be evaluated in terms of physical significance (this is equivalent to the role of regularization in matrix methods – Section 9.4). Select a computational effective and physical meaningful evaluation procedure.
5. *Adopt efficient search algorithms.* Successive estimates of the solution do not need to follow a grid-search pattern where the space of the solution $\underline{x}$ is systematically searched. Instead, the solution can be explored with a *Monte Carlo* approach by generating estimates of the solution $x^{<est>}$ with a random number generator. The Monte Carlo method is completely general and it does not get trapped in local minima. Furthermore, it provides information that can be used to determine statistical parameters.

Grid-search and Monte Carlo methods are computer intensive. Instead, the optimal solution estimate $\underline{x}^{<est>}$ can be identified following the steepest descent along the error surface until the minimum error is reached. The solution proceeds by changing one parameter x_k at the time and keeping track of variations in the error surface with respect to changes in each parameter to update gradients. Converge difficulties arise near local minima and when error surfaces are nonconvex.

10.4 TECHNIQUES FROM THE FIELD OF ARTIFICIAL INTELLIGENCE

Several techniques from the field of artificial intelligence can be applicable to the solution of inverse problems. Selected examples are explored next.

10.4.1 Artificial Neural Networks (Repetitive Problems)

Artificial neural networks (ANNs) are intended to reproduce cognitive processes by simulating the architecture of the brain. An ANN consists of layers of "neurons" that are mathematically interconnected so that they can relate input to output (Figure 10.5a). While each neuron can perform only a simple mathematical operation, their combined capability permits solution of complex problems.

A Neuron

Consider a neuron in an internal "hidden layer" (Figure 10.5b). The total input is a weighted combination of the incoming values v_i that were produced by neurons in the previous layer, according to the weights w_i of the corresponding connections:

$$\text{input} = \sum_i v_i \cdot w_i \tag{10.14}$$

Weights w_i establish the relative importance individual input values v_i have. The "activation function" determines the "neuron response"; the sigmoid function is commonly used:

$$\text{response} = \frac{1}{1 + e^{-\text{input}}} \qquad \textit{sigmoid function} \tag{10.15}$$

and the response varies between 0 and 1 for any positive or negative input. Nonlinear activation functions provide the network with the flexibility that is required to model nonlinear transformations between the input and the output.

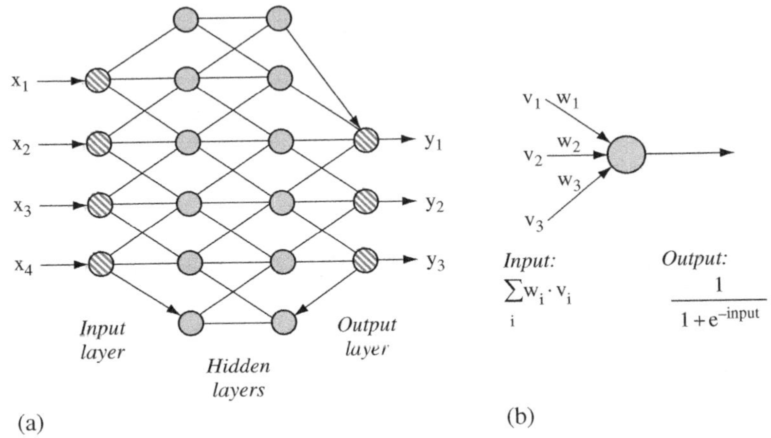

Figure 10.5 Artificial neural network. (a) The network consists of a structure of connected layers of elements or "neurons". The output and input are related through simple operations performed at each element. (b) The input of a single neuron is a linear combination of output values from the previous layer. The output of the neuron is computed with the activation function

Network Design

The activation function and the architecture of the network are selected first. The network is designed with a defined set of layers, neurons, and interconnections. The number of elements in the input vector $\underline{x}$ does not have to be the same as the number of parameters in the output $\underline{y}$. Typical networks involve hundreds of neurons that are distributed among the input and output layers and two or more hidden layers. Such a network includes thousands of connections (this is a measure of the number of unknowns).

Training

A critical step in the ANN methodology is the "assimilation" of the network to the nonlinear transformation that is intrinsically embedded in available input–output data. These data are the training set. During "network training", the weights w_i are determined for all connections. The training procedure is an inversion exercise itself: compute the output for a given input, compare with the known output, compute the residual for each parameter, and "back-propagate" the difference onto the network readjusting the weights; repeat for all case histories. Once trained, the network is used to process new cases outside the training set.

Trade-offs

ANNs are versatile. The power of an ANN is evidenced in the solution of nonlinear problems when the underlying transformation is unknown or time-variant, or when the event takes place in a noisy environment. However, the methodology provides no physical or mathematical insight into the underlying transformation.

Multiple similar cases are needed to train the network; therefore, this approach is reserved to inverse problems involved in repetitive tasks, for example the identification/characterization of objects on a conveyor belt in a production line.

The network's ability to capture the transformation and to match the training set increases with the number of hidden layers and connections. However, large networks are less reliable when addressing problems outside the scope of the training set. In other words, an ANN does not escape the trade-off between accurate fitting prior data and credible prediction of new data (Section 8.4).

10.4.2 Genetic Algorithms

Genetic algorithms combine a constrained Monte Carlo generator of new potential estimates of the solution $\underline{x}^{<est>}$ with a forward simulator to compute $\underline{y}^{<pred>}$. Although the method does not seem different from those discussed in Section 10.3, it is unique in the way it generates potential solutions $\underline{x}^{<est>}$.

In this context, the vector of unknown parameters $\underline{x}$ is the gene. Then, given a couple of initial estimates of the solution, the goal of the algorithm is to gradually enhance the estimate $\underline{x}^{<est>}$ by "reproduction" and "natural selection". There are five sequential and repetitive stages (Figure 10.6):

- The algorithm *starts* with two guessed solutions (genes $\underline{x}^{<1>}$ and $\underline{x}^{<2>}$).
- The *combination* operator generates new solution alternatives (genotypes) by randomly cutting existing solutions (parents $\underline{x}^{<1>}$ and $\underline{x}^{<2>}$) and forming new ones (offspring $\underline{x}^{<3>}$ and $\underline{x}^{<4>}$).
- As in natural genetics, *mutations* are needed to introduce genetic variety. Mutation is implemented by randomly changing a site x_k in a new solution, with some probability "p". The new value for a mutated site x_k is selected within a physically possible predefined range.
- Evolution may get trapped in local minima because many mutations must occur at specific sites in order to improve the solution. The probability of getting out of this stage is very small. To help overcome this difficulty, a *permutation* operator is added. This operator exchanges sites on the gene; for example, the

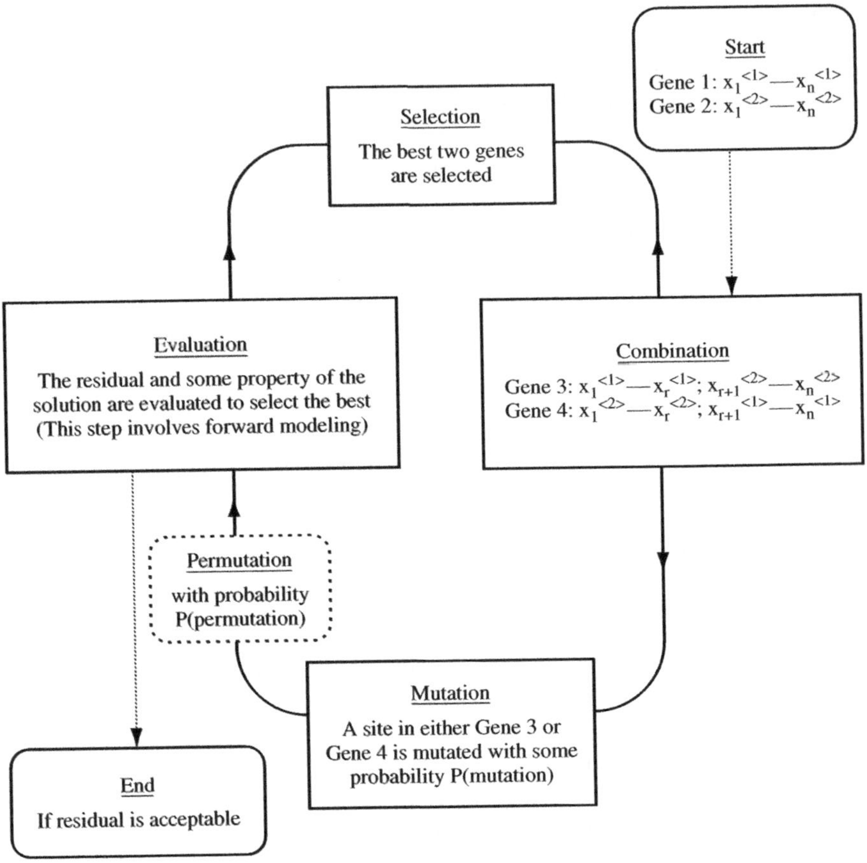

Figure 10.6 Genetic algorithm. A genetic algorithm starts with guessed solutions or "genes" $\underline{x}$. These genes are manipulated by repeating combination, mutation, permutation, evaluation, and selection, until an acceptable solution is obtained

value x_i is placed in location r and the value x_r is placed in location i. This operation is done with its own probability.

- Genes $(\underline{x}^{<1>}, \underline{x}^{<2>}, \underline{x}^{<3>}, \underline{x}^{<4>})$ are *evaluated* to identify the *fittest* ones. This operation starts by computing the values $\underline{y}^{<pred>}$ for each gene by forward simulation (the transformation may be linear or nonlinear). Then, the fitness of each gene is assessed by computing the residual $\underline{e} = \underline{y}^{<pred>} - \underline{y}^{<meas>}$ (any error definition and norm may be used). The evaluation criterion may also consider the physical meaningfulness of genes $\underline{x}^{<k>}$. Forward simulation and evaluation are the most computationally expensive steps in the algorithm.

- The fittest genes survive, and the process is repeated. Eventually, the solution $\underline{x}^{<k>}$ of the inverse problem gradually *evolves* towards the optimal solution.

In summary, the genetic algorithm approach to inverse problem solving is a semianchored Monte Carlo forward simulation where the search is conducted by hopping in the solution space $\underline{x}$ guided by the Darwinian process.

Evolution or convergence towards the optimal solution may stop when a very low mutation rate is used. However, when the mutation rate is too high, offspring are not selected and the convergence rate decreases because genetic information becomes easily corrupted; anchoring is lost, and the approach becomes the standard Monte Carlo. The probability of mutation for a given site in a gene near the optimum solution is about $p = 1/N$ (where N is the number of elements in a gene). At intermediate solutions away from optimum, other rates of mutation may be preferred. In general, the probability of permutation is smaller than the probability of mutation. Mutation and permutation probabilities are selected in an attempt to minimize the number of solutions $\underline{x}^{<est>}$ that are forward simulated and evaluated, while preventing entrapment in local minima.

Trade-offs

The solution progresses relatively fast in early iterations, even when many unknowns are involved. However, the algorithm does *not guarantee reaching the optimal solution*. Still, a "near optimal" solution may be sufficient for many inverse problems, for example when there is high uncertainty in the input parameters and in the transformation.

10.4.3 Heuristic Methods – Fuzzy Logic

Recall the problem of finding a buzzer in a dark room (Section 10.1). Our brain does not perform a formal inversion in terms of voxels or even with a parametric representation. Instead, we identify the possible position of the buzzer following simple heuristics and fuzzy logic.

Let us explore our natural ability to solve inverse problems further. Consider the study of projectile penetration in an opaque body using X-ray plates, as shown in Figure 10.7. Where is the projectile? The shadow detected on each plate is back-projected towards the source to constrain the volume where the projectile *may be*, and more importantly, where the projectile *cannot be*. Therefore, the location of the projectile becomes better defined as additional shadows are taken into consideration.

This heuristic approach can be computerized assuming that each projection $\underline{p}$ is a fuzzy set. Membership values are back-projected onto the space of the problem.

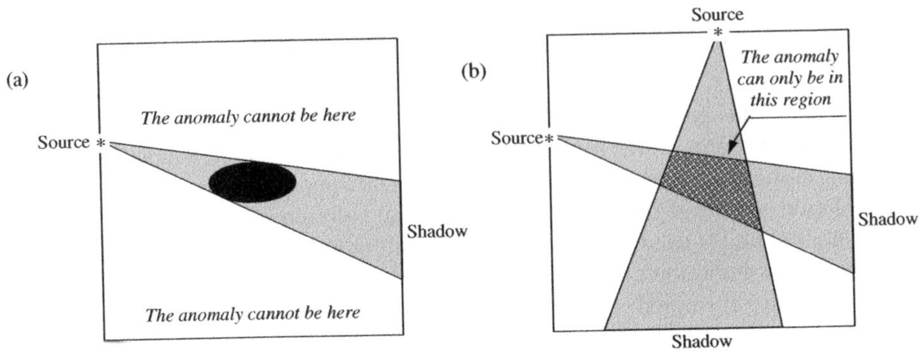

Figure 10.7 Fuzzy logic. Two flashlights illuminate the medium in different directions. (a) The presence of an anomaly causes shadows. (b) These shadows are back-projected to constrain the region in the space where the anomaly *cannot be*

The membership value of the i-th pixel μ_i to the set "background medium" is obtained as the minimum membership value of the back-propagated projections. Conversely, the membership value of the i-th pixel to the set "projectile" is obtained as the maximum of back-projected membership values. Finally, voxels in the body can be colored according to their membership values. If there are limited illumination angles, the location of the projectile would not be fully constrained and it would appear elongated in the direction of prevailing illumination. The algorithm can be expressed in the context of matrix data structures, as described in Implementation Procedure 10.4.

Implementation Procedure 10.4 Fuzzy logic for anomaly detection in tomographic imaging

Goal

To constrain the regions of the image where the anomaly cannot be.

Procedure

1. Divide the medium into a discrete set of N pixels.
2. Compute the length traveled by each of the M rays in each of the N pixels. Store these values in a matrix $\underline{\underline{h}}$ [M × N], as shown in Section 8.1.3.

3. Convert travel times to ray-average slowness (ras) for each i-th ray:

$$ras_i = \frac{t_i}{\sum_k h_{i,k}} \quad \textit{(the denominator is the length of the i-th ray)}$$

4. Compute the matrix $\underline{\underline{P}}$ ($M \times N$) of "touched pixels": divide each entry $h_{i,k}$ by the pixels size w, and round to either 0 or 1. If $h_{i,k}$ is much smaller than w, the ray touches the pixel but the travel time is not significantly affected by it.

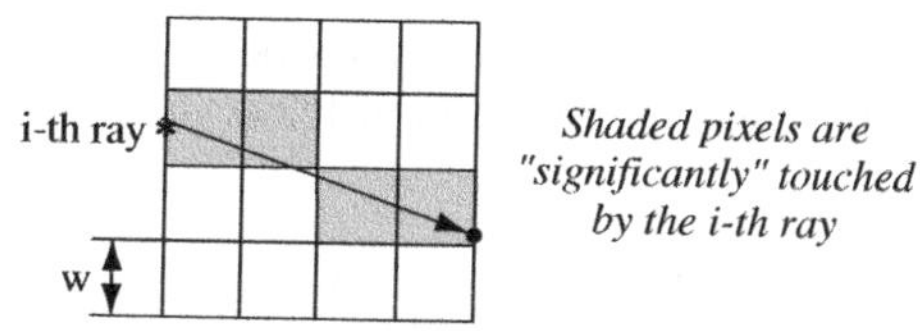

5. Back-project the ray-average slowness: replace the nonzero entries in the i-th row of $\underline{\underline{P}}$ by the corresponding average slowness ras_i computed for the i-th ray. This is the matrix $\underline{\underline{Q}}$.

6. *High slowness regions*: extract the minimum value of each column of $\underline{\underline{Q}}$

$$s_k^{<min\text{-}ave>} = \min[\underline{\underline{Q}}^{<k\ column>}]$$

7. *Low slowness regions*: extract the maximum value of each column of $\underline{\underline{Q}}$

$$s_k^{<max\text{-}ave>} = \max[\underline{\underline{Q}}^{<k\ column>}]$$

8. Plot a tomogram by coloring pixels with the computed vectors of maximum or minimum average slowness.

Trade-offs

Heuristic methods often permit the identification of salient characteristics of the solution with minimum effort; however, they fail to provide the full solution (such as the true voxel values in the example above). The result can be used to generate an initial guess $\underline{x}^{<0>}$ for other algorithms (Section 9.5.2).

10.5 SUMMARY

- Various strategies can be applied to solve inverse problems, including those outlined in this and the previous chapters and suggested as exercises at the end of this chapter.
- The selected method reflects the analyst's perception of a viable approach. It must be compatible with the problem at hand, data availability, and computer resources.
- The Monte Carlo generation of possible solutions $\underline{x}$, combined with successive forward simulations, is the most flexible inversion approach. It is also the most computer intensive. It can be applied to solve any type of inverse problem and it can involve complex physical models. Furthermore, this approach may accommodate any error definition, error norm, and additional evaluation criteria. Therefore, one can test the ability of the solution $\underline{x}^{<est>}$ to justify the data $\underline{y}^{<meas>}$ as well as the physical meaningfulness of the solution.
- Monte Carlo searches can be guided to limit the number of possible solutions that are run through the simulator and evaluation functions. Guiding criteria should prevent entrapment in local minima.
- The parametric representation of the problem in terms of a small number of unknowns leads to robust inversions.
- Efficient algorithms can be developed when the problem is carefully analyzed. Examples range from Green's functions for forward simulators, heuristic approaches, and solutions in transformed domains.
- Increased efficiency is often accompanied by lessened flexibility and more restrictive assumptions.
- The trade-off between variance and resolution is inherent to all methods.

FURTHER READING

Deming, R. and Devaney, A. J. (1996). A Filtered Back-Projection Algorithm for GPR. Journal of Environmental and Engineering Geophysics. Vol. 0, No. 2, pp. 113–124.

Goldberg, D. (1989). Genetic Algorithms in Search, Optimization, and Machine Learning. Addison-Wesley, Reading, Mass.

Gordon, R. (1974). A Tutorial on ART. IEEE Transactions on Nuclear Science. Vol. NS–21, pp. 78–93.

Herman, G. T. (1980). Image Reconstruction from Projections. Academic Press, New York.

Kak, A. C. and Slaney, M. (1988). Principles of Computerized Tomographic Imaging, IEEE Press, New York. 329 pages.

Saad, Y. (2003). Iterative Methods for Sparse Linear Systems. Society for Industrial and Applied Mathematics, Philadelphia. 528 pages.

SOLVED PROBLEMS

P10.1 *Iterative solution of system of equations.* Use ART, SIRT, and MART to solve the system of equations $\underline{y}^{<meas>} = \underline{\underline{h}} \cdot \underline{x}$, where

$$\underline{y}^{\langle meas \rangle} = \begin{bmatrix} 5 \\ 10 \\ 9 \\ 13 \end{bmatrix} \qquad \underline{\underline{h}} = \begin{bmatrix} 1 & 0 & 2 \\ 1 & 2 & 3 \\ 0 & 3 & 3 \\ 1 & 4 & 2 \end{bmatrix} \qquad \underline{x} = \begin{bmatrix} x_1 \\ x_2 \\ x_3 \end{bmatrix}$$

ART solution (Equation 10.8): The first few iterations are

s	i	$x_1^{<s+1>} = x_1 + \Delta x_1$	$x_2^{<s+1>} = x_2 + \Delta x_2$	$x_3^{<s+1>} = x_3 + \Delta x_3$	$y_i^{<meas>}$	$y_i^{<pred>} = \sum h_{i,k} \cdot x_k$	Error distribution$_k$ $\frac{h_{i,1}}{\sum (h_{i,k})^2}$	$\frac{h_{i,2}}{\sum (h_{i,k})^2}$	$\frac{h_{i,3}}{\sum (h_{i,k})^2}$	$\sum (y_i^{<meas>} - y_i^{<pred>}) \times$ error distribution$_k$ Δx_1	Δx_2	Δx_3
1	1	1.00	1.00	1.00	5	3.00	0.200	0.000	0.400	0.40	0.00	0.80
2	2	1.40	1.00	1.80	10	8.80	0.071	0.143	0.214	0.09	0.17	0.26
3	3	1.49	1.17	2.06	9	9.69	0.000	0.167	0.167	0.00	−0.11	−0.11
4	4	1.49	1.06	1.95	13	9.60	0.047	0.190	0.095	0.16	0.65	0.32
5	1	1.65	1.71	2.27	5	6.18	0.200	0.000	0.400	−0.23	0.00	−0.47
6	2	1.42	1.71	1.79	10	10.20	0.071	0.143	0.214	−0.01	−0.03	−0.05

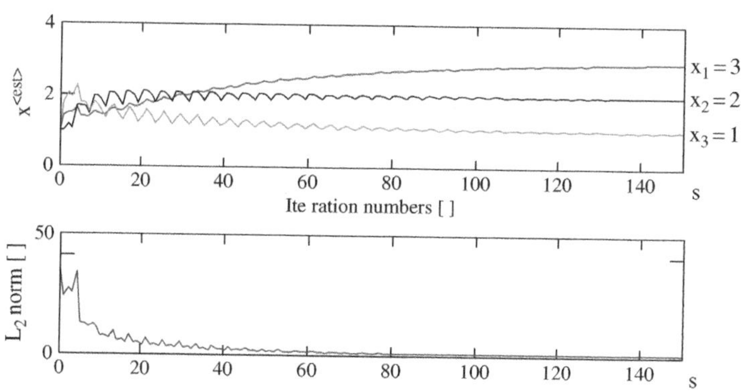

SIRT solution (Equation 10.10): $\underline{x}^{<s+1>} = \underline{x}^{<s>} + \underline{\underline{\Psi}} \cdot \underline{\underline{\Theta}}^T \cdot \underline{\underline{\Lambda}} \cdot \underline{e}^{<s>}$ where

$$\psi = \begin{bmatrix} 0.333 & 0 & 0 \\ 0 & 0.111 & 0 \\ 0 & 0 & 0.100 \end{bmatrix} \quad \Lambda = \begin{bmatrix} 0.2 & 0 & 0 & 0 \\ 0 & 0.071 & 0 & 0 \\ 0 & 0 & 0.056 & 0 \\ 0 & 0 & 0 & 0.048 \end{bmatrix} \quad \Theta = \begin{bmatrix} 1 & 0 & 4 \\ 1 & 4 & 9 \\ 0 & 9 & 9 \\ 1 & 16 & 4 \end{bmatrix}$$

s	$x_1^{<s+1>} = x_1 + \Delta x_1$	$x_2^{<s+1>} = x_2 + \Delta x_2$	$x_3^{<s+1>} = x_3 + \Delta x_3$	$y_i^{<meas>}$	$y_i^{<pred>} = \sum h_{i,k} \cdot x_k$	Error distribution $(y_i^{<meas>} - y_i^{<pred>})^T \cdot (\Delta \cdot \Theta \cdot \Psi)$		
						Δx_1	Δx_2	Δx_3
1	1.00	1.00	1.00	5, 10, 9, 13	3.00, 6.00, 6.00, 7.00	0.324	0.802	0.681
2	1.32	1.80	1.68	5, 10, 9, 13	4.69, 9.97, 10.45, 11.89	0.039	0.014	−0.024
3	1.40	1.82	1.66	5, 10, 9, 13	4.68, 9.97, 10.42, 11.94	0.039	0.012	−0.023
4	1.44	1.82	1.63	5, 10, 9, 13	4.67, 9.96, 11.39, 11.98	0.039	0.010	−0.021
5	1.48	1.84	1.61	5, 10, 9, 13	4.67, 9.96, 10.35, 12.02	0.039	0.009	−0.020
6	1.52	1.85	1.59	5, 10, 9, 13	4.67, 9.96, 10.32, 12.06	0.038	0.008	−0.019

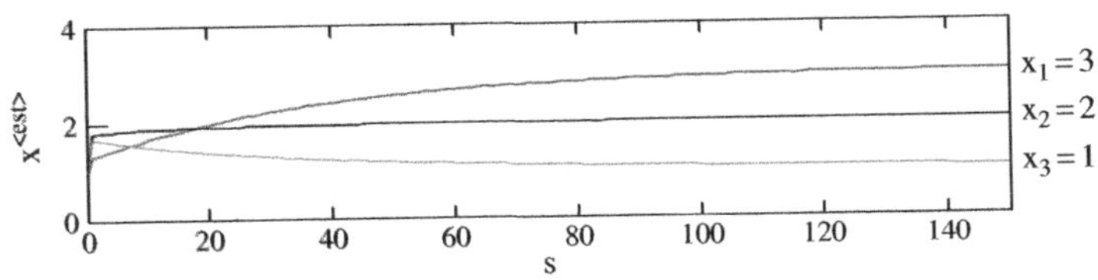

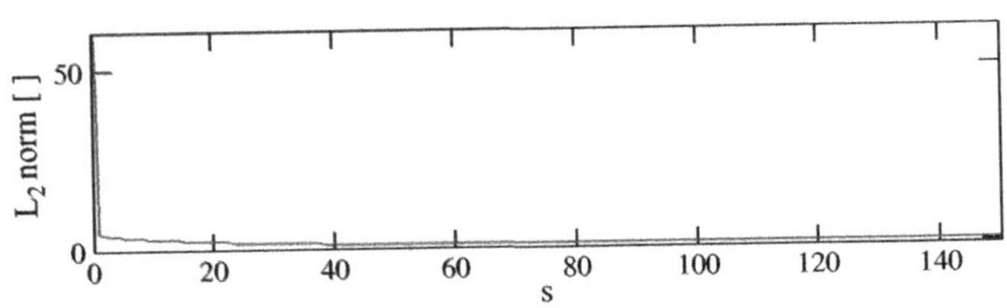

MART Solution (Equation 10.12). Update if $h_{i,k} \neq 0$. First few iterations,

s	i	$x_1^{<s>}$	$x_2^{<s>}$	$x_3^{<s>}$	$y_i^{<meas>}$	$y_i^{<pred>} = \sum_k h_{i,k} \cdot x_k$	$\text{correct}_i = \frac{y_i^{<meas>}}{y_i^{<pred>}}$	New values $x_k^{<s+1>} = (x_k^{<s>}) \cdot \text{correct}_i$		
								$x_1^{<s+1>}$	$x_2^{<s+1>}$	$x_3^{<s+1>}$
1	1	1.000	1.000	1.000	5	3	1.667	1.667	1.000	1.667
2	2	1.667	1.000	1.667	10	8.667	1.154	1.923	1.154	1.923
3	3	1.923	1.154	1.923	9	9.231	0.975	1.923	1.125	1.875
4	4	1.923	1.125	1.875	13	10.173	1.278	2.457	1.438	2.396
5	1	2.457	1.438	2.396	5	7.250	0.690	1.695	1.438	1.653
6	2	1.965	1.438	1.653	10	9.528	1.050	1.779	1.509	1.734

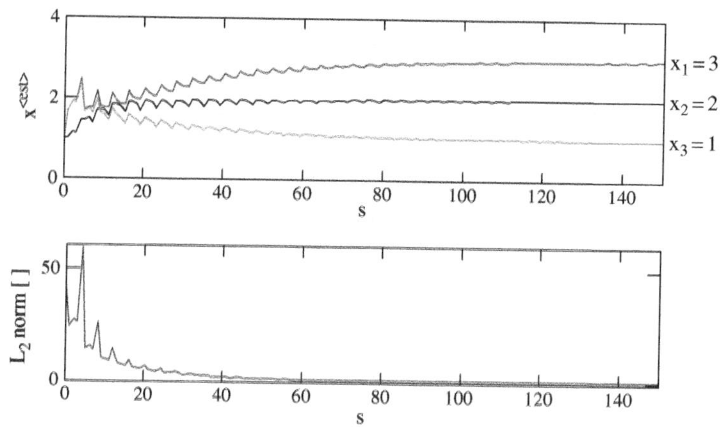

Note typical early oscillations in ART and MART algorithms.

P10.2 *Successive forward simulations.* Solve problem P10.1 using successive forward simulations and the L_2-norm.

Solution: Let us fix x_2 and x_3 and vary x_1. We find the value of x_1 that minimizes the L_2 norm, and then x_2 while keeping x_1 and x_3 fixed, and so on. Let us start at the initial guess: $x_1 = x_2 = x_3 = 2$. The following sequence of slices summarizes the search:

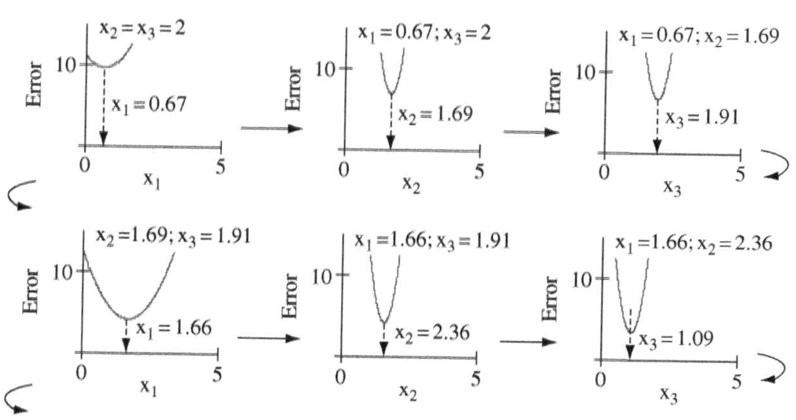

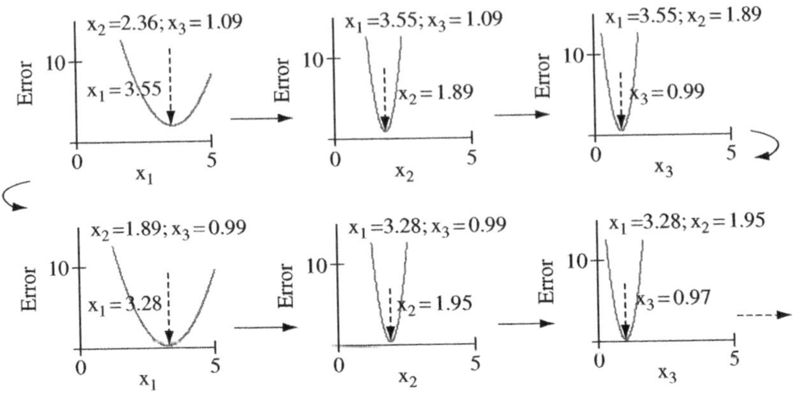

Note: The final result is $x_1 = 3$, $x_2 = 2$, $x_3 = 1$.

ADDITIONAL PROBLEMS

P10.3 *Iterative solution.* Invert the following system of equations using iterative algorithms. (a) Compare the evolution of the solution with ART and SIRT. (b) How many iterations are needed to reach 5% error? (c) Compare the results with the least squares solution (LSS). (d) Attempt the solution with MART. Discuss!

$$\begin{bmatrix} 9.8 \\ 12.1 \\ 14.2 \\ 15.7 \end{bmatrix} = \begin{bmatrix} 1 & 0 \\ 1 & 1 \\ 1 & 2 \\ 1 & 3 \end{bmatrix} \cdot \begin{bmatrix} a \\ b \end{bmatrix}$$

P10.4 *Transform methods: reducing the number of unknowns.* Study the ill-conditioning of $\underline{\underline{S}}^T \cdot \underline{\underline{h}}^T \cdot \underline{\underline{h}} \cdot \underline{\underline{S}}$ for different numbers of terms c (Equation 10.3). Modify the equation to improve frequency control. Discuss. Relate to RLSS.

P10.5 *Other methods.* Investigate other methods that can be used for the solution of inverse problems. Include: (a) linear programming and SIMPLEX algorithm, and (b) graph-search strategies. Outline the solution strategy in each case.

P10.6 *Application: parametric representation.* Consider a problem in your area of interest. Following Ockham's recommendation, cast the problem in parametric form using the smallest number of unknowns while still capturing the most important characteristics of the problem. Simulate data with noise for a given set of parameters $\underline{x}^{<true>}$. Explore the invertibility of the unknowns around $\underline{x}^{<true>}$ for different levels of noise. Plot slices of error surfaces computed with the L_1 L_2 and L_∞ norms. Recast the problem in terms of dimensionless π ratios and compare against the dimensional approach. Draw conclusions.

P10.7 *Application: solutions by forward simulations.* Consider a simple inverse problem in your area of interest. Program the forward simulator and time it. Detail the algorithm to solve the inverse problem using Monte Carlo, genetic algorithms, and ANNs.

P10.8 *Application: transformation methods.* Consider a problem in your area of interest. Can the inverse problem be inverted through a transformation? (Review the Fourier slice theorem in this chapter.)

11

Strategy for Inverse Problem Solving

Inverse problems are frequently encountered in engineering practice and scientific tasks. The solution of an inverse problem requires adequate understanding of the physics of the problem, proper experimental design, and a good grasp of the mathematics of inverse problem solving to recognize its inherent effects.

This chapter brings together knowledge gained in previous chapters to develop a comprehensive approach to inverse problem solving (review Implementation Procedure 9.3). The case of tomographic imaging is used to demonstrate concepts and methods. The experience gained from this example is readily transferable to other inverse problems in engineering and science.

11.1 STEP 1: ANALYZE THE PROBLEM

Successful inverse problem solving starts long before data inversion. In fact, the first and most important step is to develop a detailed understanding of the underlying physical processes and constraints, the measurement procedures, and inherent inversion-related difficulties. Then one must establish clear and realizable expectations and goals.

Let us analyze the inverse problem of tomographic imaging (recall Section 8.1.4). The following restrictions and difficulties must be considered.

11.1.1 Identify Physical Processes and Constraints

Tomograms can be generated with different physical processes and forms of energy. Commercially available devices include X-ray CAT scan, radar and

Discrete Signals and Inverse Problems J. C. Santamarina and D. Fratta

seismic tomography, ultrasonic imaging, positron emission tomography (PET), magnetic resonant imaging (MRI), and electrical resistivity tomography (ERT). The decision to use one form of energy determines the type of information that can be obtained from the tomographic image. For example, a tomogram generated with mechanical wave measurements captures the spatial distribution of elastic and inertial properties in the medium; on the other hand, a tomogram developed with electromagnetic wave propagation measurements reflects the spatial variability in electrical resistivity, dielectric permittivity, and magnetic permeability.

Wave propagation involves various intricate, albeit information-rich, phenomena that can be easily overlooked or misinterpreted. The most relevant complications related to the nature of wave propagation follow (see sketches in Figure 11.1):

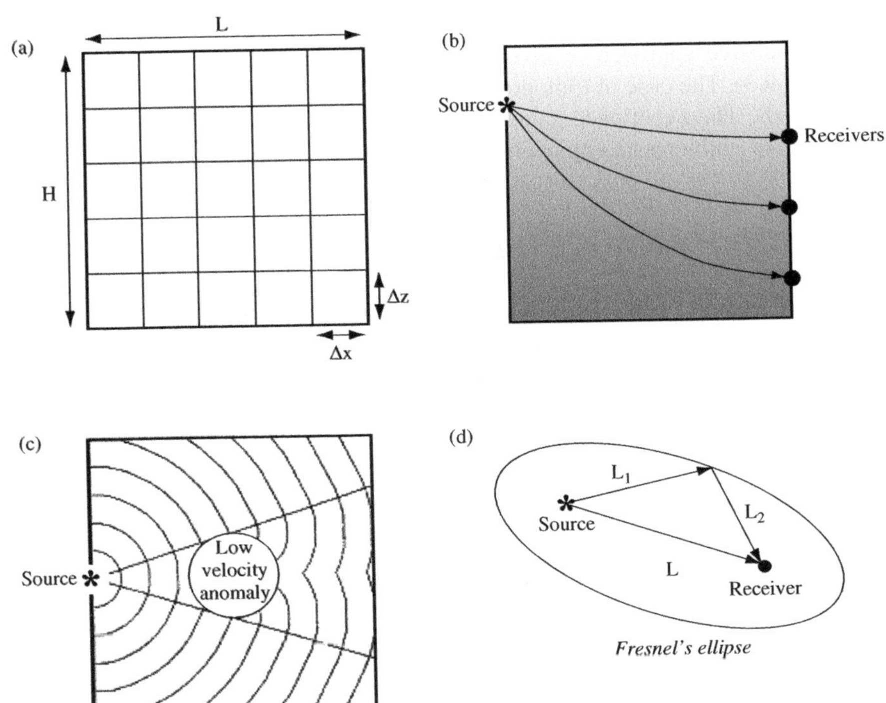

Figure 11.1 Tomographic imaging: (a) problem representation; (b–d) wave phenomena: ray bending, diffraction and Fresnel's ellipse

- *Attenuation.* Geometric spreading and material attenuation cause amplitude decay. Typically, high-frequency components are attenuated at higher rates. Low-pass material filtering incrementally rounds the wave train and biases the interpretation of signals with increasing travel distance.
- *Attenuation and noise.* Noise restricts the size of the body that can be imaged or demands higher input energy. Signal stacking and postprocessing may be required to improve the signal-to-noise ratio.
- *Trade-off: skin depth vs. resolution.* Long wavelengths are needed to penetrate large objects; however, long wavelengths provide lower spatial resolution.
- *Ray curvature.* Linear inversion presumes straight ray propagation. Spatial variability (the purpose of tomographic imaging) causes reflections, refractions, and ray bending (Figure 11.1b).
- *Anisotropy.* Materials such as wood, laminates, fiber-reinforced polymers, and rock masses can exhibit significant degree of anisotropy, which causes energy splitting or birefringence.
- *Diffraction.* Diffraction hides the presence of low-velocity anomalies (Figure 11.1c).
- *Fresnel's ellipse.* Wave propagation samples a region of the medium, not just the ray path. This region is related to the wavelength and the distance between the source and the receiver (Figure 11.1d). A "thick ray" may be preferred as a propagation model.

11.1.2 Address Measurement and Transducer-related Difficulties

Deficiencies during data gathering result in noisy data. Common testing difficulties in wave-based tomographic imaging include:

- *Source energy and frequency content.* In general, the frequency of emitted signals decreases with increasing source size and delivered energy.
- *Transducer directivity (sources and receivers).* A receiver positioned outside the radiation field of the source (and vice versa) will not detect the wanted signal.
- *Near field.* Sources and receivers in close proximity may operate in their near fields. Physical models for data interpretation are typically derived for far-field conditions and fail to explain data gathered in the near field.

- *Measurement precision* ε_t *and transducer separation.* Neighboring transducers need not be closer than $\sqrt{2 \cdot L \cdot V_{med} \cdot \varepsilon_t}$, where L is the distance across instrumented sides, V_{med} is the wave propagation velocity in the medium, and ε_t is the precision in travel time measurements.
- *Fresnel's ellipse and transducer separation.* There is no significant advantage in placing neighboring sources and transducers closer to each other than the width of the Fresnel's region: $\sqrt{2 \cdot L \cdot V_{med} \cdot \varepsilon_t + (V_{med} \cdot \varepsilon_t/4)^2}$.
- *Detectability.* The travel time in the medium, across a distance L is $t_o = L/V_{med}$. The change in travel time δt due to the presence of an inclusion size d_{inc} and velocity V_{inc} is

$$\frac{\delta t}{t_o} = \frac{d_{inc}}{L}\left(\frac{V_{med}}{V_{inc}} - 1\right)$$

  The value δt must exceed the precision in travel time measurements ε_t (averaging and error cancellation may improve this requirement).
- *Noise.* In most cases, background noise can be reduced with proper electrical, mechanical, and thermal isolation. Transducers and peripheral devices may add noise.
- *Systematic triggering error.* It cannot be corrected by stacking, but through calibration (it may be detected during data preprocessing).
- *Precision in travel time determination* ε_t. Travel time determination is enhanced in high-frequency signals. Automatic detection is fast, repetitive, and precise, but not necessarily accurate.

11.1.3 Keep in Mind Inversion-related Issues

- *Number of unknowns.* The number of unknown pixel values is $N = L \cdot H/(\Delta x \cdot \Delta z)$, where Δx and Δz define the pixel size (refer to Figure 11.1a). Even a low-resolution tomographic image made of 30×40 pixels involves $N = 1200$ unknown pixel values.
- *Number of measurements.* Assuming that transducers are mounted on each boundary pixel along vertical sides and a measurement is conducted between each source and receiver, the number of measurements is $M = (H/\Delta z)^2$. Such a transducer configuration in the 30×40 image discussed above results in $M = 40 \times 40 = 1600$ measurements or equations.
- *Available information.* A large number of rays (equations) does not necessarily imply an overdetermined condition when $M > N$. Many measurements may eventually provide the same information, effectively reducing the number of

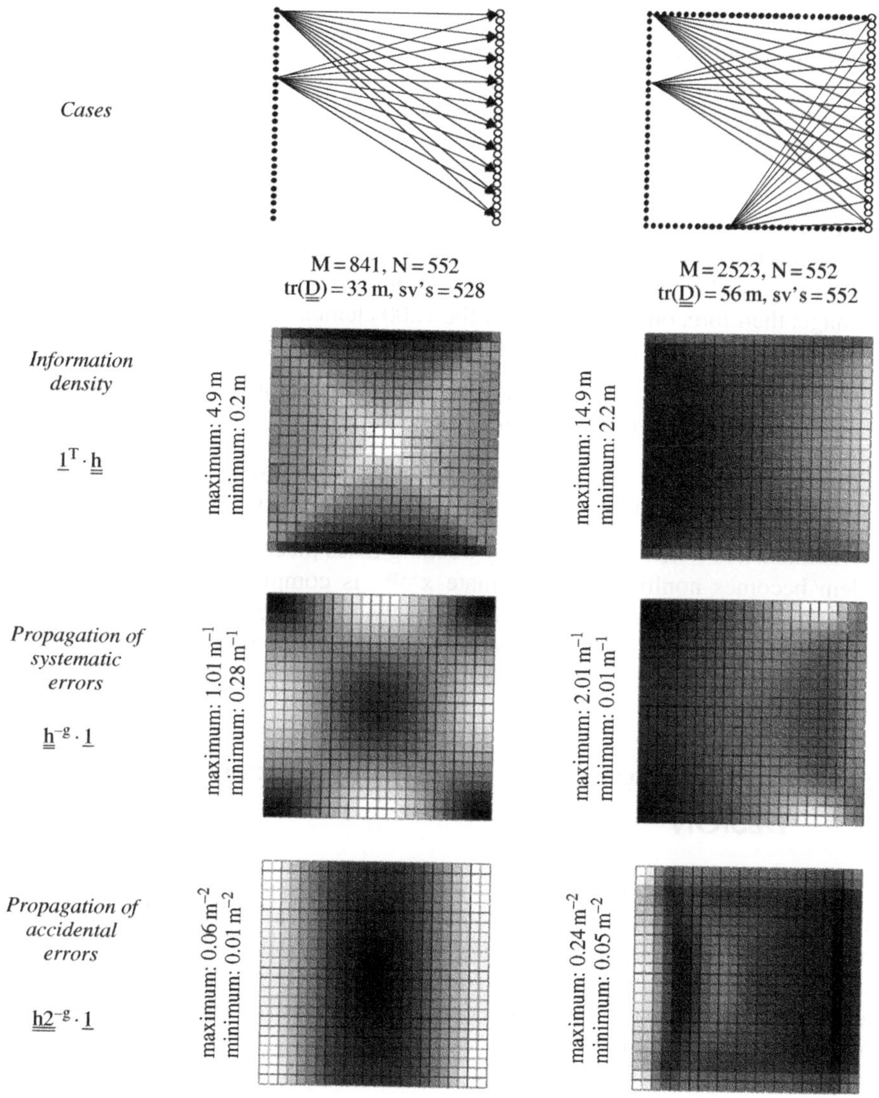

Figure 11.2 Experimental design: analysis of the transformation matrix. Generalized inverse: $\underline{\underline{h}}^{-g} = (\underline{\underline{h}}^T \cdot \underline{\underline{h}} + \lambda \cdot \underline{\underline{R}}^T \cdot \underline{\underline{R}})^{-1} \cdot \underline{\underline{h}}^T$ where $\lambda = 0.645\,\text{m}^2$. Number of singular values (sv's) correspond to condition number $\kappa > 10^4$. The quantity $\text{tr}(\underline{\underline{D}})$ is the trace of the data resolution matrix. Dark regions correspond to low values (either low information density or low error propagation)

independent observations. The availability of information can be studied with singular value decomposition (SVD), as discussed in Section 9.2. Figure 11.2 shows two configurations of sources and receivers; each ray represents a measurement. The singular values for the matrix of travel lengths $\underline{\underline{h}}$ are shown for each configuration.

- *Large and sparse matrices.* The matrix $\underline{\underline{h}}$ $[M \times N]$ in tomographic imaging problems is large and sparse (Equation 8.14). In the previous example of a low-resolution 30×40-pixel tomogram, the matrix $\underline{\underline{h}}$ is $[1600 \times 1200]$. However, each ray touches between one and two times the number of pixels across the image; therefore, only 30–60 out of the 1200 elements in each row are nonzero. Therefore, the transformation matiix $\underline{\underline{h}}$ is dccisively sparse.
- *Uneven spatial coverage.* Measurements do not sample the properties of the medium evenly, as shown in Figure 11.2.
- *Inversion parameters.* Inversion parameters (such as type and degree of regularization) should not determine the results of the inversion.
- *Nonlinearity.* When significant ray bending takes place, the tomographic problem becomes nonlinear: the estimate $\underline{x}^{<est>}$ is computed knowing $\underline{\underline{h}}$, whose entries $h_{i,k}$ are determined with the ray paths that are controlled by the spatial distribution of pixel values $\underline{x}$. The straight ray assumption applies to medical X-ray applications, but it is deficient in geotomography.

11.2 STEP 2: PAY CLOSE ATTENTION TO EXPERIMENTAL DESIGN

The viability of a solution and the attainable resolution are determined at this stage. Experimental design should address two critical aspects: distribution of measurements to attain a good coverage of the solution space, and instrumentation selection to gather high-quality data (Implementation Procedure 9.2). In addition, keep in mind that inverse problems can be data-intensive and costly; therefore, the selected test configuration should avoid unnecessary measurement duplication while preventing aliasing.

11.2.1 Design the Distribution Measurements to Attain Good Spatial Coverage

Available information, the even or uneven coverage of the solution space, the degree of ill-conditioning, and the potential for error propagation can be explored as soon as the matrix $\underline{\underline{h}}$ is formed, and before any data are gathered.

Let us apply guidelines in Implementation Procedure 9.2 to the tomographic inverse problem. Figure 11.2 shows results for two source and receiver configurations, including: (a) tabulated number of measurements, number of unknowns, and the trace of the data resolution matrix $D = \underline{\underline{h}} \cdot \underline{\underline{h}}^{-g}$; (b) a plot of sorted singular values; (c) the vector of column-sums $\underline{1}^T \cdot \underline{\underline{h}}$ presented as a 2D image to gain a preliminary assessment of spatial coverage of the solution space; and (d) the vectors $\underline{\underline{h}}^{-g} \cdot \underline{1}$ and $\underline{\underline{h2}}^{-g} \cdot \underline{1}$ that contain the row-sums in matrices $\underline{\underline{h}}^{-g}$ and $\underline{\underline{h2}}^{-g}$ (also presented as 2D images to identify pixels with highest potential for error magnification). Similar plots are generated for various test configurations until a realizable test design is identified to obtain adequate data.

Illumination anisotropy, particularly in the cross-wall configuration, will elongate the shape of inverted anomalies in the direction of prevailing illumination.

11.2.2 Design the Experiment to Obtain High-quality Data

Low-noise high-quality data are needed to reduce the effects of error magnification during inversion (Sections 9.6 and 9.8). The general tenet of experimentation "improve the test at the lowest possible level" gains even higher significance in inverse problems. Select appropriate transducers and peripheral electronics, shield them from external noise, and implement proper signal recording and processing methods. Correct measurements for the frequency response of the measurement system (review Implementation Procedures 4.1, 5.2 and 6.6).

11.3 STEP 3: GATHER HIGH-QUALITY DATA

Look at the raw data while they are being generated. Identify a suitable display that permits diagnosing test problems and even helps identify salient characteristics of the system. The simultaneous display of signals gathered at neighboring locations or time steps is particularly convenient to spot sudden changes in the system, to diagnose and remediate testing difficulties, and to identify possible outliers that can be retested.

11.4 STEP 4: PREPROCESS THE DATA

Data preprocessing refers to simple computations and graphical display strategies that are implemented to gain insight about the measurements (noise level, outliers, spatial coverage) and a priori characteristics of the solution (mean properties, spatial trends, presence of anomalies). These results facilitate the selection of the physical model, provide valuable information to guide and stabilize the inversion, and can be used to generate a viable initial guess of the solution $\underline{x}^{<0>}$. Several preprocessing strategies aimed at tomographic inversion are presented next.

11.4.1 Evaluate the Measured Data $\underline{y}^{<meas>}$

Data errors are magnified during inverse problem solving, affect the rate of convergence, and increase the presence of ghosts in the final images. Error magnification can be controlled with regularization, but it is often at the expense of resolution.

Systematic Error

A constant shift in travel time is most likely caused by the acquisition system, for example, trigger delay. This systematic error in the data can be identified by plotting travel time vs. travel length. If the medium is homogeneous and isotropic, measurements should plot on a straight line with zero time intercept; the inverse of the slope is the wave propagation velocity in the medium. A nonzero time intercept is the systematic error, and it can be removed from the data set before inversion.

Accidental Errors

Random errors in travel times are often associated with the determination of first arrivals. Accidental errors can be detected on average velocity plots. The average velocity for a ray is computed as the Pythagorean distance between the source and the receiver divided by the measured travel time.

Outliers

Gross measurement errors can be identified in average velocity plots and correspond to points that deviate few standard deviations away from the mean value. Obvious outliers should be removed from the data set. An equation is lost in $\underline{y} = \underline{\underline{h}} \cdot \underline{x}$, but the solution becomes more robust.

Case History: Kosciusko Bridge Pier

Tomographic data were obtained for a massive concrete pier underneath the Kosciusko bridge in New York City, under very noisy operating conditions. The cross-section of the pier and the location of sources and receivers are shown in Figure 11.3a. Travel time and mean ray velocity are plotted versus ray length in Figures 11.3b and c. A systematic triggering error, accidental errors, and outliers are evident.

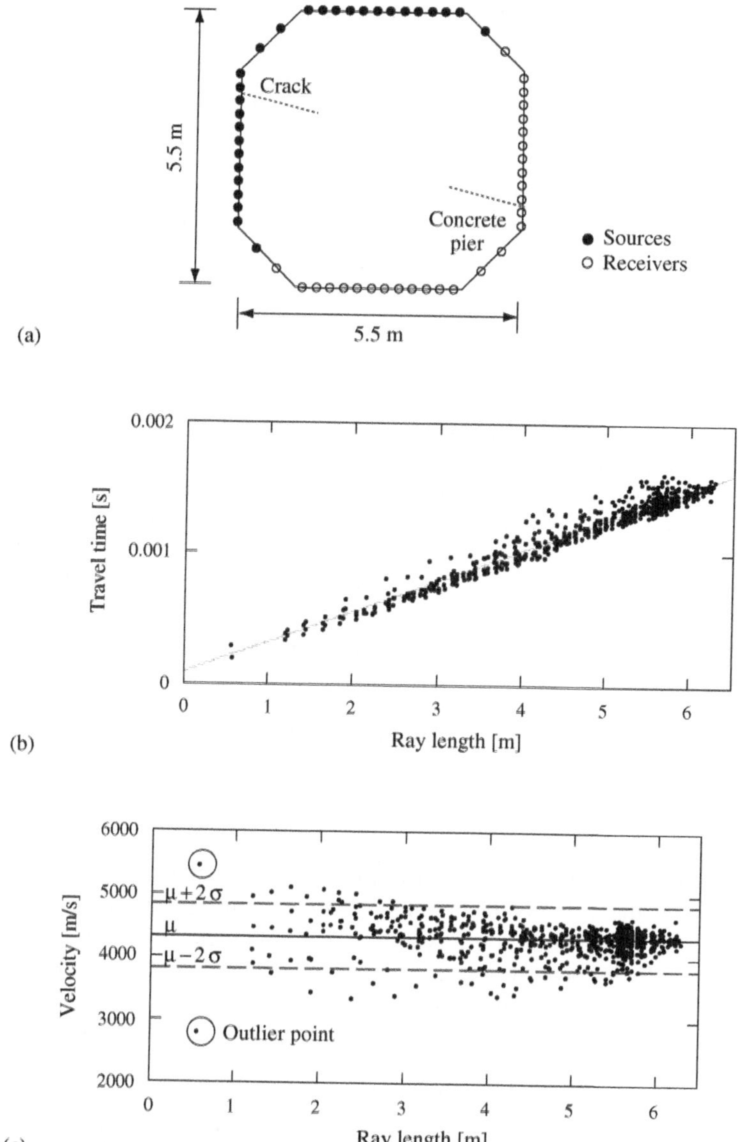

Figure 11.3 Kosciusko bridge pier – New York. Inspecting the data for systematic and accidental errors: (a) test setup; (b) systematic errors become evident in the travel time versus ray length plot; (c) average velocity versus travel length (after systematic error removal). These measurements were obtained under high ambient noise

11.4.2 Infer Outstanding Characteristics of the Solution $\underline{x}^{<est>}$

A glimpse at the characteristics of the solution $\underline{x}^{<est>}$, including general background characteristics and the presence of anomalies, can be gained by plotting projections of ray-average velocities versus position or direction. The compilation of all projections forms a 3D array called a sinogram; in parallel projections, the 3D array consists of the measured value vs. the projection direction and the sensor position. Anomalies that do not detectably affect projections or sinograms cannot be inverted.

Case History: Korean Tunnel

A cross-hole geotomographic study was conducted to detect a tunnel in Korea. Travel times were measured between two parallel boreholes. Measurements were repeated every 0.2 m for each of seven ensonification angles: +45°, +30°, +15°, 0°, −15°, −30°, and −45°, for a total of M = 1050 measurements (see sketch in Figure 11.4a). The ray-average velocity projection in Figure 11.4b shows the increase in background velocity with depth. On the other hand the variation of the ray-average velocity with ray angle in Figure 11.4c indicates global anisotropy in the host medium. All 1050 measurements are shown – the scatter reflects the variation of ray-average velocity with depth identified in Figure 11.4b.

Case History: Balloon in Air – Transillumination

A balloon filled with helium was fixed at the center of an instrumented frame in air and cross-hole travel time data were obtained using 16 sources mounted on one side of the frame, and 16 microphones mounted on the opposite side, for a total of M = 256 measurements (Figure 11.5a – the velocity of sound in air is 343 m/s; the velocity in helium is greater than air; the gas mixture and pressure inside the balloon are unknown). The complete sinogram and selected ray-average velocity projections or "shadows" are shown in Figures 11.5b and c. They clearly denote the high-velocity inclusion. Apparently accidental errors in the projections are actually coherent time shifts when all projections are seen together in the sinogram.

11.4.3 Hypothesize Physical Models that Can Explain the Data

Data preprocessing should also be implemented to gain information that can be used to select the physical model that relates the unknowns $\underline{x}$ to the measured data $\underline{y}^{<meas>}$.

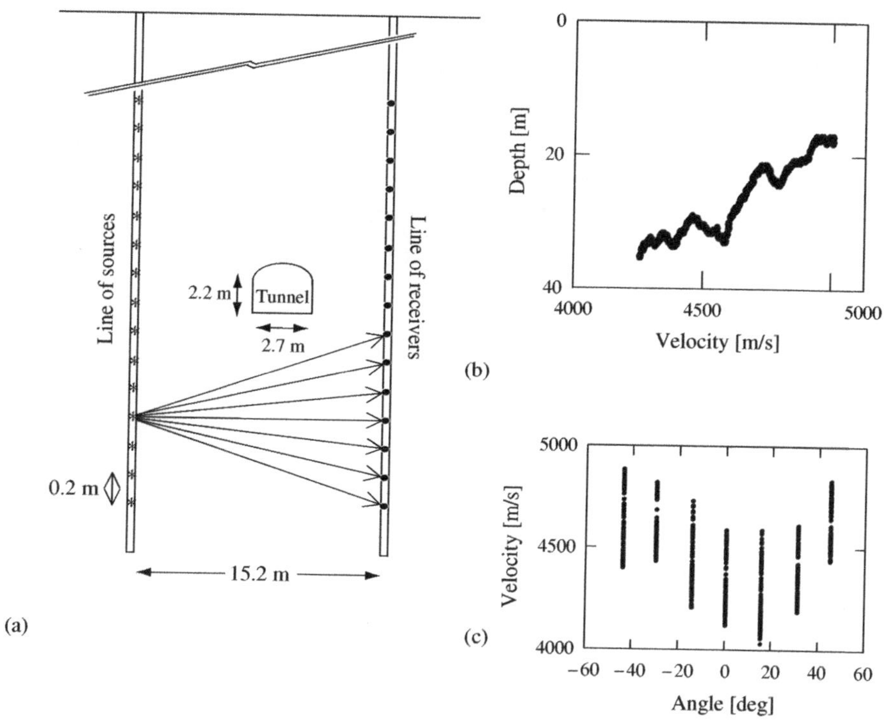

Figure 11.4 Tunnel in Korea: (a) source and receiver locations. Data were generated with seven ensonification angles: +45°, +30°, +15°, 0°, −15°, −30°, −45° (i.e. seven rays per source); (b) gradual increase in velocity with depth (ray-average velocity projection at +15°); (c) the effect of anisotropy: ray-average velocity versus ray inclination (Data courtesy of Dr R. Rechtien and Dr R. Ballard)

Case History: Concrete Monolith with Open Crack

Ultrasound transillumination data were gathered for a concrete monolith with an open crack cut across the block (Figure 11.6a). Travel time and ray-average velocity are plotted vs. ray length in Figures 11.6b and c. There are two distinct trends. Points on the linear trend in Figure 11.6b plot with constant ray-average velocity ~4700m/s in Figure 11.6c; these points correspond to measurements where the source and the receiver are both either above or below the open crack. How is energy propagating when the source and the receiver are on opposite sides of the crack? Virtually no energy goes across the open crack, and the detected signals correspond to wave propagation paths that go around the crack. The extra

System parameters:
Background medium: Air
Inclusion: Helium balloon

Medium velocity: $V_{med} = 343$ m/s
Inclusion velocity: $V_{inc} > V_{med}$
Inclusion radius: $R_{inc} = 0.23$ m
Inclusion coordinates: $x_{inc} = 0.75$ m and $y_{inc} = 0.75$ m

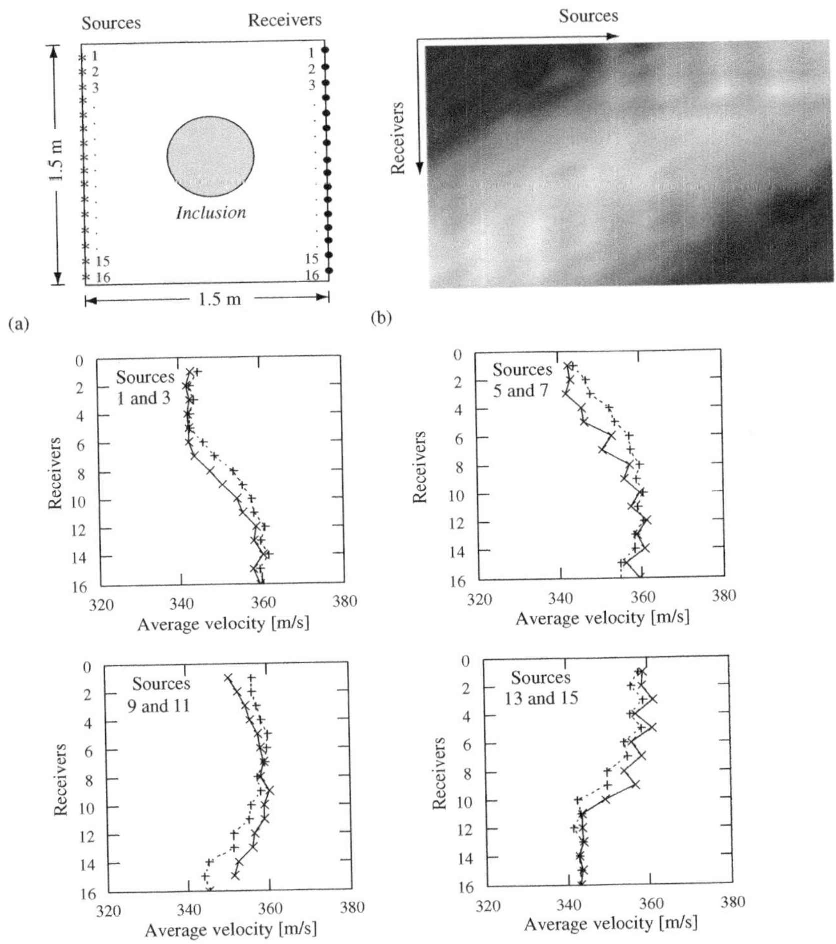

Figure 11.5 Large helium balloon in air. Transmission data: (a) test setup; (b) sinogram of ray-average velocities; (c) profiles of ray-average velocity help constrain the location of the balloon

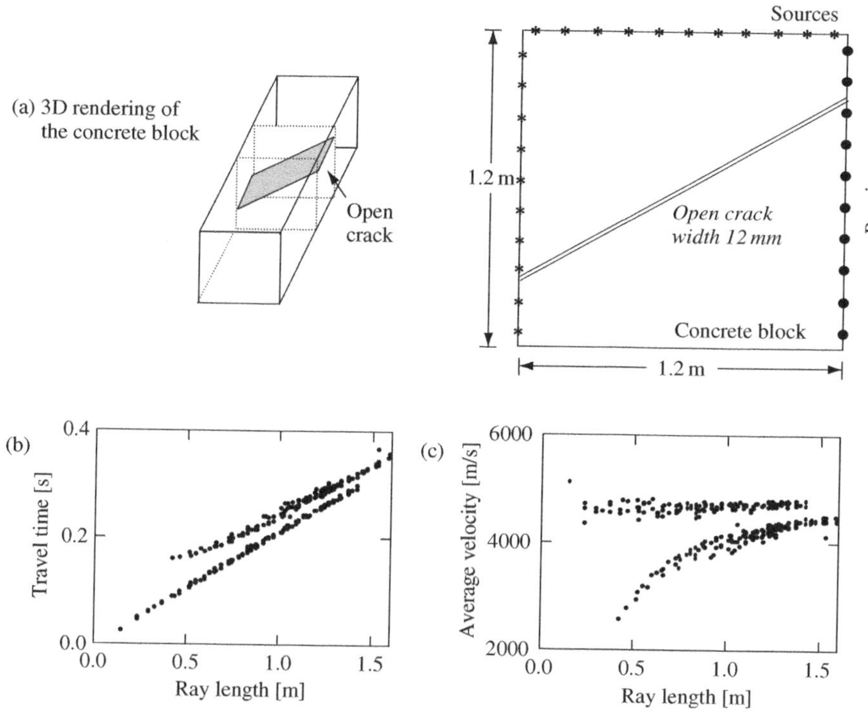

Figure 11.6 Concrete block data: (a) location of sources and receivers; (b) the travel time versus ray length plot suggests two definite propagation modes; (c) the average velocity versus ray length plot confirms that one mode of propagation takes place through a medium of constant velocity. The other trend approaches the constant velocity data as the ray length increases (Data courtesy of Ontario Hydro)

length in the out-of-plane path affects short rays more than long rays, according to the Pythagorean relation, and causes the trends observed in Figures 11.6b and c.

11.5 STEP 5: SELECT AN ADEQUATE PHYSICAL MODEL

The physical model selected to relate the measurements $\underline{y}^{<meas>}$ and the unknown parameters $\underline{x}$ must capture the essential features of the problem. An inappropriate model adds *model error* and hinders the inversion of a meaningful solution. In addition, the time required to compute the model is most important if a massive forward simulation strategy will be implemented for data inversion.

The simplest wave propagation model for travel time tomography is a straight-ray and it can be selected when ray theory applies and the spatial variation in velocity is small. The entries in the matrix $\underline{\underline{h}}$ are computed as the Pythagorean length between the intersections of the ray with the pixel boundaries. A simple example is shown in Figure 11.7. If the number of pixels is very large, the computation of accurate travel lengths loses relevance, and the Pythagorean computation can be reduced to "touched = 1" and "not touched = 0" (the row-sum of $\underline{\underline{h}}$ is then matched to the ray length).

Ray bending in heterogeneous media requires the development of efficient ray-tracing algorithms. Ray-tracing is a two-point boundary value problem: the

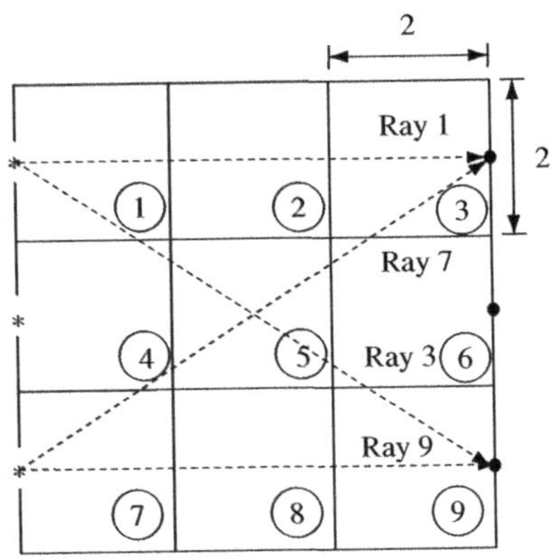

$$\underline{\underline{h}} = \begin{bmatrix} 22 & & 2 & 0 & 0 & 0 & 0 & 0 & 0 \\ \cdots & \cdots & \cdots & \cdots & \cdots & \cdots & \cdots & \cdots & \cdots \\ 1.85 & 0 & 0 & 0.55 & 2.4 & 0.55 & 0 & 0 & 18.5 \\ \cdots & \cdots & \cdots & \cdots & \cdots & \cdots & \cdots & \cdots & \cdots \\ \cdots & \cdots & \cdots & \cdots & \cdots & \cdots & \cdots & \cdots & \cdots \\ \cdots & \cdots & \cdots & \cdots & \cdots & \cdots & \cdots & \cdots & \cdots \\ 1.85 & 0 & 0 & 0.55 & 2.4 & 0.55 & 0 & 0 & 18.5 \\ \cdots & \cdots & \cdots & \cdots & \cdots & \cdots & \cdots & \cdots & \cdots \\ 0 & & 0 & 0 & 0 & 0 & 2 & 2 & 2 \end{bmatrix} \text{Rows represent rays}$$

Columns represent pixels

Figure 11.7 Ray tracing. Computing the entries of the matrix $\underline{\underline{h}}$ assuming straight rays. The rows shown in matrix $\underline{\underline{h}}$ correspond to the actual travel lengths of rays 1, 3, 7 and 9 end points of the ray are the known source and receiver positions, and the goal is to determine the ray path that satisfies *Fermat's principle of minimum travel time*. Close-form solutions for the ray path can be derived for simple velocity fields; an example is summarized in Table 11.1.

Table 11.1 Ray paths in a heterogeneous and anisotropic velocity field

Velocity field. It is defined as:

Vertical wave velocity V_v: linear with depth q: $V_v(q) = a + b \cdot q$

Constant anisotropy between V_v and V_h: $c = \frac{V_v(q)}{V_h(q)}$

Elliptical variation of velocity with ray angle q′: $V(q, q') = V_v(q) \cdot \sqrt{\frac{1+q'^2}{c^2+q'^2}}$

The a, b, and c parameters define the wave velocity field V(q, q′).

Ray path: The source and the receiver are at horizontal positions $p^{<R>}$ and $p^{<S>}$, and their vertical positions are such that the vertical velocities are $V_v^{<R>}$ and $V_v^{<S>}$ respectively. The depth q of the ray at position $p^{<R>} < p < p^{<S>}$ is

$$q = \sqrt{\left(\frac{V_v^{<S>}}{b}\right)^2 + (p - p^{<S>}) \cdot \left[\frac{\left(V_v^{<R>2} - V_v^{<S>2}\right)}{b^2 \cdot \left(p^{<R>} - p^{<S>}\right)} + c^2 \cdot (p^{<R>} - p)\right]} - \frac{a}{b}$$

this is the ray path.

Travel time. A differential of the ray length "dℓ" is $d\ell = dp \cdot \sqrt{1 + (q')^2}$, where the slope q′ of the ray at position p is obtained by differentiating the ray path,

$$q' = \left[\frac{\left(V_v^{<R>2} - V_v^{<S>2}\right)}{b^2 \cdot \left(p^{<R>} - p^{<S>}\right)} + c^2 \cdot \left(p^{<R>} + p^{<S>} - 2 \cdot p\right)\right] \frac{b}{2 \cdot (b \cdot q + a)}$$

Finally, the travel time is obtained by numerical integration along the ray path:

$$t = \int_S^R \frac{d\ell}{V(q, q')} = \int_S^R \frac{\sqrt{1+q'^2}\,dp}{V(q, q')} \approx \sum_{p^{<R>}}^{p^{<S>}} \frac{\sqrt{c^2+q'^2}}{(a+bq)} \Delta p$$

Examples:

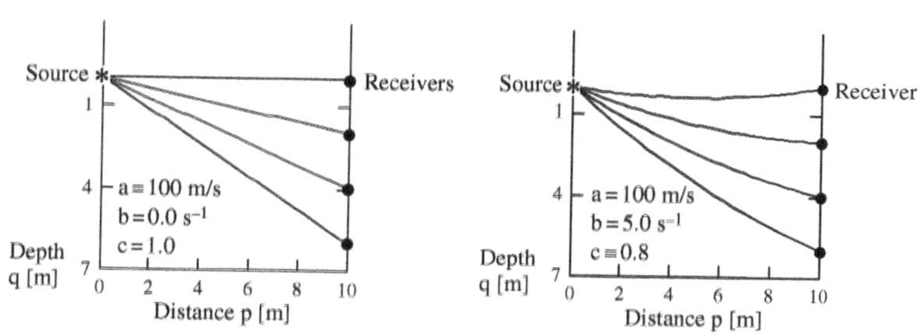

Note: Derived using calculus of variation – collaboration with M. Cesare.

Diffraction takes place when the wavelength approaches the size of the anomalies. In this case, full-wave solutions are preferred. Diffraction around low-velocity anomalies "heals" the wave front and hinders their tomographic detection (Figure 11.1c).

When diffraction or ray bending take place, the selection of a straight ray model for tomographic inversion results in poor quality images due to the amplification of model error.

11.6 STEP 6: EXPLORE DIFFERENT INVERSION METHODS

Invert the data using different inversion methods. Guided by Ockham's criterion, attempt to reduce the number of unknowns. Consider the parametric representation of the problem and invert the data by forward simulations. Then, explore less constrained representations, for example within the framework of matrix-based inversion. For repetitive problems, run multiple forward simulations and assemble a library of "solved cases" that can be used to identify an initial guess by data matching. Do not hesitate to explore other inversion strategies that may result from a detailed mathematical analysis of the problem or even heuristic criteria. The result $\underline{x}^{<est>}$ should reflect a balance between justifying the data $\underline{y}^{<meas>}$ (low error norm) and the physical meaningfulness of the solution $\underline{x}^{<est>}$.

11.6.1 Heuristic Methods

Heuristic inversion procedures are demonstrated next for data gathered in transillumination and reflection modes.

Case History: Steel Pipe in Air – Echolocation

Echolocation is extensively used by bats and dolphins, and in applications such as radar, sonar, and ultrasonic nondestructive material evaluation; the underlying physical concepts led to ultrasound imaging in medical diagnosis. Laboratory data were gathered using a hollow steel cylinder as an anomaly in air; short sound signals were emitted with a speaker and microphones detected the reflections. The source and receiver positions for each measurement are the foci of an ellipse that constrains the possible location of the reflecting anomaly: *the length of the string that is used to draw the ellipse is the velocity of the medium times the measured travel time*. Ellipses for all measurements are shown in Figure 11.8 for three different positions of the anomaly. The true location of the anomaly is also shown. This is the most rudimentary form of a geophysical technique known as migration! The graphical solution readily shows uncertainty in the inversion of R_{inc} (see Solved Problems at the end of this chapter for more details).

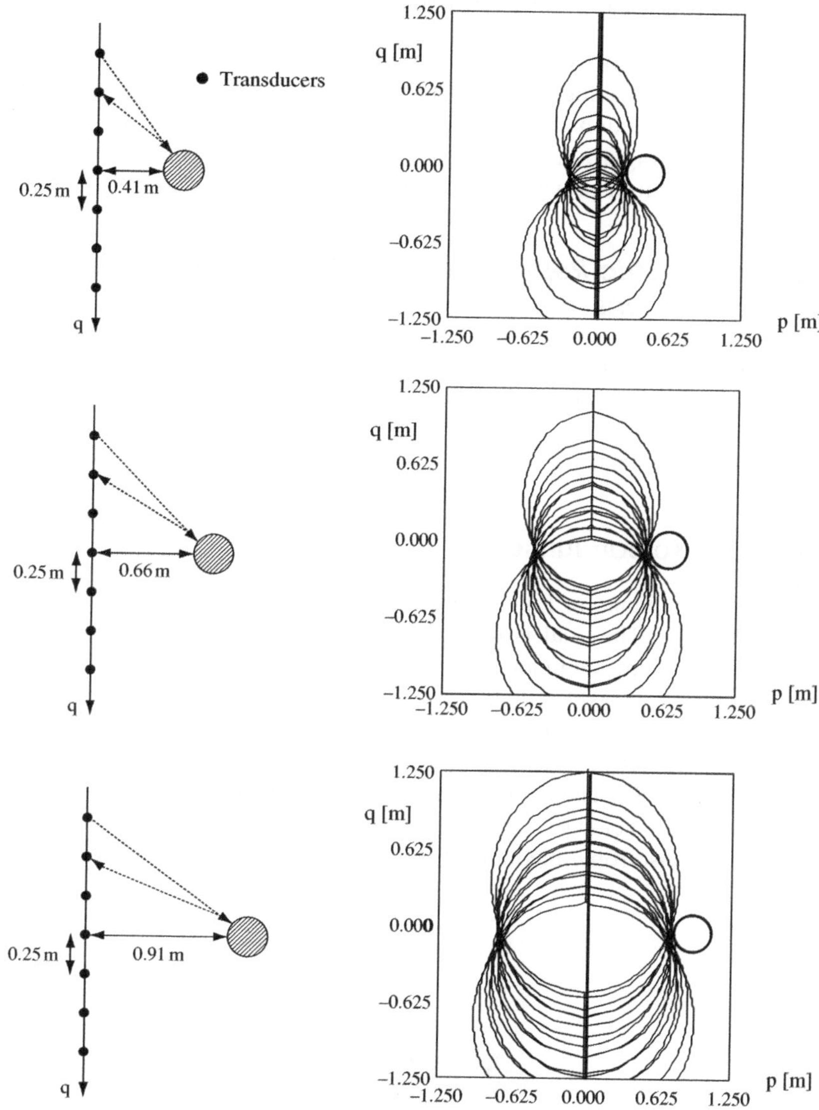

Figure 11.8 Constraining the position and size of the reflector with ellipses. The foci of each ellipse are at the source and the receiver locations for the corresponding measurement (along the centerline). Note the enhanced delineation of the first anomaly which is closest to the string of transducers (Data courtesy of S. Sloka)

Case History: Balloon in Air – Transillumination

The 16 ray-average velocity projections (Figure 11.5) are back-projected to constrain the location of the high-velocity anomaly following the fuzzy-logic procedure introduced in Section 10.4 (Implementation Procedure 10.4). The resulting image presented in Figure 11.9 clearly denotes the anomaly, which appears elongated in the direction of prevailing illumination.

11.6.2 Parametric Representation – Successive Forward Simulations

Data for three case histories are inverted next. The first example is the Korean tunnel and it is used to invert for the velocity field of the host medium. The other two examples address the detection of an anomaly using either reflection or transillumination data.

Case History: Korean Tunnel

The 1050 measurements in Figure 11.4 are analyzed using the close-form solution in Table 11.1. The goal is to identify the parameters of the velocity field by successive forward simulations guided by the L_1 and L_2 error norms. Four possible media are considered: homogeneous–isotropic ($a \neq 0, b = 0, c = 1.0$), homogeneous–anisotropic ($a \neq 0, b = 0, c \neq 1.0$), vertically heterogeneous and isotropic ($a \neq 0, b \neq 0, c = 1.0$), and vertically heterogeneous and anisotropic ($a \neq 0, b \neq 0, c \neq 1.0$). The comparisons between calculated and measured travel times and inverted velocity parameters are summarized in Figure 11.10. The heterogeneous–anisotropic medium fits the data with the least residual; although

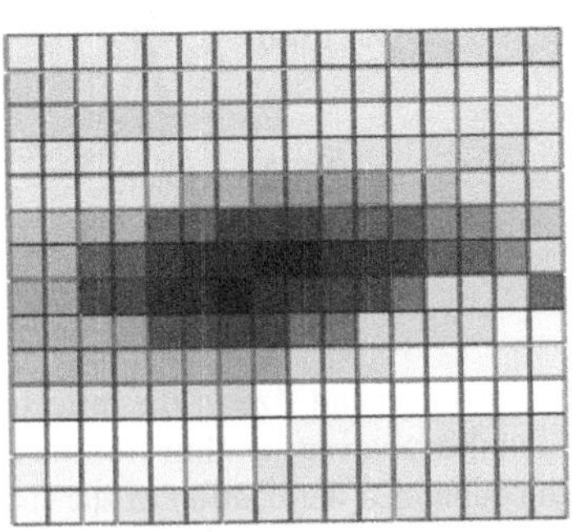

Figure 11.9 Tomographic study with helium ballon (laboratory data – Figure 11.5). Image generated with fuzzy-logic solution

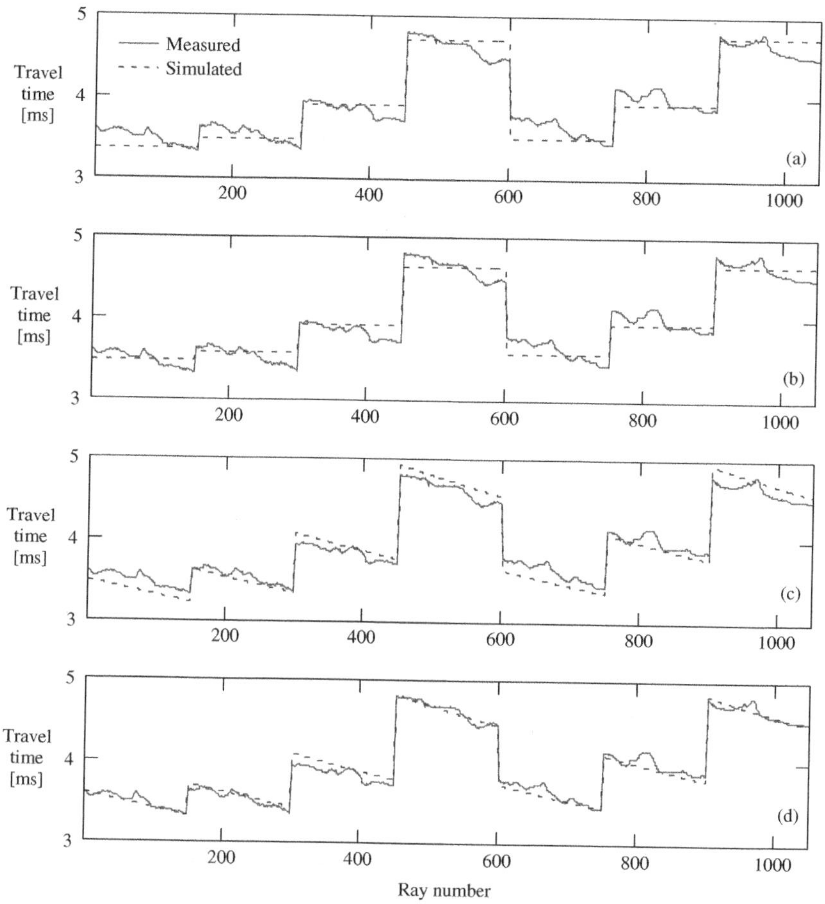

Assumed medium	a	b	c	Squares error	Absolute error
(a) Homogeneous – isotropic	4510	0	1.00	3.49	2.79
(b) Homogeneous – anisotropic	4890	0	1.12	2.78	2.35
(c) Heterogeneous – isotropic	3270	12.0	1.00	2.73	2.42
(d) Heterogeneous – anisotropic	3560	13.0	1.12	1.62	1.26

Figure 11.10 Tunnel in Korea – assessing the host medium (refer to Figure 11.4). Inversion by successive forward simulations. Assumed model: close-form solution for the ray path in vertically heterogeneous, anisotropic media (Table 11.1). Plots show travel time versus ray number. (Note the seven data sets for different illumination angles; there are N = 1050 measurements.) The continuous and dotted lines correspond to measured and predicted travel times, respectively

this material model has more degrees of freedom, data preprocessing results in Figure 11.4 clearly support it.

Case History: Steel Pipe in Air – Echolocation

The inverse problem in Figure 11.8 is cast in terms of four unknown parameters: the inclusion position and size p_{inc}, q_{inc}, size R_{inc} and the velocity of the medium V_{med}. A straight ray model is used for forward simulation, and convergence is guided with the L_2 norm. The optimal solution for each of the three tests is summarized in Table 11.2 (additional insight is gained by analyzing results in the solved problem at the end of this chapter). The distance to the anomaly from the string of transducers q_{inc} is resolved better than the anomaly position p_{inc} parallel to the string of transducers. The anomaly size is poorly resolved, and there is a strong interplay between the size R_{inc} and the distance q_{inc} so that the value that is resolved best is $q_{inc}-R_{inc}$. A careful analysis of the heuristic solution in Figure 11.8 elucidates these observations.

Table 11.2 Inversion of reflection data by successive forward simulations

Field setup	Solution	V_{med}	p_{inc}	q_{inc}	R_{inc}	$q_{inc}-R_{inc}$
Case 1	Experiment	343	0.80	0.41	0.16	0.25
	Inverted (L_2)	381	0.86	0.29	0.02	0.27
Case 2	Experiment	343	0.80	0.66	0.16	0.50
	Inverted (L_2)	385	0.84	0.56	0.01	0.55
Case 3	Experiment	343	0.80	0.91	0.16	0.75
	Inverted (L_2)	389	0.82	0.85	0.11	0.74

Note:
Test setups are shown in Figure 11.8.
Inversions are based on travel times – no additional data.
Convergence is driven to minimize L_2 norm.
Data and model errors enhance the trade-off between q_{inc} and R_{inc}.
The point on the anomaly closest to the line of transducers is $(q_{inc}-R_{inc})$ away.

Case History: Anomaly in Air – Transillumination Data

The parametric representation involves five unknowns: the velocity of the medium V_{med}, and the properties of the inclusion including position p_{inc} and q_{inc}, size R_{inc} and velocity V_{inc}. A straight-ray forward simulator is used first. Slices of the error surfaces obtained with L_1, L_2, and L_∞ norms are presented in Figure 11.11.

Two important observations follow from these results. First, the resolvability of the vertical position of the inclusion q_{inc} parallel to the instrumented sides is

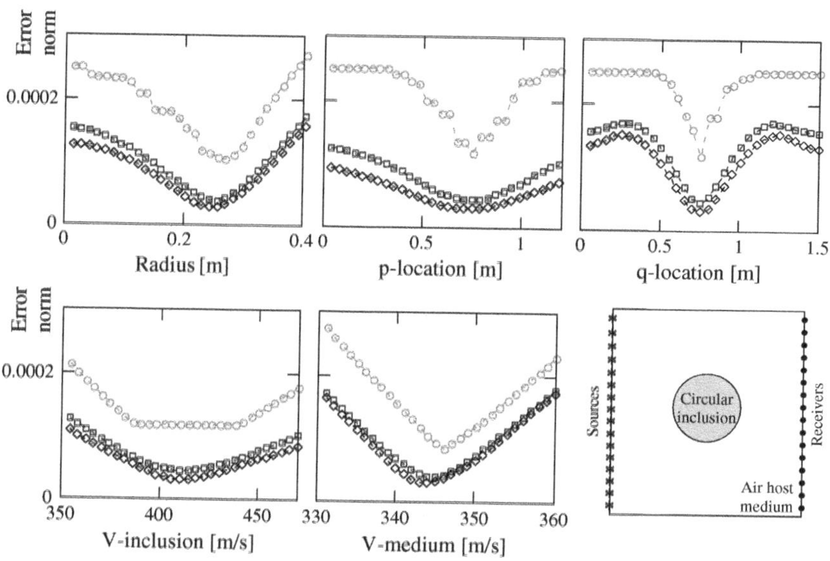

Figure 11.11 Inversion of transmission data by successive forward simulations. Parametric representation. Helium ballon in air (details in Figure 11.5): P_{inc}, q_{inc}, R_{inc}, V_{inc}, V_{med}. Slices of the error surface near optimum. (◇) L_1 norm; (□) L_2 norm; (o) L_∞ norm. (Data courtesy of A. Reed)

significantly better than for the horizontal position p_{inc}. The incorrect location of the anomaly in the horizontal direction affects mostly the same rays as the true location (this is equivalent to the elongation of the anomaly in the direction of illumination observed in Figure 11.9). By contrast, the incorrect q-location affects a significant number of rays: originally untouched rays become touched by the inclusion, and several of the rays that traverse the inclusion in the true location are not touched in the new assumed position.

Second, L_1 and L_2 error surfaces are nonconvex in the q-direction, and inversion may diverge from the minimum. Why does this occur? As discussed above, when the anomaly is displaced upward, many rays will be affected, either because they used to traverse the anomaly or because they did not. But if the anomaly is considered totally outside the region, only those rays that traverse the anomaly in its true position contribute to the residual, and the total misfit measured by the L_1 and L_2 norms decreases. However, the L_∞ norm is only concerned with the worst residual for any ray, so it remains convex. This confirms that the L_∞ norm is insensitive to uneven information density, as noted in Section 8.3.

The minimum value of the error surface is a measure of model and data errors (Section 8.6). Errors raise the error surface, reduce convergence gradients, round the error surface near the minimum, and hinder the unequivocal identification of the unknowns. Data noise may also cause local minima.

The straight-ray model predicts a significantly larger inclusion size (Figure 11.11) because the inclusion is a high-velocity anomaly and acts as a divergent lens. The inclusion size is correctly predicted with curved rays (not presented here).

11.6.3 Matrix-based Inversion

Case History: Balloon in Air – Transillumination

The data are inverted using the regularized least-squares solution. The selected regularization criterion is the minimization of variability. Therefore, the regularization matrix is constructed with the Laplacian kernel and imaginary boundary points satisfy zero gradient across the boundary. Results for different regularization coefficients λ are summarized in Figure 11.12. To facilitate the comparison, tomograms are thresholded at a mean measured velocity of 370 m/s. The following observations can be made:

- Normalized errors $(y_i^{<meas>} - y_i^{<just>})/y_i^{<meas>}$ decrease towards zero when the inversion is underregularized (lowλ). Therefore, the data are better justified when λ is low.
- The spread in pixel values is small when a smoothness criterion is imposed and the inversion becomes overregularized (highλ). Eventually, a featureless image is obtained when a very high regularization coefficient is used.
- Data and model errors are high. (Results shown in the figure are computed with straight rays.) The spread in prediction errors remains $\pm 1\%$ even as the problem becomes ill-conditioned for low λ values.

Clearly, the inversion cannot be data-driven only. Instead, the characteristics of the solution must be taken into consideration as well.

11.6.4 Investigate Other Inversion Methods

Physical insight and mathematical analysis may help identify exceptional inversion strategies besides those explored in this book. The solution of tomographic imaging in the frequency domain using the Fourier slice theorem is an excellent example (Section 10.1). Its extension to the diffraction regime provides further evidence of the benefits that insightful inversion approaches can have when combined with a detailed analysis of the problem (Fourier diffraction theorem).

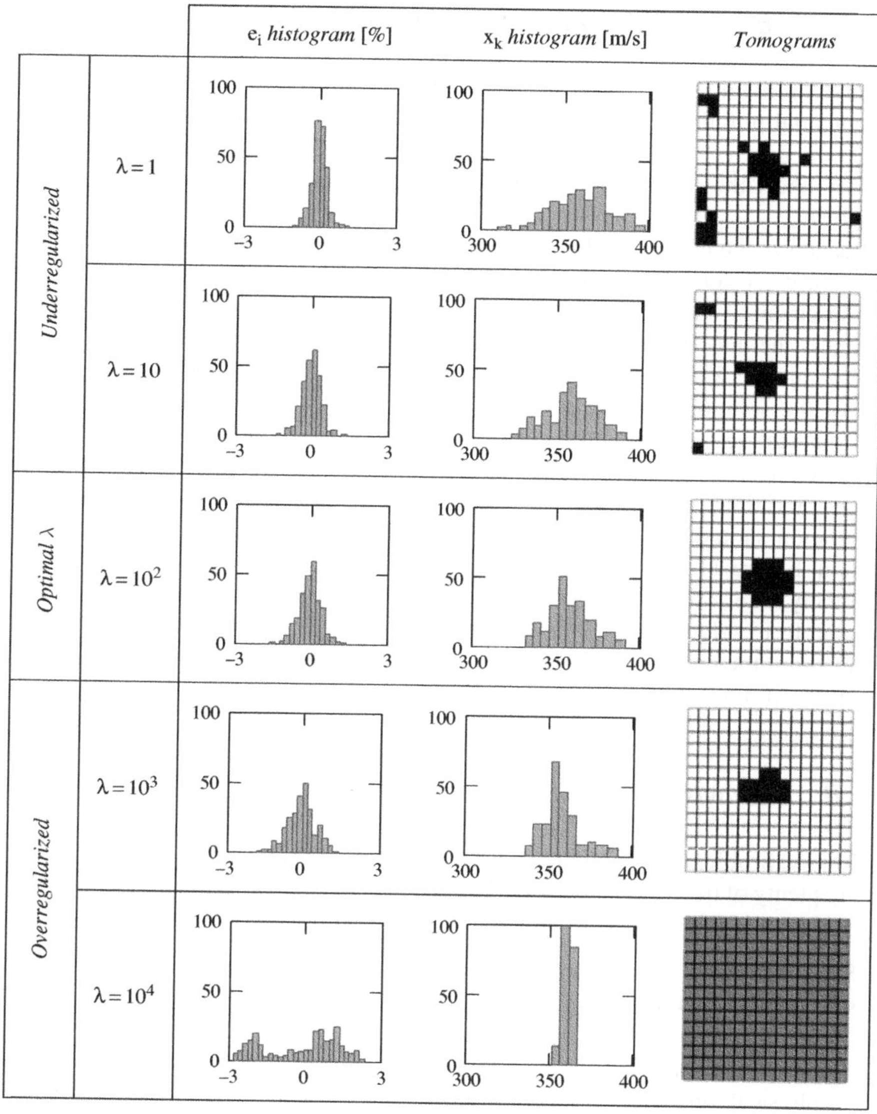

Figure 11.12 Regularized least squares solution – helium ballon. Tomograms are thresholded at 370 m/s. Notice the trade-off between data justification (e_i histogram) and image quality (histogram of pixel values and tomograms)

11.7 STEP 7: ANALYZE THE FINAL SOLUTION

Inverse problem solving may appear simple at first glance, but there are plenty of potential pitfalls along the way. The solution may be completely wrong even when the data $\underline{y}^{<meas>}$ are well justified and the residuals are small. Indeed, this is very likely the case in ill-conditioned and underregularized problems. *Remain skeptical!*

Reanalyze the procedure that was followed to obtain the measurements: did you measure what you think you measured, or are measurements determined by the measurement system (instrumentation and distribution of information density)? Reassess the underlying physical processes assumed for inversion in light of the results that were obtained. Consider all information at your disposal.

Plot the solution estimate $\underline{x}^{<est>}$ against the following vectors: column-sums $\underline{1}^T \cdot \underline{\underline{h}}$ indicative of information content, row-sums $\underline{\underline{h}}^{-g} \cdot \underline{1}$ indicative of systematic error propagation, and row-sums $\underline{\underline{h2}}^{-g} \cdot \underline{1}$ indicative of accidental error magnification. Scrutinize any correlation. In the case of tomographic images, no clear correlation should be observed between tomograms in Figure 11.12 and the 2D plots in Figure 11.2.

Finally, the well-solved inverse problem can convey unprecedented information, from subatomic phenomena, to the core of the earth and distant galaxies. In all cases, the physics of the problem rather than numerical or computer nuances should lead the way.

11.8 SUMMARY

- Inverse problem solving may appear deceptively simple at first; however, there are plenty of traps along the way. To stay on course, retain a clear understanding of the problem at all times and stay in touch with its physical reality.
- Successful inverse problem solving starts before data collection. The following steps provide a robust framework for the solution of inverse problems:

  1. *Analyze the problem.* Develop an acute understanding of the underlying physical processes and constraints, measurement and transducer-related issues and inherent inversion difficulties. Establish clear and realizable expectations and goals.
  2. *Pay close attention to experimental design.* The viability of a solution is determined at this stage. Design the distribution of measurements to attain a proper coverage of the solution space, and select transducer and electronics to gather high-quality data.

3. *Gather high-quality data.*
4. *Preprocess the data* to assess their quality, to gain a glimpse of the solution, and to hypothesize physical models. Data preprocessing permits identification and removal of obvious outliers and provides valuable information that is used to guide and stabilize the inversion.
5. *Select an adequate physical model* that properly captures all essential aspects of the problem. Fast model computation is crucial if the inverse problem is solved by successive forward simulations.
6. *Invert the data using different inversion methods.* Consider a parametric representation of the problem combined with successive forward simulations, as well as less constrained discrete representations in the context of matrix-based inversion strategies. Do not hesitate to explore other inversion strategies that may result from a detailed mathematical analysis of the problem or heuristic criteria.
7. *Analyze the physical meaning of the solution.*

- Discrete signal processing and inverse problem solving combine with today's digital technology to create exceptional opportunities for the development of previously unthinkable engineering solutions and to probe unsolved scientific questions with innovative approaches. Just . . . image!

SOLVED PROBLEMS

P11.1 *Study with simulated transmission data: parametric representation.* Consider the helium balloon problem in Figure 11.5. Simulate noiseless data assuming a straight-ray propagation model. Then, explore the error surfaces corresponding to L_1, L_2 and L_∞ error norms.
Solution: Assumed model parameters: velocity of the host medium $V_{med} = 343 m/s$, velocity of the helium balloon $V_{inc} = 410 m/s$, radius of the balloon $R_{inc} = 0.23 m$, and the coordinates of the balloon $p_{inc} = 0.75 m$ and $q_{inc} = 0.75 m$. Travel times are computed. Then, the simulated travel times are inverted as if they have been measured $\underline{t}^{<meas>}$. Slices of the error surfaces across optimum are generated as follows: (1) perturb one model parameter at the time; (2) compute the travel time $t_i^{<pred>}$ and the residual $e_i = (t_i^{<meas>} - t_i^{<pred>})$ for all N rays, and (4) evaluate the norm of the residual. Results are plotted next (L_1 and L_2 norms are divided by the number of rays N to obtain the "average residual error per ray"):

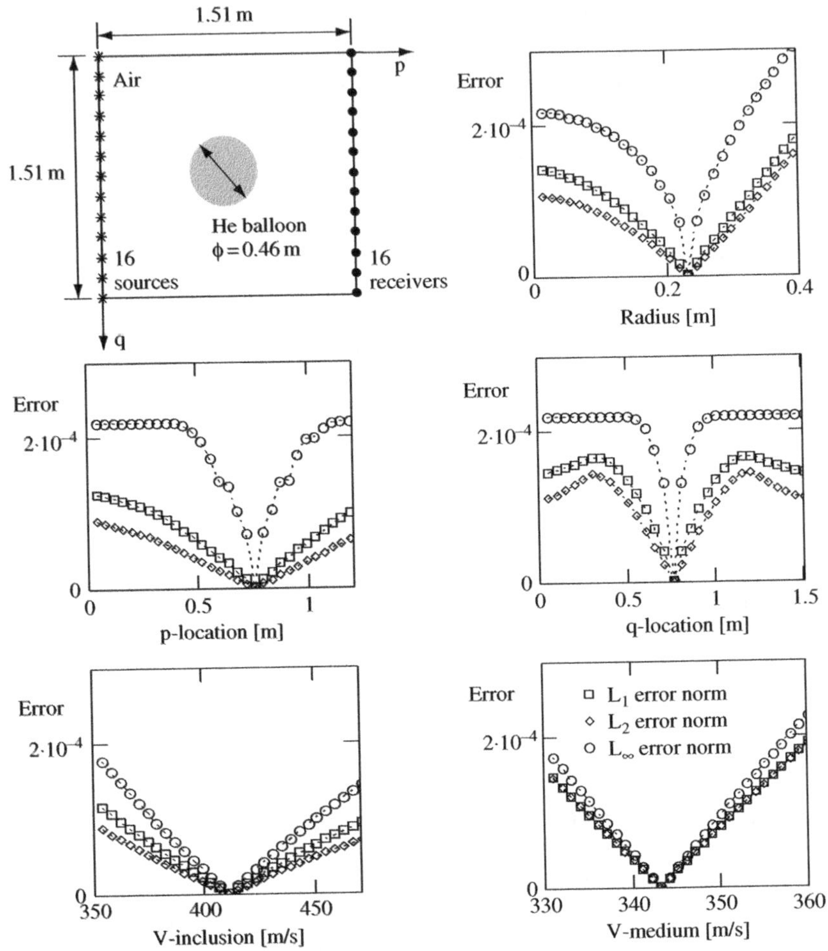

Results allow the following observations to be drawn (compare to Figure 11.11 – review text):

- The cross-section of the error surface along different variables shows different gradients. The L_∞ norm presents the highest gradients for all five parameters. In the absence of model and measurement error, all norms reach zero at the optimum.
- The velocity of the host medium affects all rays, and for most of their length. Hence, the convergence of the background velocity is very steep.

- The resolvability of the vertical position of the inclusion q_{inc} parallel to the instrumented boreholes is significantly better than for the horizontal position p_{inc}. The incorrect location of the anomaly in the horizontal direction mostly affects the same rays when the inclusion is horizontally shifted. By contrast, the incorrect q-location affects a significant number of rays.
- The striking feature is that the L_1 and L_2 error surfaces are nonconvex. This is critical to inversion because algorithms may converge to local minima, thus rendering inadequate tomographic images (see discussion in text).

Repeat the exercise adding systematic and then accidental errors, and outliers to the "measurements".

P11.2 *Study with simulated reflection data: parametric representation.* Recall the reflection problem in Figure 11.8. Simulate noiseless data for the following test configurations, assuming a straight-ray propagation model, and explore the L_2 error surface in each case.

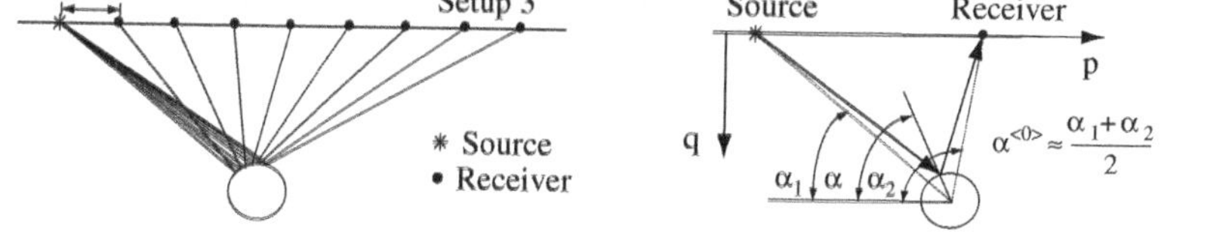

Solution: Travel times are computed by identifying the point on the reflecting surface that renders the minimum travel time. It can be shown that the reflection point is approximately located at an angle $\alpha \approx (\alpha_1 + \alpha_2)/2$ (see sketch above). The L_2-norm error surface is studied near the optimum, following the methodology in Problem P11.1. Slices of the error surfaces are shown next:

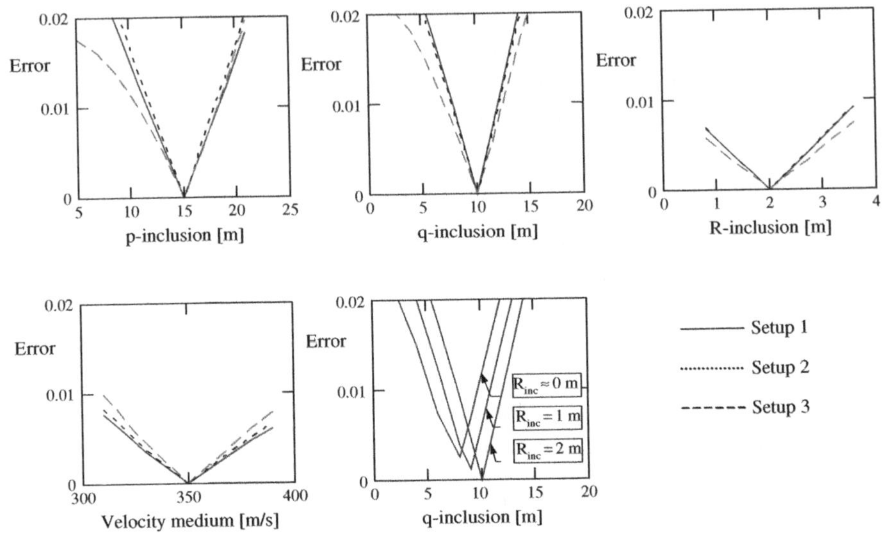

Results indicate that there is no major difference in invertibility given the data from the different field test setups. The invertibility of the size of the anomaly is poor compared to the position of the anomaly. Furthermore, there is a strong interplay between the size R_{inc} and the depth q_{inc} of the anomaly, as shown in the last figure. Therefore, it is difficult to conclude using travel time alone whether the reflector is a distant large anomaly or a closer anomaly of smaller size. The value that is best resolved is the distance from the line of transducers to the closest point in the anomaly, $q_{inc} - R_{inc}$. Compare these results and observations and those presented in Table 11.2 and related text.

ADDITIONAL PROBLEMS

P11.3 *Graphical solution*. Back-project the average velocity shadows plotted in Figure 11.5 to delineate the position of the helium balloon.

P11.4 *Transformation matrix*. Complete the matrix of travel lengths in Figure 11.7.

P11.5 *Attenuation tomography*. Express the attenuation relation in standard matrix form so that the field of material attenuation can be inverted from a set of amplitude measurements (correct measurements for geometric spreading first).

Attenuation equation: $A(k) = A_0 \cdot \left(\frac{r_o}{r_k}\right)^n \cdot e^{-\alpha \cdot r_k}$

P11.6 *Experimental design.* Design a tomographic experiment to identify anomalies in a block size 2 m × 2 m. Expect a wave velocity between 4000 and 5000 m/s. Follow the step-by-step procedure outlined in this chapter; simulate data and explore different simulation strategies.

P11.7 *Application of tomographic imaging.* Travel time data are gathered for three different locations of a single helium balloon in air. A total of 49 measurements are obtained in each case with seven sources and seven receivers (Table P11.1).

- Preprocess the data to determine the characteristics of the background medium. Assess accidental and systematic errors in the data.
- Plot average velocity shadows. Constrain the position of the anomaly using the fuzzy logic technique.
- Capture the problem in parametric form, and solve by successive forward simulations.
- Then use a pixel-based representation and solve with LSS, DLSS, RLSS, and SVDS. Identify optimal damping and regularization coefficients and the optimal number of singular values. Add an initial guess obtained from previous studies.

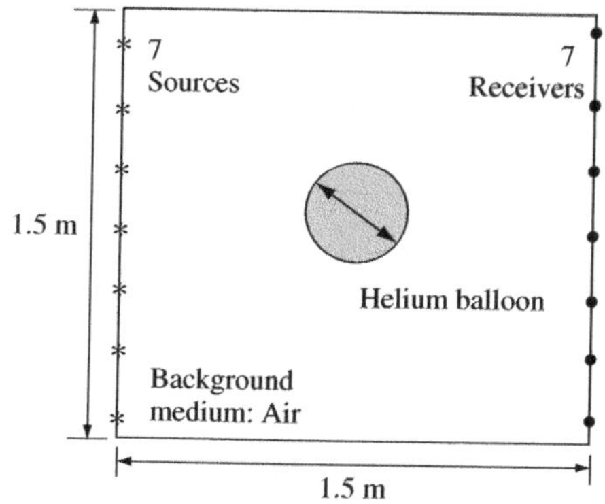

Table P11.1 Travel time tomographic data for the imaging of a helium ballon

Source positions [m]		Receiver positions [m]		Travel times [ms]		
p	q	p	q	Case 1	Case 2	Case 3
0	0.0762	1.5320	0.0762	4.38	4.52	4.38
0	0.0762	1.5320	0.3048	4.42	4.56	4.42
0	0.0762	1.5320	0.5334	4.54	4.58	4.58
0	0.0762	1.5320	0.7620	4.80	4.74	4.79
0	0.0762	1.5320	0.9906	5.06	5.06	5.06
0	0.0762	1.5320	1.2190	5.42	5.50	5.38
0	0.0762	1.5320	1.4480	5.78	5.90	5.76
0	0.3048	1.5320	0.0762	4.50	4.54	4.46
0	0.0348	1.5320	0.3048	4.42	4.30	4.40
0	0.3048	1.5320	0.5334	4.50	4.18	4.42
0	0.3048	1.5320	0.7620	4.62	4.32	4.54
0	0.3048	1.5320	0.9906	4.84	4.60	4.76
0	0.3048	1.5320	1.2190	5.10	5.00	5.04
0	0.3048	1.5320	1.4480	5.36	5.56	5.44
0	0.5334	1.5320	0.0762	4.68	4.50	4.58
0	0.5334	1.5320	0.3048	4.52	4.24	4.40
0	0.5334	1.5320	0.5334	4.46	4.12	4.34
0	0.5334	1.5320	0.7620	4.32	4.28	4.39
0	0.5334	1.5320	0.9906	4.34	4.54	4.50
0	0.5334	1.5320	1.2190	4.52	4.82	4.76
0	0.5334	1.5320	1.4480	4.92	5.24	5.08
0	0.7620	1.5320	0.0762	4.90	4.62	4.82
0	0.7620	1.5320	0.3048	4.46	4.38	4.56
0	0.7620	1.5320	0.5334	4.20	4.24	4.36
0	0.7620	1.5320	0.7620	4.14	4.36	4.30
0	0.7620	1.5320	0.9906	4.22	4.46	4.38
0	0.7620	1.5320	1.2190	4.42	4.62	4.56
0	0.7620	1.5320	1.4480	4.78	4.90	4.84
0	0.9906	1.5320	0.0762	4.84	4.92	5.14
0	0.9906	1.5320	0.3048	4.56	4.68	4.78
0	0.9906	1.5320	0.5334	4.38	4.56	4.52
0	0.9906	1.5320	0.7620	4.26	4.52	4.42
0	0.9906	1.5320	0.9906	4.32	4.48	4.38
0	0.9906	1.5320	1.2190	4.44	4.50	4.44
0	0.9906	1.5320	1.4480	4.58	4.60	4.60
0	1.2190	1.5320	0.0762	5.30	5.32	5.42
0	1.2190	1.5320	0.3048	5.00	5.08	5.04
0	1.2190	1.5320	0.5334	4.74	4.92	4.74
0	1.2190	1.5320	0.7620	4.60	4.70	4.56
Source positions [m]		Receiver positions [m]		Travel times [ms]		
p	q	p	q	Case 1	Case 2	Case 3
0	1.2190	1.5320	0.9906	4.46	4.56	4.46
0	1.2190	1.5320	1.2190	4.42	4.48	4.42
0	1.2190	1.5320	1.4480	4.44	4.52	4.46
0	1.4480	1.5320	0.0762	5.86	5.80	5.86
0	1.4480	1.5320	0.3048	5.46	5.54	5.46
0	1.4480	1.5320	0.5334	5.10	5.16	5.12
0	1.4480	1.5320	0.7620	4.82	4.88	4.82
0	1.4480	1.5320	0.9906	4.58	4.68	4.58
0	1.4480	1.5320	1.2190	4.44	4.52	4.46
0	1.4480	1.5320	1.4480	4.36	4.42	4.40

P11.8 *Application: gravity anomalies.* Consider the problem of the local gravity field caused by a 0.5 m^3 gold nugget cube hidden 1.5 m under a sandy beach.

- Study the problem. Evaluate its physical characteristics, measurement issues, transducer difficulties.
- Design the experiment to gather adequate data, consider the spatial and temporal distribution of data and procedures to obtain high-quality data.
- Simulate data for the gravity anomaly (search for Bouguer equations).
- Develop versatile and insightful strategies for data preprocessing to gain information about the data (including biases, error level, outliers) and a priori characteristics of the solution.
- Identify the most convenient inversion method; include if possible regularization criteria and other additional information.
- Prepare guidelines for the physical interpretation of the final results.

P11.8 *Application: strategy for inverse problem solving in your field of interest.* Consider a problem in your field of interest and approach its solution as follows:

- *Analyze the problem in detail*: physical processes and constraints, measurement and transducer-related issues. Establish clear goals.
- *Design the experiment.* Include: transducers, electronics, and the temporal and/or spatial distribution of measurements. Identify realizable and economically feasible configurations that provide the best possible spatial coverage.
- *Gather high-quality data.* If you do not have access to data yet, simulate data with a realistic model. Explore the effect of data errors by adding a constant shift, random noise, and outliers.
- *Develop insightful preprocessing strategies* to assess data quality, to gain a glimpse of the solution, and to hypothesize physical models.
- *Select an adequate physical model* that properly captures all essential aspects of the problem. Fast model computation is crucial if the inverse problem is solved by successive forward simulations.
- *Invert the data using different problem representations and inversion methods.* Explore other inversion strategies that may result from a detailed mathematical analysis of the problem or heuristic criteria.
- *Identify guidelines for the physical interpretation of the solution.*

Index

Discrete Signals and Inverse Problems J. C. Santamarina and D. Fratta

Printed and bound by CPI Group (UK) Ltd, Croydon, CR0 4YY
16/09/2026
14963379-0005